## Plant Conservation

The rate of species and natural habitat loss across our planet is steadily accelerating. This book argues that existing practices of plant conservation are inadequate and firmly supports the placement of ecological restoration at the cornerstone of biodiversity conservation. The author unifies different aspects of conservation into one coherent concept, including natural area protection, *ex situ* conservation, and *in situ* interventions, through either population management or ecological restoration. Assisted colonization, experimentation, and utilization of threatened plant species are raised as crucial elements in restoration, with partly novel ecosystems being among its major target areas. Covering a wide spectrum of plant conservation examples, and offering practical methodologies alongside the theoretical context, this is a vital resource for students, research scientists, and practitioners in conservation biology and restoration ecology.

SERGEI VOLIS is Professor of Biology at Kunming Institute of Botany, China. He has authored many publications investigating genetic and demographic population processes, plant adaptations to local environments, initial stages of speciation, and species phylogeography. His major research interest is plant conservation in its theoretical and applied aspects, an interest which goes back to his years of studying ecology in Simpheropol State University in the former Soviet Union and his PhD project in Ben-Gurion University of the Negev, Israel.

# Plant Conservati

## The Role of Habitat Restoration

SERGEI VOLIS
Kunming Institute of Botany, Chinese Academy of Sciences

CAMBRIDGE
UNIVERSITY PRESS

# CAMBRIDGE
## UNIVERSITY PRESS

University Printing House, Cambridge CB2 8BS, United Kingdom

One Liberty Plaza, 20th Floor, New York, NY 10006, USA

477 Williamstown Road, Port Melbourne, VIC 3207, Australia

314–321, 3rd Floor, Plot 3, Splendor Forum, Jasola District Centre,
New Delhi – 110025, India

79 Anson Road, #06-04/06, Singapore 079906

Cambridge University Press is part of the University of Cambridge.

It furthers the University's mission by disseminating knowledge in the pursuit of
education, learning, and research at the highest international levels of excellence.

www.cambridge.org
Information on this title: www.cambridge.org/9781108480376
DOI: 10.1017/9781108648677

© Sergei Volis 2019

First published 2019

Printed in the United Kingdom by TJ International Ltd, Padstow Cornwall

A catalogue record for this publication is available from the British Library.

Library of Congress Cataloging-in-Publication Data
Names: Volis, Sergei, author.
Title: Plant conservation : the role of habitat restoration / Sergei Volis.
Description: New York, NY: Cambridge University Press, 2019.
Identifiers: LCCN 2018039107 | ISBN 9781108480376 (hardback) |
ISBN 9781108727334 (paperback)
Subjects: LCSH: Plant conservation. |
Restoration ecology. | Biodiversity conservation.
Classification: LCC QK86.A1V65 2018 | DDC 333.95/316–dc23
LC record available at https://lccn.loc.gov/2018039107

ISBN 978-1-108-48037-6 Hardback
ISBN 978-1-108-72733-4 Paperback

*"Despite the increase in conservation planning and initiatives in the past few decades, the evidence that habitats are being lost or degraded and species increasingly threatened or even extinguished is beyond doubt, and current trajectories suggest that this will continue unless some innovative approaches are adopted"*

Heywood 2016

# Contents

Color plates can be found between pages 208 and 209.

# Foreword

In this thought-provoking and richly documented book, Sergei Volis lays out our human dilemma in stark but realistic terms, and describes in great detail the meaning of our time, the Anthropocene, for the possibilities of conserving as many as possible of the more than 400 000 species of vascular plants that inhabit this planet with us. They not only live with us, but they and their metabolic activities, historical and current, enable us to live here and for our lives to continue. Against this background, Volis lays out his concept of the most effective way to conserve plant diversity, through the careful management of restored biological communities. He discusses every step of this process in scholarly and practical detail, providing exhaustive citations of the pertinent literature and showing how they contribute to our understanding of the whole approach to conservation that he presents. Consequently, the book will serve as a valuable resource for restoration biology or conservation however practiced, and will clearly lead us on to even better ways of dealing with the chaos and destruction that we have caused in the world around us.

Considering that the global human population has grown from about 2.5 billion people in 1950 to more than 7.6 billion now, with the increase in levels of consumption expanding far more than the tripling that a population increase of this magnitude implies, it is no wonder that the necessary conditions for the preservation of plants, or of ourselves, have also been greatly altered. In fact, it is estimated that we are currently using about 175% of the sustainable productivity that the Earth can supply on an ongoing basis (www.footprintnetwork.org), a condition that clearly bodes ill for every living thing. Estimates that our population will grow to 9.8 billion over the next 32 years, while the population of sub-Saharan Africa more than doubles from its current 1 billion people, together with the observation that our greed for ever-increasing levels of consumption seems, if anything, to be accelerating, make the future for conservation seem even darker than it does at present. It is small wonder, given these observations, that a quarter of all tropical forests, the richest repository of life on Earth, has been cut since the ratification of the Convention on Biological Diversity a quarter century ago.

Plants are so critical for our own survival, though, that we should be working as hard as we can to find ways to conserve them, in as much of their diversity as we possibly can. Reserves and protected areas are certainly not working well; they are used by people for many purposes, and few governments allocate sufficient funds even to make a decent effort to maintain them. This difficulty reveals what any thoughtful person would understand, which is that we have basically reshaped the surface of the Earth and must now find ways to fit the parts back together in order to make the whole sustainable. Global climate change is proceeding rapidly in the face of our inadequate efforts to curtail it, and will change the nature of all biological communities progressively while we strive to save them and the organisms in them. Alterations in the design and maintenance of protected areas could allow them to function more effectively, but there would still be serious gaps in what would survive.

In our efforts to conserve higher plants, we have one advantage over our efforts to preserve most other groups of organisms: properly studied, and with optimal conditions maintained, their seeds and tissues can be stored virtually indefinitely at low temperatures and still remain capable of regeneration under appropriate conditions. Plants can also be grown and maintained in botanical gardens and similar facilities, but then, obviously, environment stresses such as climate change and pollution can affect their ability to keep on growing in a particular place. Moreover, there is rarely the room or the funds to make possible the maintenance of populations sufficiently large to include a viable sample of their genetic diversity. In seed banks, the stored samples can readily accommodate the genetic diversity of the natural populations from which they were drawn, but the inevitable question remains: where could they ever be planted and then maintained in our rapidly changing world? Nevertheless, such *ex situ* conservation remains important and should be pursued actively. As American tropical biologist Daniel Janzen has said, "If you don't save them now, you can't save them later."

Sergei Volis argues convincingly through this fascinating volume that for the most effective conservation of plant species, it will be necessary to work with and often modify natural communities, structuring them to meet the needs of conservation in the most effective ways possible. Where *restoration ecology* as generally understood attempts to return natural communities to their original states, other strategies differ. Over the last couple of decades, the idea of building *novel ecosystems*, with whatever composition suits human needs, has gained popularity. Thus the general concept of restoration ecology is definitely compatible with conservation, whereas that idea has no place in the idea of novel ecosystems solely for the benefit of humans. In effect, novel ecosystems are an extension of the agriculture that our ancestors have been practicing for some 11 500 years, and which now, with grazing, occupies more than a third of the Earth's surface.

Volis devotes much of this book to documenting the ways to build seminatural communities explicitly designed for species preservation, a novel conservation strategy that he terms the *habitat restoration paradigm*. The design and maintenance of such communities is complex, depending for best results on the proper application of quantities of information made available in large datasets. The most seriously endangered species in particular regions should be given priority and included whenever possible in the remnants of communities where they once occurred. The role of succession in the maintenance of many imperiled species of organisms is so important that it needs to be carefully built into conservation restoration. Invasive species often, although not always, need to be removed. The range of genotypes in the species being preserved is of key importance, just as when seed samples are conserved in seed banks, and the populations of the species being conserved need to be maintained at high enough levels to enable them to be genetically resilient. Assisted migration to areas where a species might currently thrive is discussed in detail and presented as a part of building conservation communities. It certainly does appear to be one necessary aspect of conserving species in our rapidly changing world. A useful strategy that Volis terms *quasi in situ* conservation and documents in some detail calls for the establishment of new, genetically diverse populations of imperiled species in waste lands or other places where few other kinds of plants will grow. In conserving the species, it likewise is a particularly effective way to build up seeds for use in repopulating.

For the implementation of conservation plans like those presented here, the legislation leading to the preservation of natural and quasi-natural areas will need to be modified widely to include all modifications to be made. Without such modifications, we shall, in most areas, simply be waiting for the communities and the species in them to disappear, perhaps comfortable in the pious hope that we have. In this truly remarkable book, Sergei Volis has given us many options for ways in which we might choose to avoid such a fate, along with a well-presented analysis of the literature and a rich mine of thoughts from which we shall be able to progress to even better ways of conserving as much as we can of the organisms with which we share our troubled planet.

Peter H. Raven
*President Emeritus,*
*Missouri Botanical Garden,*
*St. Louis, Missouri, USA*

# Preface

The major motivation for this book was that the current practices of conserving biodiversity are generally inadequate and unlikely to stop ongoing environmental degradation and species loss; plant conservation needs a new paradigm. The aim of the book is to highlight the limitations of existing approaches, introduce a new concept, and link the latter with ecological theory and available practices. The reader will judge how successful the author was in this venture.

The text will hopefully fulfill two tasks. One is to provide a long-needed text on plant conservation that is useful for conservation practitioners. The second is to unify, within a coherent concept, many aspects of plant conservation, such as natural area protection, preservation of wild germplasm in seed banks and living collections, and *in situ* interventions through either population management or ecological restoration.

The proposed approach has a solid foundation in the existing fields of population and community ecology, conservation biology, and restoration ecology, and is especially applicable in regions that have many threatened species within habitats that are threatened themselves, with both species and the whole ecosystem requiring immediate action. To ease the application of the concept in real settings, detailed methodological guidelines are provided. They are based on the latest theoretical developments and vast experience gained by conservation and restoration practitioners in the last several decades.

The book is divided into seven chapters. The first chapter introduces plant conservation as a discipline, summarizes existing theory and practices, and discusses approaches to dealing with the accelerated loss of species and habitats. The second chapter presents the author's concept of plant conservation. The principles on which the concept is based are relevant to the restoration of entire ecosystems and the recovery of particular plant species and populations. The major principles of the concept introduced in the second chapter are explained in detail throughout Chapters 3 and 4. Chapter 3 is devoted to the recovery of threatened plant species, and Chapter 4 presents and synthesizes the topics most relevant to the restoration of threatened species habitat. Chapter 5 provides a toolkit for the restoration of forest land. Chapter 6 includes a series of sections covering application of the methodology described in Chapters 4 and 5 for the restoration of particular systems, such as logged forests, abandoned agricultural fields and pastures, and forest plantations. Each section provides a concise but comprehensive coverage of a specific topic, cites the essential classic literature and up-to-date research in the field, provides examples of how the relevant methodology is currently being applied, and discusses how it can be applied in agreement with the proposed concept. The closing chapter of the book links the theory with practice, identifying areas of emergency for restoration, and presenting case studies of conservation-oriented restoration, envisioned, planned, or partly implemented.

The book is intended for graduate students, research scientists, and practitioners from the fields of plant conservation biology and restoration ecology. This book would not have been possible without the encouragement and moral support of Dr. Peter Raven.

# 1  Introduction

1.1  WHAT PLANT CONSERVATION IS ABOUT AND
WHY BIODIVERSITY SHOULD BE CONSERVED

Defining plant conservation as a discipline is simple. It is part of an applied science called conservation biology, specifically focusing on plants. Thus, the majority of theoretical developments of conservation biology apply to plant conservation and are utilized in it.

Conservation biology, in turn, is a young, multidisciplinary science, which provides the principles and tools to deal with the crisis confronting biological diversity, and which is fundamentally different from other branches of science in several aspects (Soulé 1985). One is that, because of its concern with imminent threat and extinction, it is under severe time constraints and in constant need of actions. The second is its holistic nature, making it a synthesis of a variety of other disciplines, namely population and community ecology, population genetics, biogeography, landscape ecology, environmental management, and economics. Another important aspect of conservation biology, at least until recently, was its moral obligation, i.e., the implicit assumption that it is morally wrong for our species to drive other species to extinction. This aspect, however, is now challenged by those who call themselves "new conservation biologists" (Kareiva and Marvier 2007, 2012; Daily *et al.* 2009; Marvier and Kareiva 2014) and neo-Marxist social scientists (Fletcher 2010; Büscher and Dressler 2012; Büscher *et al.* 2017), who share the view that biotic diversity does not have an intrinsic value independent of providing humans with goods and services. Apologists for these movements, which according to Büscher *et al.* (2017) "are more democratic, equitable and humane" than traditional conservationists, adopt the position that nonhumans can be morally relevant only to the extent that they affect human well-being, and the latter must be given priority in any conservation efforts. This issue is discussed further in Section 1.3, but a few points must be made clear. Virtually every text on the socioeconomic aspects of nature conservation repeats a common mantra that land acquisition for conservation means missed economic opportunities for a society. A truth, however, is that under rapid and accelerating human population growth the "economy first" or "feeding poor first" principles will inevitably result in: (1) every patch of land eventually undergoing one or another form of anthropogenic transformation that will be extremely difficult or impossible to reverse; (2) many fragile ecosystems becoming bare lands useless for any human activity; and (3) a mass extinction of animals and plants. Is this what we want?

Another line of reasoning for rejecting the priority of the "economy first" principle in decision-making regarding biodiversity is provided by Kormos and Zimmerman (2014). Analyzing existing practices of commercial logging, the authors came to the conclusion that "industrial logging of tropical hardwoods from natural forests is biologically unsustainable under virtually any scenario that approximates financial viability" and therefore "logging in these forests will end in the not-too-distant future." Their following rhetoric question is "Sooner or later protection will be necessary to prevent conversion: why not keep tropical forests intact, a much easier task when there are few or no roads (Laurance *et al.* 2009), rather than seeking to protect them after they have been degraded?"

Instead of sacrificing wilderness and biodiversity to economic requirements, we need to deal with and solve the root cause of the conservation crises: a growth in human population and per capita consumption, a trend that shows no signs of slowing down (Barnosky *et al.* 2017).

## 1.2   THE OLD CONCEPT

The emergence of the discipline of conservation biology is demarcated by the publication of the book *Conservation Biology: An Evolutionary-Ecological Perspective* in 1980 (Soulé and Wilcox 1980). The papers in this book, which employ the principles of the theory of island biogeography and population biology, explicitly address issues of the "decay of biological diversity" and "the rampant pace of habitat destruction."

A detailed overview of the history of conservation biology is beyond the scope of this book. However, it is important to mention the major steps in the development of the discipline. The first is the island biogeography theory (MacArthur and Wilson 1967), which established a causal link between the size of the area, distance from the source of propagules, and species richness, a central principle for explaining the effects of habitat fragmentation. The second, coined by Jared Diamond, known as the SLOSS dilemma (single large or several small) laid the foundations for the planning of protected areas. Large reserves are better than small ones; reserves closer together are better than those far apart; reserves grouped and linked together are better than those that are separated and arranged in a straight line; and round reserves are better than elongated reserves (Diamond 1975).

Subsequent contributions were made by population biologists investigating demographic and genetic processes in populations. Mark Shaffer coined the concept of minimum viable population (MVP), the minimum size of a population to survive, with a specified probability and a specified number of generations (Shaffer 1981). He was also the first to define four types of stochastic causes that can drive small populations to extinction: environmental, demographic, and genetic stochasticity, and natural catastrophes. Similarly, Ian Franklin analyzed the minimum size of a population to avoid negative effects of inbreeding depression and genetic drift, and derived what is known as "Rule 50/500": 50 individuals are required to avoid inbreeding and 500 individuals to guarantee adaptability (Franklin 1980).

This research direction of investigating processes that take place in small populations was dubbed the "small population paradigm" by Caughley (1994). Because this paradigm concerns population smallness and because it is based on the biological characteristics of populations, it can give robust predictions. The small population paradigm dominated the conservation literature of the 1980s (Frankel and Soulé 1981; Schonewald-Cox et al. 1983; Soulé 1986, 1987). Despite its strengths (well-established theoretical grounds, ease of computer simulations, and relative ease of experimentation), this paradigm turned out to be less useful in real conservation settings (Caughley 1994). The "declining population paradigm," according to Caughley (1994), in contrast to the "small population paradigm," focuses on detecting population declines and their external causes. This paradigm does not have such a strong theoretical basis as the small population paradigm, but is more applicable in many situations. The 1990s witnessed rapid development of this paradigm into a sub-discipline of conservation biology called population viability analysis (PVA). Later, it became evident that the small population paradigm and the declining population paradigm, summarized in Table 1.1, can tackle the same problem in complementary ways (Beissinger 2002), and both are incorporated in modern PVA (Boyce 2002). Although less numerous than we would expect, there are quite a few examples of how PVA enabled estimation of MVP and identified optimal population and habitat management strategies to sustain it (e.g., Drechsler et al. 1999; Oostermeijer 2000; Hunt 2001; Quintana-Ascencio et al. 2003; Volis et al. 2005; Maschinski et al. 2006). A review of the use of PVA in recovery planning for plant species listed under the US Endangered Species Act revealed 223 publications describing 280 PVAs for 246 species (Zeigler et al. 2013).

Other important theoretical developments that greatly contributed to the evolution of conservation biology into a truly applied science include metapopulation theory (Gilpin and Hanski 1991; Gotelli 1991; Hanski and Simberloff 1997), ecological niche modeling (Stockwell and Peters 1999; Hirzel et al. 2002; Phillips

Table 1.1 *Comparison of the proposed concept (habitat restoration paradigm) with the two existing conservation paradigms*

| | Small population paradigm | Declining population paradigm | Habitat restoration paradigm |
|---|---|---|---|
| Major concern | Small population size | Population decline | Degradation of a habitat |
| Focus | Within-population processes | Identification of external threats to a population | Causes and effects of habitat degradation |
| Solutions | Protection and population management | Protection and removal of a threat by a population or habitat management | Restoration of a degraded habitat |
| Emphasized actions | Actions increasing within-population genetic variation and population growth, population augmentation | Optimal disturbance regime management, pest, disease, and invasive species control, reintroduction, and augmentation | Appropriate species choice in restoration plantings, addressing plant–animal interactions, assisted migration |

*et al.* 2006), and the development of algorithms for reserve selection (Kirkpatrick 1983; Margules *et al.* 1988; Nicholls and Margules 1993).

As a result of the development of conservation biology in the last 50 years, we have seen impressive achievements in understanding processes that occur in intact versus human-affected habitats and populations, and an explosion in the variety of conservation tools.

At the same time, there have been vast investments during this period in nature conservation both financially and in terms of the extent of protected areas. For example, the total budgets of the Worldwide Fund for Nature were over US$350 million in 2001, the Nature Conservancy were over US$300 million in 2002, and Conservation International were US$92 million in 2004 (Robinson 2006). The protected areas, according to the 2014 United Nations List of Protected Areas, cover approximately 15.4% of the Earth's terrestrial surface. However, it is now clear that mere designation of protected areas, which has been the primary approach to conserving biodiversity, will fail to protect biodiversity (e.g., Brashares *et al.* 2001; Tang *et al.* 2010; Gardner 2011; Clark *et al.* 2013; Leisher *et al.* 2013). It is also clear that there is a weak link between theoretical developments in conservation biology and implementation of this knowledge in real-world situations. The reason is that conservation biology research conducted over the last 50 years has provided us with quite a good understanding of how natural ecosystems operate under either no or minimal to moderate human impact. But human population growth and expansion are so quick, and their effects on nature so devastating, that existing conservation practices have become ineffective and obsolete. With respect to the conservation of plants, what are these conservation practices? Briefly they include:

1. assessments of biodiversity summarized in the International Union for Conservation of Nature (IUCN) species categorization and lists of threatened species;
2. global and regional prioritization of species, habitats, and areas for conservation;
3. establishment of protected areas preserving natural habitats and species that are under risk of extinction;
4. no intervention in strictly protected areas; minor interventions in less strictly protected areas usually limited to control of invasive species and prescribed burning;
5. preservation of threatened species in *ex situ* seed banks and botanic garden living collections with minimal coordination between *ex situ* and *in situ* actions;

6. reinforcement or reintroduction of endangered species usually conducted at single or very few locations;
7. focus on conservation plans for single species rather than on groups of species or species assemblages.

The realization that the existing methods do not work and cannot put a halt to the rapid disappearance of nature requires new conceptual thinking in conservation and a search for novel approaches. In my view, the theoretical tools to address new challenges already exist. What is needed is to use them in a new way, adapted to the modern-day reality of the Anthropocene.

## 1.3   NEW CHALLENGES AND TWO ALTERNATIVE SOLUTIONS

Although some anthropogenic disturbances may have no, little, or even positive impacts on natural ecosystems, the vast majority are negative or extremely negative (Figure 1.1). Among the main threats to biodiversity (habitat loss, overexploitation, spread of exotic species, environmental change), habitat loss and associated fragmentation are the most serious ones. Rates of landscape modification and habitat fragmentation became so dramatic in the twenty-first century that few, if any, ecosystems remain untouched by the impact of human activity. Loss of natural habitat reached such critical levels (Figure 1.2) that we will never know how many species we have driven to extinction even before they have been described. The combined effects of invasions, altered disturbance regimes, changing climate, species loss, and ecosystem degradation has, at times, exceeded the ability of ecosystems to maintain their structure and function. While the ecosystem effects of individual drivers can usually be predicted, their combinations introduce a lot of uncertainty and complexity. Not surprisingly, historically authentic, coevolved biotic assemblages rapidly disappear, being replaced by new combinations of species living under environmental conditions that have no historical analogs.

Witnessing rapid loss of biodiversity, despite all the conservation efforts, and realizing that all the above anthropogenic effects

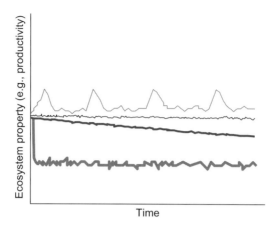

FIGURE 1.1 Four possible ecosystem trajectories under anthropogenic disturbance. Line width corresponds to the probability of occurrence: no change (black), periodic successional changes (green), gradual directional reversible change (blue), and sudden irreversible change (red). *A black and white version of this figure will appear in some formats. For the color version, please refer to the plate section.*

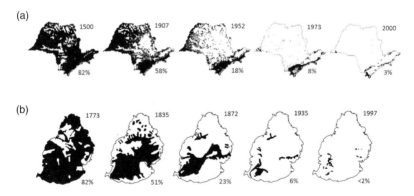

FIGURE 1.2 Two examples of fragmentation and disappearance of indigenous forest. (a) Atlantic coastal forest in São Paulo state of Brazil from 1500 to 2000 (after Oedekoven 1980, modified from Burkey 1997). (b) Mauritius island from 1773 to 1997 (modified from Florens 2013). The percentages denote the extent of remaining native forest cover.

move ecosystems outside of their historical range of variability should we accept the changes passively or keep trying to revert the altered habitats to historical conditions? Knowledge of historical conditions is essential in both conservation and restoration to identify reference states. However, historical records are either nonexistent or fragmentary and ambiguous. Nowadays, global climate change, the shift from a static to a dynamic view of ecological communities, the ambiguities of the past, and uncertainties of the future make the use of historical reference increasingly problematic and impractical.

This crisis of reference baselines is, to a large extent, responsible for a dichotomy of two global views of nature conservation, being forward and backward looking (Alagona *et al.* 2012). One wing of the conservation community, so-called "new environmentalists" or "new conservationists," declares that the whole baseline concept is obsolete, and suggests abandoning history, focusing not on the past but the future. The other wing is trying to refine or redefine the reference concept, often placing the baseline in the deep, distant past, adopting so-called "rewilding" or "resurrection ecology."

The forward-looking view proposes dropping the term "restoration ecology," seeing its historical focus as inappropriate, and wishes to replace it with terms emphasizing the new focus on intervention and creation of communities that have no historic analogs. These terms include "intervention ecology," "reconciliation ecology," "win–win ecology," and "futuristic restoration" (Choi 2004, 2007; Allison 2007; Halle 2007; Choi *et al.* 2008; Hobbs *et al.* 2011), with the goals of the latter being to repair or reinstate the key ecosystem services. This would be done through the creation of "novel ecosystems" (Hobbs *et al.* 2013) that may contain new combinations of species as a result of deliberate or inadvertent introduction, anthropogenic disturbance, changes in land use, pollution, or rapid climate change. The concept of the "novel ecosystem" goes back to the paper of Chapin and Starfield (1997) describing Arctic tundra transitioning to boreal grassland steppe under an altered climate and fire regime. This term has since been applied to a diverse range of ecosystems, e.g., mixed exotic–native forests established on degraded lands in many tropical regions, tussock grasslands that replaced forests in New Zealand or stands of exotic pines that replaced fynbos in South Africa (Lugo and Helmer 2004; Hobbs *et al.* 2006). All these ecosystems have one thing in common – they have no analogs in the past ecosystems, and are expected to dominate in the future. While many conservationists consider this an ecological disaster, the "new conservationists" view this change optimistically, as an opportunity to build "a new, more positive and forward-looking environmentalism" (Marris *et al.* 2012) that will secure ecosystem goods and services rather than species from extinction. The latter movement tries to shift the emphasis from preserving biodiversity *per se* to creation and management "of those natural systems that benefit the widest number of people, especially the poor" (Kareiva and Marvier 2007; Kareiva *et al.* 2012). This concept, welcoming free migration of any species across the globe and utilization of the last remaining pristine habitats until their complete disappearance, gets considerable support from the public, mostly a lay audience understanding much more about social rather than life or environmental sciences (Marvier and Wong 2012). Not surprisingly, we are witnessing the rapid proliferation of these views in the literature (e.g., Marris 2011; Theodoropoulos 2013; Kirksey 2015; Pearce 2015; Emmett and Nye 2017). Clearly, this movement has no sentiment for preserving species that do not fulfill an important functional role in novel "natural systems." Although advocates of the novel ecosystem concept state that it does not abandon the conservation of historic habitats and endangered species as central goals of natural area management (Kueffer *et al.* 2013), those people whose activities interfere with traditional nature conservation perceive these novel ecosystems in exactly that way (examples can be found in Simberloff *et al.* 2015).

Thus, while one of the two extreme views of conservation and restoration treats ecosystems as static and looks backward for a reference, the other one celebrates completely transformed landscapes and looks forward to ecological novelty. But, is there a third way of having as an ultimate goal to preserve biodiversity and for the latter compromising backward and forward views? It would be a way that accepts the important role of historical knowledge in conservation and restoration but which "seeks neither to recreate an imagined past nor to abandon history entirely" (Alagona *et al.* 2012). In my view, a third way exists, and I will try to convince the reader in the following chapters.

## I.4  A PROPOSED CONCEPT TO TACKLE NEW CHALLENGES

In the mid-twentieth century, no one could have imagined the extent of nature destruction that would happen in the following half century. The belief that the nature reserves and national parks could properly preserve biodiversity has started to vanish. The steadily accelerating loss of species and natural habitats in the last century has been an alarm call for many conservation biologists that existing practices of plant conservation are inadequate and new approaches are necessary. This inadequacy was recognized by Donald Falk in the late 1980s when he called for widening the scope of conservation approaches and their better integration. In Falk's view, there was a need to develop an array of conservation methods which would be complementary and mutually reinforcing (Falk 1987, 1990a, b).

Several attempts to reconcile conservation practices with the new realities of worldwide environmental degradation have since been made. All these concepts and ideas emphasized restoration of the degraded habitats as being crucial for conserving biodiversity in the twenty-first century (Dobson *et al.* 1997; Young 2000; Burney and Burney 2007; Wiens and Hobbs 2015). Young (2000) even predicted that ecological restoration would be the long-term future of conservation biology.

The idea of incorporating ecological restoration into the majority of conservation plans is a real leap forward and is the cornerstone of the concept that I would like to present as the third way mentioned in the previous section. The future will show whether this idea can become a new paradigm for plant conservation (Table 1.1).

As almost all natural habitats have already undergone some level of degradation due to human impact, virtually no pristine habitats still exist. Therefore, conservation of existing biodiversity cannot rely solely on protecting these habitats. This is a dead end. Instead, conservation should undertake a much more creative and flexible approach in which nature's regenerative capacity is enforced by efficient and complex management practices. In the new conservation concept presented and discussed in this book, interventions, assisted colonization, and experimentation are the crucial components. However, in contrast to the "new conservationists'" view, they have nothing to do with benefiting local human population ecosystem services. Similarly, in this concept, apparent inadequacies of the current conservation methodology (i.e., the low effectiveness of threatened species legislation and protection in nature reserves, and low success of plant reintroductions) require critical evaluation. Though, this is not to downplay the role of protected areas at the expense of botanical gardens and germplasm banks.

The concept presented in this book adopts the idea of creating partly novel (i.e., having species compositions that differ from historical analogs) ecosystems, but it is not a Trojan horse for traditional conservation. On the contrary, it can be adopted as an opportunity for the conceptually new conservation of threatened species and their habitats, which was always the focus of traditional conservation.

# 2 The Concept's Major Principles

The proposed concept does not cast doubt on the vital importance of protected areas in biodiversity conservation. Without protection from logging, grazing, poaching, and other activities that are detrimental to the natural habitat, mass species extinction is inevitable. However, passive conservation through strict protection that forbids any modifications of the protected habitat is not a viable strategy because populations of many threatened species have a low chance of recovery unless some crucial interventions are made. These interventions must either maintain the dynamic processes of ecosystems, such as succession, or release dispersal and establishment limitations responsible for the commonly observed lack of regeneration in many populations of threatened species, even in reserves and national parks. The interventions can target one or a few species (e.g., introduction of conspecifics) or the whole ecosystem (e.g., introduction of a suite of species that are functionally important for an ecosystem, thinning of vegetation, or creation of deadwood). On the other hand, if the habitat is unprotected, the interventions will make no sense. The restored but unprotected habitat will almost certainly once again become a victim of human destructive utilization.

Thus, I see interventions as crucial for protected areas to fulfill their goal of biodiversity conservation and these interventions should include many more options than removal of invasive species or disease/pest control, including creation of canopy gaps, thinning of pioneer species in favor of late-successional tree species, liberation of juveniles of threatened species from competing vegetation, and various forms of translocation. In translocation, not only are the reinforcement of existing populations and the reintroduction of extirpated species crucial, but also assisted colonization, i.e., introduction of threatened species beyond their current range. For assisted colonization to be successful, what is vital is not only a good knowledge of a species realized and potential niches, and possible effects of climate changes, but also wide-scale experimentation. Interventions and translocation of threatened species performed as replicated over space experiments in protected areas are the major management principles of the proposed concept.

A focus on improving a degraded habitat that is the cause of population decline rather than on improving the population itself through various forms of intervention makes the proposed concept similar to a particular form of ecological restoration called "restoration *sensu stricto*" (Aronson *et al.* 1993; Jackson *et al.* 1995). The aim of this is not reclamation of severely or completely degraded habitats, where the whole ecosystem must be recreated or altered to a desirable state, but restoring desired species composition, structure, and processes in a degraded but still functioning ecosystem. What makes the proposed concept different from restoration *sensu stricto* is the wide utilization of threatened plant species in restoration. This idea has the following logic. On the one hand, threatened species will have a future only in the restored habitats. On the other hand, if they belong to the functionally important plant category, a category needed to restore the ecosystem integrity, why not use them in the form of seedlings or saplings for ecological restoration? An area to be restored can be subject to either assisted establishment or assisted colonization depending on whether the target threatened species is still present there or there are no past records of its occurrence. Introduced plants can revitalize an existing population by improving availability of mates and pollinators' visitation, or decreasing genetic drift and thus its negative effects on offspring performance. In addition to this, the plants introduced into both the currently occupied and unoccupied but suitable habitat can improve the whole ecosystem and conditions for other species (vegetation cover, canopy structure, trophic web, mutualistic interactions) if the threatened species has a functionally important role in the eco-system (which many threatened tree and shrub species do). Decline in abundance or the disappearance of species more vulnerable to habitat degradation and fragmentation often result in the dominance of just a few species that are most tolerant to anthropogenic disturbances. As dominance is related inversely to species diversity, thinning of the dominant species in favor of threatened species can be a way to preserve both high species diversity and the threatened species. Liberating suppressed juveniles or replacement planting can be espe-cially relevant for those threatened species that are poor competitors and have been outcompeted by aggressive fast-growing and dispersal-efficient dominant species.

If we accept the idea that the threatened species can in principle be used in restoration, to bring this idea to a practical use we need a clear understanding of which of them can be used for restoration, under what circumstances, and in what way. Matching these species to a particular location must be based on accepting the widespread phenomenon of local adaptation accounting for predicted climate change effects. Production of the threatened species outplants in suf-ficient quantities and of sufficiently high genetic variation required for successful introduction can be done only through use of plant germplasm maintained and propagated in dedicated *ex situ* living collections. In addition, a site to be restored using the threatened species must have legal status preventing unauthorized anthropo-genic disturbance but permitting pre- and postplanting management interventions. These basic principles of the proposed concept will be covered in more detail below.

## 2.2   ACTIVE INTERVENTION

The current practices of conserving biodiversity *in situ* rely too much on protecting the area, i.e., on passive conservation (Heywood 2016, 2017). This approach appears to be generally inadequate and unlikely to stop the ongoing species loss; a much stronger emphasis on inter-vention is needed (Hobbs *et al.* 2011; Heywood 2016).

Active intervention is needed when dealing with ecosystems in which natural disturbance regularly returns the ecosystem to the earlier successional stage. Periodic disturbances are essential for early-successional plants and animals, as well as for maintaining ecosystem functioning and the overall biotic diversity. For many of the world's ecosystems, disturbances such as fires, floods, droughts, storms, and herbivory are natural components. Therefore, for the successful management of an ecosystem, it is important to recognize its historical disturbance regime and how this will continue to influence the ecosystem under ongoing or future changes. The goal of disturbance management should be to restore ecosystem processes (e.g., energy flow, nutrient, or water cycling) and to keep the ecosystem at the desired succession stage with the desired community structure and composition.

Regularly occurring fire is important for maintaining the open nature of such vegetation types as grasslands, shrublands, and open forests. In these environments, fire removes the thick litter and scrub layer and alters the microclimate and nutrient content of surface soil, enhancing productivity. Many plant species have specific adaptations to survive in a post fire environment, but are displaced by fire-intolerant competitors in areas protected from fire. Fire suppression causes fuel accumulation thus increasing the risk of high-intensity, more destructive fires, and a higher probability of disease and insect outbreaks. A typical management substitute for wildfires is prescribed burning. Using prescribed burning, it is possible to suppress non-native or invasive species, remove dead tree debris and leaf litter, control diseases and parasites, reduce the risk of severe wildfires, and enhance ecosystem productivity and diversity. The frequency of required prescribed burns varies among ecosystems and, in general, is higher in areas with higher annual precipitation. To control sprouting woody plants (e.g., oak or elm), in comparison with nonsprouters (e.g., juniper), more frequent fires are necessary. Similarly, more frequent burning is needed in tallgrass versus shortgrass prairies.

Herbivory is another important disturbance agent. Native rangeland plants have adapted to grazing by developing extensive root systems and rapid resprouting. Once the European and North American grasslands and open forests were grazed by herbivorous wildlife, but today, livestock have replaced native herbivores in many rangeland ecosystems. Like fire, grazing prevents the establishment of woody plants in favor of grasses and forbs in open vegetation types. For example, Smart et al. (1985) found that in African grassland excluding elephants was even more detrimental than fire suppression, and led to the rapid encroachment of woody *Acacia sieberiana* and loss of many species, including the original grassland dominants. Similarly, grazing prevents, or reverses the succession of African savanna to woodland. Rangeland ecosystems are adapted to a certain level of grazing, but not to overgrazing. Under both extremes, overgrazing or no grazing, a grassland will change into another plant community. If it is overgrazed, the destroyed vegetation will be replaced by a species-poor ruderal plant community. Conversely, under no grazing, the plant community will be invaded by woody plants and eventually become a forest. Given the impact of grazing on community structure, it has been recognized as a management tool in conservation applications (Hopkins and Wainwright 1989; Plassmann et al. 2010).

Conservation efforts can use the above two disturbance regimes, as well as others, in the reestablishment of ecosystem dynamics. Examples include increased flooding dynamics in riparian ecosystems (Moerke and Lamberti 2004), prescribed burning in boreal forests (Fule et al. 2004), long-term grazing in species-rich sand dunes (Plassmann et al. 2010), hand-cutting of woody vegetation in oak savannas (Reinhardt et al. 2017), mowing of wetlands and wet meadows (Kolos and Banaszuk 2013), and mowing and fertilization in nutrient-poor grasslands (Pechackova et al. 2010). Thus, disturbance can be used as a management tool to modify ecosystem dynamics for conservation/restoration purposes, for instance, to restart the succession, reassemble a community, or remove obstacles to species establishment.

The purpose of removing obstacles to species establishment and recruitment is especially relevant for threatened species. Among endangered plants there are many pioneer tree species that critically depend on disturbance. Many conifers, and especially threatened ones, require periodic disturbance for seed germination and seedling/sapling development. The disturbance agents for conifers include fire for giant sequoia *Sequoiadendron giganteum* (Swetnam 1993), windstorms for Douglas fir *Pseudotsuga menziesii* (Franklin *et al.* 2002), fire and windstorms for kauri *Agathis australis* (Steward and Beveridge 2010), and floods for baldcypress *Taxodium distichum* (Keim *et al.* 2006).

To understand the importance of disturbance as a process for maintaining an ecosystem in which endangered/relict pioneer species are an important component, giant sequoia is a good example. Adaptation of the life history of giant sequoia (*S. giganteum*) to a mixed-severity disturbance regime is described in York *et al.* (2010). Regeneration in this species occurs after a fire severe enough to kill multiple overstory trees and scorch the lower crowns of the largest sequoia trees. The following colonization is facilitated by cone serotiny, releasing huge numbers of seeds in pulses, and bare soils. Under adequate soil moisture conditions, seeds germinate and seedlings rapidly establish. In the following phase of competition with other species, canopy openings greater than 0.1 ha allow *S. giganteum* to outgrow all associated conifer species for at least the first 7 years. In the mixed conifer forests of the Sierra Nevada during the presettlement period, fires occurred at intervals ranging from 8 to 20 years (Stephens *et al.* 2007), giving a competitive advantage to rapidly growing *S. giganteum* saplings and young trees. In addition to these high-frequency, moderate-severity disturbances, high-intensity fires inevitably occurred at longer intervals. Unlike the other local conifers, individual *S. giganteum* can survive higher-intensity fires due to their ability to withstand very high levels of crown scorch, their thick and nonresinous bark, high crown base, and the production of epicormic sprouts. Those individuals that do persist through fire-caused disturbances quickly capture resources made available by the disturbance (light and soil nutrients) and start to dominate the canopy (and probably the below-ground space).

This persistence/release mechanism involving moderate or local high-severity disturbances may occur numerous times over the life span of long-lived species, leading to their dominant canopy position. Cessation of these naturally occurring processes (e.g., fire suppression) that started in the early 1900s in the United States have altered forests in which fires were naturally occurring (Stephens and Ruth 2005) and significantly reduced or completely halted regeneration of fire-dependent species. These forests need active intervention if past processes, structures, states, or trajectories are to be reinstated.

Nowadays, active intervention is also a necessity for ecosystems with minor or no periodical disturbance. There are two reasons for this. First, intervention in the form of removal and control of invasive species is required in many well-preserved protected areas because aggressive invaders are capable of encroaching and displacing native plants even in intact pristine habitats (Rose and Hermanutz 2004; Foxcroft et al. 2014; Braun et al. 2016; Florens et al. 2016; Bellingham et al. 2018). In 2007, according to a Global Invasive Species Programme report (De Poorter 2007), invasive alien species were recorded as a threat in 487 plant areas (PAs). Alien invasive species are perceived by the majority of managers of PAs in Europe as the second greatest threat to their areas after habitat loss (Pyšek et al. 2014), and more than half of US national park managers view them as a moderate or major concern (Randall 2011). Recognizing this threat, IUCN created the Invasive Species Specialist Group (ISSG, www.issg.org), one of five thematic specialist groups organized under the auspices of the Species Survival Commission (SSC). Managing invasive alien plants in PAs varies from limited-scale operations such as early detection and elimination or containment of nascent foci to total eradication in severely invaded large areas (Simberloff 2014; Tu and Robison 2014). For many PAs, it is not the risk of disturbing soil, canopy, or understory that is a concern, but that there are often too many invasive species and populations to control and never enough resources to manage all of them.

Accepting the idea of deliberate removal of certain plants even in a strictly protected ecosystem to preserve its health, we should take the next step of accepting as necessary the removal of both alien and native plants in certain cases for the same goal. The reason for this is wide-scale anthropogenic disturbance. Only a small number of nature reserves protect primeval rather than secondary forests. Virtually all ecosystems preserved in nature reserves have undergone some degree of human-induced changes in the past that disrupted species interactions and ecological processes (e.g., Chapman *et al.* 2010). In addition, exploitation, fragmentation, and environmental degradation have reduced the population sizes of many species below the viability threshold. The still existing in an area individuals indicate that the survival of the species at a specific location is principally possible, but only if the factors contributing to a population decline are identified and eliminated. Elimination of these factors in the majority of situations is impossible without active interventions (e.g., introduction of nurse plants, thinning of competing species, or rewilding). In many cases, suppression or removal of aggressive invasive species and outplanting of native species will succeed only if practiced simultaneously (Samways *et al.* 2010b; Griffin-Noyes 2012; Ammondt *et al.* 2013).

## 2.3   CONSIDERING LOCAL ADAPTATION

Historically, the concept of adaptive, habitat-related intraspecific variation goes back to Turesson (1922) who studied populations of several herbaceous species in transplant common garden experiments, and introduced the terms "genecology" and "ecotype," the latter being defined as a genetically distinct population adapted to specific environmental conditions. Later, Clausen *et al.* (1940, 1948) extended the study of adaptive population differentiation by using climatically different sites over a range of altitudes. Subsequent research has shown that ecotypic variation in plants is a widespread phenomenon and there is now a large body of literature on local adaptation and natural selection in plant populations (e.g., Linhart and Grant 1996; Briggs and Walters 1997; Kawecki and Ebert 2004; Leimu and Fisher 2008; Hereford 2009; Blanquart *et al.* 2013).

Because identification of appropriate sources of planting material is a key question in ecological restoration and conservation programs, an issue of local adaptation, i.e., a tradeoff in performance of plants across environments, is of great importance in both plant conservation and restoration ecology. Maladaptation to site conditions may result in high mortality, reduced growth, or poor seed set of introduced plants. A good example of maladaptation to local conditions which appeared years after planting is provided by Johnson *et al.* (2004). *Pseudotsuga menziesii* provenances introduced into Oregon, United States, in 1915, performed well until 1955, and then were hit by an unusual and prolonged cold period; while local sources survived, the nonlocals were either badly damaged or killed.

Spatially structured intraspecific phenotypic variation can result from either spatially heterogeneous selection causing adaptation of populations to local environmental conditions or limited gene flow (Endler 1977; Bradshaw 1984; Linhart and Grant 1996; Rasanen and Hendry 2008). If the observed spatial structure of phenotypic variation is due to genetic drift, a sampling design should capture as much species variation as possible, though this variation is neutral and not related to the performance of the plants across environments. If, however, local selection is important, recognition of spatial distributions of locally adapted phenotypes is of crucial importance for delimiting the geographic zones within which the locally collected seeds can be used for planting.

Below are considerations pertinent to whether it is likely that local is or is not best.

1. Local adaptation is expected under spatial but not temporal environmental heterogeneity (although exceptions exist such as bet-hedging germination strategy in temporally varying environments)

(Garcia-Ramos and Kirkpatrick 1997; Kirkpatrick and Barton 1997; Doebeli and Dieckmann 2003), and the probability of local adaptation increases with spatial scale of variation. Therefore specialists are expected in spatially varying but stable environments, and generalists are expected in frequently disturbed habitats (Kassen 2002).

2. Probability of local adaptation increases with decreasing dispersal rate, and therefore is less likely for species with higher seed and pollen dispersal distance. Local adaptation can be impeded by gene flow swamping potentially beneficial alleles (Kirkpatrick and Barton 1997; Bridle and Vines 2007), but also constrained by genetic drift and a lack of genetic variation for adaptively important traits that may happen under no gene flow (Holt and Gomulkiewicz 1997; Barton 2001; Kawecki and Ebert 2004; Alleaume-Benharira *et al.* 2006). Genetically depauperate populations will have reduced adaptive potential and can also be locally maladapted. Therefore, habitat fragmentation leading to small effective population sizes and low population connectivity will make local adaptation less likely (Broadhurst *et al.* 2008; Weeks *et al.* 2011).

3. Locals can be expected to be well adapted under pristine conditions, but adaptation is less likely when the environment has been modified (Jones 2013). Local adaptation can be conditioned by the species distant past and "genetic memory" (traits conferring adaptation to new climatic conditions) may be recalled when past climate conditions recur) (Montalvo *et al.* 1997). However, genetic memory is irrelevant when environmental conditions change in an unprecedented manner (Sgro *et al.* 2011). If human impacts and/or climate changes create novel conditions, there can be no preadapted genotypes anywhere (Williams and Jackson 2007), although for the widespread generalist species this is less likely and the superior nonlocal populations can exist across the species range (Johnson *et al.* 2004).

Abiotic conditions usually vary across the species range, as well as biotic interactions, and there is a general trend of increase in intensity of abiotic stress and/or interspecific competition toward the limits of species realized niches, with lower availability of suitable habitats, which results in lower and more fluctuating population densities, smaller population sizes, and reduced population connectivity than in the species interior (Hengeveld and Haeck 1982; Lawton 1993; Gaston 2003; Vucetich and Waite 2003). Because the populations at the species edge experience the most extreme conditions for the species, a very important question is: what is the conservation utility of these populations? Answering this question requires knowledge about whether the peripheral populations are locally adapted. There are two alternative views on predominant evolutionary factors that shape the genetic make-up of peripheral populations. One view is that peripheral populations may lack selectively important alleles as a result of their higher spatial isolation and smaller sizes leading to greater impact of genetic drift (Eckert *et al.* 2008). The alternative view is that peripheral populations do possess, due to strong selection under conditions marginal for the species conditions, locally adapted alleles that can be important for the species long-term survival (Lesica and Allendorf 1995).

Evidence suggests that in many species the peripheral populations are locally adapted (Davis and Shaw 2001; Volis *et al.* 2002, 2015, 2016a; Griffith and Watson 2006) and the more limited the gene flow from the species core, the more likely it is that they will be locally adapted (Mimura and Aitken 2010). Mimura and Aitken (2010) found that geographically isolated populations of Sitka spruce (*Picea sitchensis*) are better adapted to their environment than peripheral populations from a continuous species range. Thus, for populations on the edge of a species range, local is very likely to be best.

Experimental testing for local adaptation is done through transplantation experiments, as schematically shown in Figure 2.1. To provide reliable recommendations for conservation/restoration application these experiments should (1) maximize the number of populations, sites, life-history stages, and generations assessed; (2) use population growth rate (λ) as a response variable, or, if this is not feasible, test for fitness differences in response to selective agents at specific life-history stages concurrently, rather than sequentially; and (3) perform transplantation over environmental gradients/clines and multiple scales (Gibson *et al.* 2016). For species with no

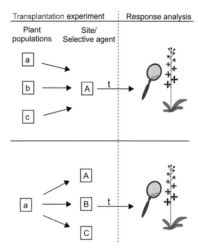

FIGURE 2.1 Schematic graph of local adaptation experiments (modified from Gibson *et al.* 2016). Lower-case letters (a–c) indicate plant populations; upper-case letters (A–C) indicate sites or selective agents; and t indicates time after the beginning of the experiment. The gray-shaded colors, underlying boxes (a–c, A–C) represent an environmental or geographic gradient. Ideally, material from multiple plant populations should be crossed with either multiple sites or selective agents.

experimental data on local adaptation, it will be safe to assume that plants in different parts of a range are locally adapted if a species is distributed along a steep environmental gradient or in different climatic/soil environments, has intraspecific forms, or exhibits reproductive isolation.

## 2.4   CONSIDERING CLIMATE CHANGE EFFECTS

Because local adaptation is common and is expected to result in a fitness tradeoff between local and nonlocal environments, decision-making in restoration and reintroduction programs about seed source traditionally focused on local provenancing (Keller *et al.* 2000; McKay *et al.* 2005; O'Brien *et al.* 2007). This is well justified for stable and spatially heterogeneous environments, but it may not work under rapidly occurring climate change because the strongest evidence for local adaptation is provided by cases of population adaptations to climate (Howe *et al.* 2003; Johnson *et al.* 2004; Savolainen *et al.* 2007). Thus the "local is best" sourcing practice needs adjustments to account for future climate change (Broadhurst *et al.* 2008; Sgro *et al.* 2011). The advantages of using local versus nonlocal sources are partly negated by changing climatic conditions.

FIGURE 2.2  Pictures illustrating global climate change effects. Boulder Glacier (Montana's Glacier National Park, USA) in 1932 (a) and in 2005 (b). Grinnell Glacier (Glacier National Park, USA) from 1938 to 2006 (c). Photos by US Geological Survey.

Multiple lines of evidence suggest widespread impacts of global warming on species and ecosystem processes, threatening the continued persistence of many plant populations (Thuiller *et al.* 2005; Grimm *et al.* 2013; Staudinger *et al.* 2013). Because of the rapidly changing temperature and precipitation patterns (Figure 2.2), species that are locally adapted to the current climatic conditions will have to adapt to novel conditions, or shift ranges to match the changing climate, in order to escape local or even global extinction (Parmesan 2006; Aitken *et al.* 2008; Kelly and Goulden 2008). The new ranges of many endangered species will be outside currently protected areas (Hannah *et al.* 2007; Heller and Zavaleta 2009). While many widespread and highly fecund species will just shift their ranges in response to climate change, and in the worst case will suffer adaptational lag for several generations, the situation will be much worse for those species that have low seed dispersal distance, infrequent seed production, or low fecundity. For these species, environmental conditions that change too rapidly will interrupt a continuous process of local adaptation making the residents less fit. For many, the change in climate may exceed the species innate capacity to adapt. In this situation the source material from geographically and ecologically distant populations may have adaptations more closely matching the new environmental settings. At least for some species, movement of genotypes to cooler and wetter conditions should be considered as an option (Carter 1996; Andalo *et al.* 2005).

The expected species range shifts are the contraction of the equatorward edge and the expansion of the poleward edge of the species suitable area (Parmesan *et al.* 1999; Hickling *et al.* 2005). The expanding and contracting parts of the species distribution are termed the "leading edge" and "trailing edge," respectively. Warming climate will promote adaptation and migration rather than cause any stress to the populations at the leading edge, because gene flow from the range core would introduce alleles preadapted to the warmer climate (Davis and Shaw 2001). Conversely, the populations at the trailing edge will experience a reduction in fitness due to gene swamping from the range core, and the further maladaptation of the immigrant alleles under the changed (warmer) climatic conditions. Therefore, the management implications for these two edges would differ substantially. For the leading edge, the most appropriate focus of conservation planning would be on reserve design, habitat protection, and restoration (Heller and Zavaleta 2009). The populations from the leading edge are most likely to be preadapted for ability to colonize because they include genotypes that were successful during population expansion after past climate changes (e.g., glacier retreat) (Cwynar and Macdonald 1987; Thomas *et al.* 2001). For this reason, they can be a source of the most suitable genotypes to be introduced beyond the current species range. In contrast, the trailing edge should be targeted for intensive sampling to preserve *ex situ* genetic diversity unique and most threatened by climate change (Hampe and Petit 2005).

The need to incorporate climate change considerations into nature conservation policy and management (Schmitz *et al.* 2015) has resulted in a suite of strategies concerning the assessment of species or habitat vulnerability to climate change (Heller and Zavaleta 2009; Mawdsley *et al.* 2009). Two major approaches are vulnerability indices ranking species in their vulnerability to climate change, and species distribution modeling predicting future versus current areas suitable for species (Rowland *et al.* 2011). Vulnerability is a function of factors that are extrinsic to a species such as its exposure to regional climate change and local habitat effects; and factors that are intrinsic to a species and determine its sensitivity to climate changes (Williams *et al.* 2008; Rowland *et al.* 2011). The latter include life history, and physiological and other traits that govern, either directly or indirectly (through biotic interactions), a species response to abiotic habitat conditions, including climatic conditions.

An example of a simple and widely used vulnerability index is the Climate Change Vulnerability Index (Young *et al.* 2012). This index evaluates species characteristics that determine climate sensitivity (e.g., physiological attributes, dispersal mechanisms, pollinator specificity, and geologic substrate fidelity), as well as climate exposure information to predict species vulnerability to climate change. As a result of analysis, a species is assigned one of six vulnerability scores: extremely vulnerable (EV), highly vulnerable (HV), moderately vulnerable (MV), presumed stable (PS), increase likely (IL), and insufficient evidence (IE). Species ranking using vulnerability indices is important for prioritizing species for conservation actions, and provides insights into why – and due to which life-history characteristics – certain plants are more sensitive to climate change than others (Rowland *et al.* 2011).

Species climatic tolerances determine to a large extent the range of environments which a species can occupy. This range, called environmental niche space, can be evaluated through species distribution modeling (SDM) (also known as ecological or environmental niche modeling or climate envelope modeling). SDM involves the use of spatially explicit environmental and species occurrence data through a geographic information system (GIS) to predict current or future areas suitable for species to grow (known as "suitable areas") (Guisan and Thuiller 2005; Franklin 2009; Guisan *et al.* 2013). Future climate projections employ mathematical models of the general circulation of the planetary atmosphere or ocean (GCMs) and emissions scenarios (i.e., how much greenhouse gas will be released into the atmosphere by humans).

Species distribution models use indicator variables (e.g., minimum and maximum temperatures) to set thresholds of climatic suitability for a given species, and then identify geographic areas where the modeled climatic suitability is above a selected threshold. The thresholds of climate suitability are generated by relating current locations of species occurrences to contemporary climatic values for these locations. Models are validated through a split of the species occurrence data and separate use of the two datasets for predicting suitable habitat area under current climate and comparing predictions with known occurrence data. Well-supported models can then be used to identify those parts of the current species range where climatic extremes will shift beyond the identified tolerance limits and assess the probability of species or population extinction under different climate change scenarios (Elith and Leathwick 2009).

Most of the available SDM models do not take into consideration species biological capacity for colonization, core–peripheral clines in population density and genetic diversity, differences between leading and trailing edge populations, intensity of gene flow, and degree of local adaptation in different parts of the species range. In most species, SDM accounting for local adaptation will apparently reveal much smaller sizes of climatic envelopes for individual populations than the climate envelope of the species as a whole. Different plant populations often have different ecological and evolutionary history which defines their trajectory of response to the climate change, and species with strong intraspecific genetic differentiation, e.g., Douglas fir (*Pseudotsuga menziesii*), display a variety of population-specific growth responses to climate change. Analysis of a growth response of Douglas fir populations to changes in climate (using the mean temperature of the coldest month as the most sensitive indicator of climate) revealed that these responses were seed source dependent. Transferred plants of all origins performed better in locations with warmer winters, but cold climate populations achieved their maximum growth at lower temperatures than warm climate populations (Leites *et al.* 2012).

Ideally, assessment of climate impact should be done using both approaches as complementary and cross-validating (Still *et al.* 2015), and this indeed has been done in a number of studies (e.g., Anacker *et al.* 2013). However, in practice, the appropriate approach for a particular species assessment method will be one for which the necessary information exists. For species with little published information on life history and environmental requirements, vulnerability assessment will be problematic. Likewise, for a species with only a few spatial occurrences, predictive distribution modeling will be unreliable. Another important aspect is the scale at which assessment is done. Vulnerability indices suit regional planning well due to their relative nature, but otherwise have limited utility. Additionally, because their output is not spatial, these indices cannot inform where a species might be most vulnerable to climate change within the management unit. SDM, on the other hand, is not limited by regional boundaries, can be used in the context of the broad species distribution, and can provide detailed maps of climatic suitability at the required spatial resolution.

A species distribution response to climate change can be determined by comparing the present and future suitable area size, calculating whether these areas overlap, and assessing a change in climatic suitability at known occurrence locations (Hijmans and Graham 2006; Schwartz *et al.* 2006; Maggini *et al.* 2014; Still *et al.*

2015). Range overlap is a useful measure to identify areas where the habitat remains suitable over time. In comparison to the change in range or range overlap, a change in suitability score does not require setting a threshold value. In addition, the suitability score is based solely on habitat suitability at known occurrence locations and does not require consideration of predicted areas of suitable habitat that are outside the current range. These three measures may show different patterns and, for this reason, Maggini *et al.* (2014) proposed integrating different SDM-based indicators of vulnerability to climate change via a composite "SDM score" index.

Under rapidly occurring environmental change, such as global climate change, the ability to respond through adaptive change, known as adaptive potential, in addition to the issue of local adaptation, must be given full consideration. The higher the adaptive potential, i.e., extent of genetic variation, the higher the chances that a population or species will survive through adaptation to new environmental conditions. Species that inhabit a variety of distinct environments or that are distributed along environmental clines should have higher adaptive variation than those with a narrow distribution confined to a particular habitat type. If an environmental gradient is present within a species range, its steepness is positively related to the probability of adaptive variation (Moritz 2002; Vandergast *et al.* 2008) and therefore to the species adaptive potential. The evolutionary adaptive potential of particular populations is also important. Key questions are what is the adaptive potential of populations at the species range edge and how different is this from the adaptive potential of those in the range interior? The answers have obvious practical implications because peripheral populations are assumed to have a crucial role in a species response to rapid environmental change (Etterson and Shaw 2001; Parmesan 2006; Bell and Gonzalez 2011).

Very few studies have tested the adaptive potential of edge versus core populations. Pujol and Pannell (2008) found a reduced potential to respond to selection after expansion and lower neutral genetic variation in edge populations. However, in a study by

Volis *et al.* (2016a), a reduction in neutral genetic variation was not associated with a concomitant decrease in adaptive quantitative trait variation of the edge populations. In that study, the edge populations demonstrated sufficient variation in adaptively important traits, despite both low effective population size and gene flow from the species interior.

Several studies have demonstrated that gene flow among edge populations can improve their adaptive potential (Lavergne and Molofsky 2007; Sexton *et al.* 2011; Volis 2011). Potential positive effects of this gene exchange include creation of new combinations of alleles with enhanced fitness (Volis 2011) and purging of deleterious mutations that lead to inbreeding depression (Pujol *et al.* 2009; Facon *et al.* 2011).

In predicting species range shifts and developing species conservation programs, it is important to take into account that some areas within a species range can be less susceptible to the change in climate than others. Refugia are locations where species have persisted through periods of climatic extremes for longer than in other areas due to more favorable conditions, and are usually the areas of the species maximum range contraction (Stewart *et al.* 2010). Such sites where species could find shelter during periods of regionally adverse climate represent habitats with relative eco-climatic stability (Fjeldsa and Lovett 1997; Hewitt 2000; Tzedakis *et al.* 2002; Médail and Diadema 2009), and therefore are thought to have been critical for maintaining biodiversity through the glacial–interglacial Quaternary climate fluctuations. Once climatic conditions became more favorable (i.e., in the Holocene), species with good dispersal ability and a wide potential ecological niche were able to migrate out of refugia and colonize formerly glaciated or inhospitable areas. However, the range of species that did not possess these characteristics (e.g., many Tertiary relics) would remain restricted to their sanctuaries (Linder 2001; Tribsch 2004) adding to the high value of these locations for protection. Conservation biology is currently highly interested in identifying potential climate refugia for species at risk from ongoing climate change (Shoo *et al.* 2011; Groves *et al.* 2012; Olson *et al.* 2012). For species that have persisted in long-term refugia in the preceding glacial–interglacial cycle, their existence suggests that, at least in some locations, they can tolerate major climate change (Hampe and Petit 2005), and that these locations are more likely than other areas to remain as refugia in the future (Gavin *et al.* 2014). These areas can be identified by a combination of paleoclimatic and fossil data, phylogeography, and SDM, and can be included as high priority areas in the conservation plans (e.g., Klein *et al.* 2009).

## 2.5  ASSISTED COLONIZATION

Assisted colonization can be equated with other terms used synonymously in the literature, such as facilitated adaptation, assisted migration, assisted range expansion, and managed re- or translocation (Hunter 2007; McLachlan *et al.* 2007; Hoegh-Guldberg *et al.* 2008; Hayward 2009; Heller and Zavaleta 2009; Richardson *et al.* 2009; Vitt *et al.* 2010; Aitken and Whitlock 2013; Havens *et al.* 2015). Assisted colonization has been defined as "the purposeful movement of species to facilitate or mimic natural range expansion, as a direct management response to climate change" (Vitt *et al.* 2010), and later as "the purposeful movement of individuals or propagules of a species to facilitate or mimic natural range expansion or long-distance gene flow within the current range, as a direct management response to climate change" (Havens *et al.* 2015). This latest definition reflects a broadening of the term, as any propagule movement beyond typical gene flow distances. This concept attempts to enrich the introduced gene pool, traditionally focusing on local genetic diversity, with genes from nonlocal populations, for example those presumably adapted to projected climate changes (Havens *et al.* 2015). The main idea of assisted colonization is movement of a target outside its current range with one of the following motivations: (1) maintain genetic diversity, (2) protect species from extinction, (3) mimic dispersal interrupted by human-created habitat barriers, (4) support the recipient ecosystem functionality and biodiversity, or (5) maintain a population used in natural resource extraction (Schwartz *et al.* 2012; Lunt *et al.* 2013). Actually, reintroduction and assisted colonization differ only in defining an operational species range, and in many cases we know too little about the species historic range to distinguish it from the current range as currently unsuitable. For this reason, there is little ecological justification to separate the two (Dalrymple *et al.* 2012; Vitt *et al.* 2016).

A strong argument in support of assisted colonization is that conventional conservation strategies will not provide sufficient protection from rapidly occurring environmental change (Vitt *et al.* 2010, 2016; Thomas 2011). For species with small populations scattered in fragmented landscapes and those for which distribution is confined to rare habitat types, the anticipated range shift can be beyond their tolerance limits. Plant communities found on patches of unusual bedrock and soil types, such as serpentine, gypsum, limestone, and dolomite, contain many endemic species and make large contributions to regional floristic diversity. For example, in California, 35% of the state's 1742 rare plant species grow on special substrates (Skinner and Pavlik 1994). Many of the world's rare plant species are similar edaphic endemics and they are known to be extremely vulnerable to alterations of their habitats (Shultz 1993; Skinner and Pavlik 1994; Briggs and Leigh 1996; Kelso *et al.* 1996; Sivinski and Knight 1996).

Narrowly distributed species that could thrive elsewhere but which cannot disperse across hostile anthropogenic or natural surrounding areas, or for which dispersal rate is too slow to contend with rapid climate change are among the most important potential targets for assisted colonization (Thomas 2011; Vitt *et al.* 2016). Many species whose climate-defined range is predicted to disappear due to climate change might still be able to survive if they could disperse to the areas with more suitable conditions. Thus, assisted colonization can be seen as the artificially increased dispersal capacity of endangered species. A good example is the Joshua tree (*Yucca brevifolia*) inhabiting the North American Mojave Desert.

A significant species range shift is predicted in the near future as a result of global warming, with only a few populations within the current range remaining viable. However, their fossils have revealed that dispersal of the species in response to increased aridity that happened ~11 700 years ago occurred at a rate of only ~1 to 2 m/year, which would be too slow to allow the natural expansion into new climatically suitable areas (Cole *et al.* 2011). In this situation a solution is relocation/assisted colonization (Figure 2.3).

Recruitment limitations and availability of suitable microsites for recruitment are two factors that, in addition to low seed dispersal distance and physical barriers for dispersal, limit range shifts in plants. Plants can expand their ranges only if climate change creates new microsites suitable for recruitment (Eriksson and Ehrlen 1992; Turnbull *et al.* 2000) and sufficient numbers of seeds arrive at these microsites (Primack and Miao 1992; Ehrlen and Eriksson 2000; Myers and Harms 2009). Therefore, rates of species range shifts in response to climate change will depend on what factors constrain recruitment: those that limit production and arrival of seeds to suitable microsites, or those that limit availability of suitable microsites for recruitment (Jones and del Moral 2009; Dullinger and Huelber 2011). According to the available literature, most species are seed limited and up to 50% of species current potential ranges are not filled due to factors limiting seed availability (e.g., Turnbull *et al.* 2000; Clark *et al.* 2007), and seed availability tends to decline toward species range edges either because of low abundance or reduced fecundity. If seed limitation is caused by low plant abundance, then range expansions can be slow because it will be possible only after newly establishing individuals at the leading range edge reach reproductive maturity, reproduce, and increase in abundance. Conversely, if seed limitation is due to low fecundity at the range edge, and fecundity increases with warming climates, then range shifts can be quite rapid (Case and Taper 2000). Kroiss and HilleRisLambers (2015) found that decline in seed availability toward species range edges is primarily due to low parent tree abundance rather than declining fecundity.

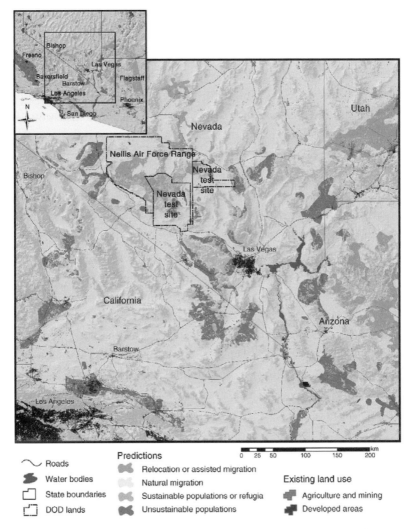

FIGURE 2.3 Assisted migration (= colonization) as a solution to the effect of climate change on Joshua tree (*Yucca brevifolia*) (from Cole *et al.* 2011). Four different colors designate areas where SDM predicts existing populations will become unsustainable (red), remain viable (orange), migrate naturally and persist (yellow), or conditions are suitable for assisted colonization in protected areas (green). DOD stands for the US Department of Defense. *A black and white version of this figure will appear in some formats. For the color version, please refer to the plate section.*

In long-lived plants, such as trees, range expansions in response to climate change are thus likely to lag, as long as generation times preclude rapid increases in plant density.

Low availability of microsites suitable for recruitment can severely limit a species ability to colonize new areas (Clark *et al.* 1999; Caspersen and Saprunoff 2005). The latter can be due not only to abiotic (e.g., climatic) factors, such as soil moisture and snowpack duration, but also to negative (competition, predation) and positive (facilitation) biotic interactions (Case and Taper 2000; Adler and HilleRisLambers 2008; Price and Kirkpatrick 2009). Seed production and germination rates are often not only highly variable across years but also uncorrelated, implying that successful recruitment requiring both availability of seeds and suitable germination conditions can be a rare event (Kroiss and HilleRisLambers 2015).

Implementation of assisted colonization is retarded by lack of clear guidelines on how to do it safely and effectively (Vitt *et al.* 2016). Opponents of translocating species claim that there is a high associated risk that these species will become invasive in their new ranges. However, most historically translocated species have remained rare within recipient regions and did not eliminate native species (Thomas 2011).

The available data proving or disproving the feasibility of assisted colonization are scarce. There are a few large transplant experiments that moved plants beyond the current species range. Schreiber *et al.* (2013) moved *Populus tremuloides* provenance 2300 km northwest from Minnesota to northeast British Columbia and did not find decreased survival or growth. Similarly, in a study on three species of Mexican pine (*Pinus devoniana, Pinus leiophylla,* and *Pinus pseudostrobus*) testing seedling survivorship after a shift in altitude 300 and 450 m upward, survival over 3 years was greater than 75% for 300 m and 56% for 450 m (Castellanos-Acuna *et al.* 2015). The Florida-endemic *Torreya taxifolia* was successfully translocated to the Appalachians to safeguard this critically endangered tree from fungal pathogens in its native range (Thomas 2011). Similarly, giant sequoia (*Sequoiadendron giganteum*) has been successfully planted outside its historical range (York *et al.* 2007). For three co-occurring species of *Lomatium*, perennial herbs, the experimental populations created outside the geographic range had equal or better survivorship compared to those within the current distribution (Marsico and Hellmann 2009).

The experience of New Zealand conservationists should also be mentioned. They have developed a successful strategy of establishing endangered species on offshore islands where invasive species are absent or eliminated, regardless of whether there are records of the species presence on that specific island (Clout and Craig 1995; Towns 2002). The rationale for doing so was that the islands to which the endangered species were translocated had similar conditions to those where the species used to occur, and that they might once have occurred there naturally.

To summarize, threatened species with small populations, highly fragmented ranges, long generation cycles, limited adaptation and migration capacities, and regeneration problems, or those suffering declines due to introduced pests, diseases, or invasive competitors should be first priority candidates for assisted colonization (Gallagher *et al.* 2015). This can be recommended from both the perspective of the habitat being restored and the species being conserved, within and beyond their current known range (Vitt *et al.* 2016).

## 2.6 EXPERIMENTATION

Accepting that active intervention is a necessary part of modern plant conservation, we must recognize experimentation as a major tool for a number of reasons.

Experimentation is needed for the efficient management of protected areas. For some areas, especially those that harbor critically endangered species with no viable relocation options, it would be desirable to apply a variety of experimental treatments among and within protected areas, so that some areas remain untouched, while others are managed. Comparison of the outcomes will make it possible to identify the best treatment to facilitate the transition of ecosystems along desired trajectories (Radeloff *et al.* 2015).

Management of protected areas may include creation of favorable microsites for the target species. Some habitats, e.g., calcareous grasslands or desertified shrublands, suffer from microsite limitation (Shachak *et al.* 1998; Wagner *et al.* 2016); however, artificial creation of such microsites involves a lot of uncertainty because the optimal levels of required intervention differ among species. Some species prefer drier microsites with strongly reduced competition, whereas others, whose seedlings are more vulnerable to desiccation, may prefer wetter sites. Thus, different types of disturbance of varying intensity need to be tested for working out the optimum required to create suitable conditions for the establishment of introduced species.

Experimentation is also needed to mitigate climate change effects. Studies of species–climate relations suggest that the amount and periodicity of precipitation and temperature–precipitation interactions determine species potential niche (e.g., Rehfeldt *et al.* 2006), and SDM can help to predict suitable locations for the species under both current and future climate. However, only experimentation will allow us to identify those locations where the species can establish viable populations.

Responses to climate are usually species specific, producing, as a consequence of range shifts and competition, complex and difficult-to-predict novel species combinations. As a result, species with previously nonoverlapping ranges under new conditions will reassemble into novel communities and ecosystems (Williams and Jackson 2007; Hobbs *et al.* 2009; Gilman *et al.* 2010). While many of these new ecosystems will be unsuitable for imperiled species, in some of them they may find a new home. For those threatened species for which habitat has been altered, reference habitat does not exist, and abiotic/biotic requirements are poorly understood, there is no alternative to experimental translocation (Roncal *et al.* 2012;

Menges *et al.* 2016; Vitt *et al.* 2016; Volis 2016c; Wendelberger and Maschinski 2016). Such introduction over a range of environments rather than within the historical distribution or single "preferred habitat" should increase the chances of successful establishment.

Experimental translocation is analogous to the creation of small founder populations in a habitat new for the species (Aitken *et al.* 2008). If, in addition to hospitable climate conditions, the biotic species requirements in the new habitat are met, these populations will become reproductively capable and regenerating. They can absorb long-distance pollen flow, which will make them genetically variable. If these founder populations get established and persist for sufficient time, they can become nuclei for range expansion. In the course of time, these populations have a reasonable chance of becoming (1) sufficiently large and genetically variable to be viable and (2) locally adapted (Petit *et al.* 2003; Mimura and Aitken 2007).

At a lower geographic scale, successful introduction requires knowledge of the species regeneration niche requirements and preferred microhabitats. The latter can be reliably determined only through experimental introduction across a range of microhabitats and monitoring of the introduced plants for a sufficient period of time (e.g., Volis *et al.* 2010; Menges *et al.* 2016; Wendelberger and Maschinski 2016).

Large-scale restoration projects, particularly when using rare and threatened species with limited seed availability, should always be preceded by experiments investigating species- and treatment-specific responses (Cabin *et al.* 2002; Brudvig *et al.* 2017). The importance of experimentation in restoration is recognized by "systemic experimental restoration," which is the establishment of mosaics of replicated treatments within mosaics of habitats that are purported to support metapopulation dynamics and population colonization within altered landscapes (Howe and Martinez-Garza 2014). This approach is an alternative to the commonly applied "best available practice" of establishing a single combination of a limited number of plant species. Created in this way, plant communities are expected to differ in species composition among the introduction microsites, mostly in the presence and abundance of rare species, and be potential sources of colonization for each other. The experimental approach is highly relevant to restoration where the goal is to rehabilitate existing habitats or create new habitats for threatened species because introductions of such species in general have a low chance of success (Maunder 1992; Seddon *et al.* 2007; Godefroid *et al.* 2011; Dalrymple *et al.* 2012; Drayton and Primack 2012). Thus, broadening the list of species introduced in different combinations and treatments (Howe and Martinez-Garza 2014) and replicating introduced populations over time and space (Guerrant 1996) is a way to maximize the likelihood of reintroduction success. In restoration, given high variability in seedling survival over even small spatial scales (Gomez-Aparicio *et al.* 2005a; Garrido *et al.* 2007; Albrecht and McCue 2010; Bontrager *et al.* 2014; Inman-Narahari *et al.* 2014), trials conducted at multiple sites are becoming more and more popular (Butterfield 1996; Calvo-Alvarado *et al.* 2007; Wishnie *et al.* 2007; Zahawi and Holl 2009; Holl *et al.* 2011; Alvarez-Aquino and Williams-Linera 2012; Roman-Danobeytia *et al.* 2012; Encino-Ruiz *et al.* 2013; Martinez-Garza *et al.* 2013; Yang *et al.* 2013).

## 2.7    RESTORATION OF THREATENED SPECIES HABITATS

We are witnessing the almost complete disappearance of original habitats in many countries, with even nature reserves preserving ecosystems that underwent different levels of anthropogenic disturbance. For example, in South Asia a quarter of the land inside protected areas is classified as human modified (Clark *et al.* 2013). Because this disturbance has altered once existing species assemblages and biotic interactions, many extant populations of threatened plant species do not regenerate naturally even in the most strictly protected areas. Humans can destroy the regeneration niche by making unsuitable either the reproductive, dispersal, or recruitment niches, e.g., by preventing germination in altered soil or light conditions, or by extirpating pollinators/seed dispersers and thus making seed production/dispersal impossible. While short-living species in these new conditions will soon disappear, longer-living species can temporarily persist as nonrecruiting adults ("living dead" *sensu* Janzen 2001). Although not without exceptions, relict species are long-lived plants with very localized distributions and are usually represented by small and isolated populations with limited or completely absent recruitment. For example, in the Mediterranean mountains of southern Spain, regeneration in relict populations of *Pinus sylvestris* is prevented by high levels of seed predation by rodents and birds (Castro *et al.* 1999) and of *Frangula alnus* subsp. *baetica* by browsing from introduced game animals (Hampe and Arroyo 2002), while populations of *Taxus baccata* do not regenerate because they are strongly dependent on currently lacking or too scarce understory shrubs that facilitate the germination of seeds and establishment of seedlings (Garcia *et al.* 2000). A relict tree, *Zelkova abelicea*, endemic to the island of Crete, has very poor regeneration because of browsing and trampling of seedlings and saplings by goats (Kozlowski *et al.* 2012). In China, even in protected areas, regeneration of many relict species is prevented by activities of local human residents, such as clearing understory vegetation to promote the growth and spread of bamboo stands (*Davidia involucrata*; Qian *et al.* 2018) or cultivation in the understory of cash plants (*Metasequoia glyptostroboides*; Tang *et al.* 2011).

For this type of species, there are three possible solutions, as outlined by Young *et al.* (2005): (1) We can restore the once existing population's environmental conditions to repair the lost links in the recruitment chain and give the population the chance of survival (Figure 2.4). (2) We can extend the dispersal niche by promoting the species dispersal into the environments where that species had not been previously recorded (Figure 2.4). (3) We can create more living dead nonrecruiting populations by planting individuals in sites where seeds cannot germinate or seedlings develop, but saplings and adults can thrive (Figure 2.4) (Young *et al.* 2005). Unfortunately, although it

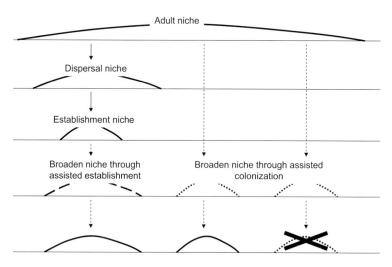

FIGURE 2.4 Ontogenetic niche and the possible effects of restoration activities (from Young 2005 with changes). The x-axis is a gradient of an environmental parameter that defines niche space. Seeds can disperse to more sites than are suitable for establishment. Conversely, dispersal limitations can prevent the arrival of seeds at some suitable sites. Restoration activities may broaden the recruitment niche through assisted establishment (dashed lines) or the dispersal niche through assisted colonization (dotted lines). Assisted colonization can result in establishment with regeneration or without it; in this case, colonization cannot be considered successful.

has never been a goal, we succeed in the latter much more than in the former two options. Many introductions of rare and endangered plant species have resulted in the establishment of the introduced individuals but without producing new generations (Godefroid *et al.* 2011; Dalrymple *et al.* 2012). This should be an alarm call that apparently focusing only on a focal species in conservation projects rarely works and needs a shift to another conservation paradigm.

Delayed extinction applies not only to living fossils, relict, and other long-lived plant species. Short to substantial delay but inevitable local extinction can occur with any species following habitat loss or degradation if the current landscape structure and habitat quality do not permit its long-term survival. A species can initially survive alteration or fragmentation of its habitat but later becomes extinct, even under no further habitat deterioration, if a threshold in habitat quality, area, and connectivity is crossed (Hanski and Ovaskainen 2002). This time delay in extinction is called *relaxation time* (Diamond 1972) and the phenomenon that declining populations will eventually go extinct in fragmented or degraded habitats has been described as *extinction debt* (Tilman *et al.* 1994). At the community level, this phenomenon predicts a decline in species richness as a result of local species extinction following alteration of the habitat (Tilman *et al.* 1994; Tilman and Lehman 1997). Empirical support for extinction debt across different habitat types and taxonomic groups (including species with fast turnover times) is constantly growing (Vellend *et al.* 2006; Bommarco *et al.* 2014; Gonzalez-Varo *et al.* 2015; Otsu *et al.* 2017; Otto *et al.* 2017).

The widespread phenomenon of extinction debt has important implications for the conservation of biodiversity and restoration. First, it suggests that, under present habitat loss and fragmentation, populations of many species are on a deterministic path to extinction even without any further habitat deterioration. Second, the transient persistence of species in altered and fragmented habitats may cause underestimation of the number of threatened species in the area and quality of a habitat needed to conserve these taxa, and overestimation of the species richness that the altered habitat can support in the long term (Hanski and Ovaskainen 2002). Third, the extinction debt can be paid by either allowing the species to go extinct or by restoring the habitat, before the species disappeared, in such a way that the threshold conditions are met again. As long as a species that is predicted to become extinct still persists, there is time for crucial conservation interventions leading to the improvement in habitat quality and connectivity (Kuussaari *et al.* 2009; Wearn *et al.* 2012; Hylander and Ehrlen 2013; Olivier *et al.* 2013; Newmark *et al.* 2017). On the other hand, assisted colonization of other habitats can prevent the species global extinction even if the local extinction debt is paid. A meta-analysis of the mortality of plants introduced outside versus inside the historic range showed remarkably good plant performance (expressed as a mortality risk ratio, RR) in the former ($RR_{\text{sites outside historic range}}$ = 0.177, $CI_{0.95}$ = 0.053, 0.588, $n$ = 7; $RR_{\text{previously extant sites}}$ = 0.827, $CI_{0.95}$ = 0.646, 1.059, $n$ = 23; $RR_{\text{sites within historic range}}$ = 0.665, $CI_{0.95}$ = 0.578, 0.763, $n$ = 99) (Dalrymple *et al.* 2012). Thus, large-scale restoration and assisted colonization should be viewed as complementary crucial aspects of the conservation of biodiversity allowing a significant number of species to recover. The combination of such approaches as analysis of spatial distribution and (meta)population viability modeling can make possible, on the one hand, estimation of the amount, type, and spatial locations of the crucial habitat, and, on the other hand, the appropriate population maintenance, or habitat restoration interventions that will most favor species persistence in the long term.

## 2.8   UTILIZATION OF THREATENED SPECIES IN RESTORATION

Plant species can be at risk because of their initial natural rarity (i.e., inherently adapted to low abundance), as a result of human disturbance, or both, and it can be difficult to determine the contribution of these two factors with certainty (Weller 1994). Among the currently rare species, the proportion of inherently rare species, i.e., those having small populations everywhere, appears to be small (Rabinowitz *et al.* 1986) and most of the currently rare species are likely to be "anthropogenic rarities" (Fiedler and Laven 1996) or "new rares" (Huenneke 1991; Oostermeijer *et al.* 2003). These species might have been common in the past but nowadays have become threatened as a result of higher vulnerability, in comparison with other species, to the alteration of once existing habitats and biotic interactions. Species in which decline is not due to intrinsic causes such as extreme habitat specialization, but to extrinsic causes (e.g., invasion of non-native species, livestock grazing, fire suppression, or land conversion), may turn out to be useful for restoration of altered or partly degraded habitats. Moreover, uncertainty about the cause of rareness for many threatened species, but with a high probability that they are "new rare" species, allows substantial broadening of lists of candidate species for habitat restoration with threatened species. Some of these species, after introduction into numerous apparently suitable locations within their potential distribution range, can become not only sporadic, but subdominant, or even dominant species in some of the restored ecosystems.

Comparisons with congeneric common species show comparable, albeit lower, fruit and seed set, ovule quality, and germination rates (e.g., endangered *Manglietia grandis* versus nonendangered *Manglietia hookeri*, Fu *et al.* 2009; or extremely rare *Amsinckia grandiflora* versus common *Amsinckia tessellata*, Carlsen *et al.* 2002). Similarly, a comparison of rare with common congeners shows comparable, although more irregular, distributions of individuals among size classes in rare species, apparently due to more fluctuating reproductive output and recruitment into adult size classes (Byers and Meagher 1997; Kelly *et al.* 2001).

A key, or at least a clue, to identification of threatened species having good potential for habitat restoration is their viable population demographic structure, in which all life-cycle stages and size categories are present and the size–frequency distribution approaches an inverse J-shaped curve. This population structure with relatively high abundance of juveniles relative to adults is characteristic of healthy, nondeclining populations (Poorter *et al.* 1996; Condit *et al.* 1998; Feeley *et al.* 2007). Good natural regeneration is an especially important criterion, and at least some threatened species satisfy this requirement. For example, in a study by Chien *et al.* (2008), all six threatened subtropical tree species showed vigorous natural regeneration in a protected environment. In several studies, rare and endangered tropical tree species exhibited high germination and seedling survival rate, and the density of young individuals was comparable in logged and unlogged forests (Guarino and Scariot 2012; Ngo and Hölscher 2014). All examined populations of the endangered trees *Craigia yunnanensis* (Gao *et al.* 2010) and *Aquilaria crassna* (Jensen and Meilby 2012) had a vigorous demographic structure. Endemic to Nusakambangan Island of Indonesia and critically endangered due to illegal logging, the tree *Dipterocarpus littoralis* has very few mature trees but a continuous recruitment (Robiansyah and Davy 2015).

In some threatened species regeneration can be apparent. Lack of individuals in some size classes is not necessarily an indication of repeated regeneration failure as a result of altered biotic and abiotic environment, but can be due to direct anthropogenic effects (e.g., seed or seedling collecting, grazing, or logging) (Gomez-Aparicio *et al.* 2005b; Souza 2007). Once the anthropogenic pressure is relieved, the true viable population demographic structure becomes evident.

The reasons for limited usage of threatened species in conservation projects include the requirement for large quantities of seed and a lack of knowledge of the species reproductive biology and efficient methods of propagation and planting. Knowledge about propagation by seeds or cuttings, ideal individual size, and planting time are vital for successful establishment. Although still very limited, this knowledge is steadily accumulating for rare and endangered species (e.g., Iturriaga *et al.* 1994; Adjers and Otsamo 1996; Sakai *et al.* 2002; Danthu *et al.* 2008; Moreira *et al.* 2009; De Motta 2010; Kay *et al.* 2011; Martins *et al.* 2011; Ratnamhin *et al.* 2011; Koch and Kollmann 2012; Castellanos-Castro and Bonfil 2013; Gratzfeld *et al.* 2015; Lu *et al.* 2016). Once the necessary knowledge is acquired and protocols are available, comparable common species cost per seedling will make restoration practitioners more likely to incorporate rare and threatened species into their plans (Rodrigues *et al.* 2011), because availability of seedlings rather than the cost of a seedling per added species is an obstacle to planting high-diversity species pools (Aronson *et al.* 2011).

Determination of locations suitable for a threatened species locations requires good knowledge of its historical range (when available), climatic, and edaphic preferences, and interactions with other species. Once environmental requirements of the target species are known, suitable locations for each species within the target area can be

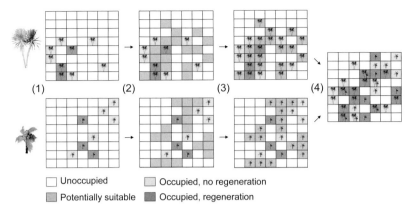

Unoccupied

Potentially suitable

Occupied, no regeneration

Occupied, regeneration

FIGURE 2.5 Necessary steps of conservation-oriented restoration using threatened plants: (1) mapping and population demographic survey; (2) species distribution modeling and analysis of regeneration niche; (3) multiple site introduction trials; (4) monitoring of establishment. The introduction trials may result in no establishment, establishment without regeneration, and establishment with regeneration. Only the introductions that resulted in establishment with regeneration can be considered successful. *A black and white version of this figure will appear in some formats. For the color version, please refer to the plate section.*

identified and mapped using SDM, and, after maps are superimposed, the species assemblages can be defined for any spatial scale (Figure 2.5). Identification of suitable sites for the target threatened species can be based on either environmental variables (used in SDM) or on species co-occurrence. A study by Baumberger *et al.* (2012) showed that the predictive power of the model based on co-occurring plant species was similar to that of the model based on environmental variables but the latter model showed more false positives than the former one. Thus, the presence of frequently co-occurring indicator species can be used to narrow down the range of suitable translocation sites predicted by SDM.

Some endangered species depend heavily on disturbance, such as some large-sized and light-demanding tropical conifers (e.g., *Araucaria angustifolia*, Souza 2007) or angiosperms (e.g., *Afzelia xylocarpa*, Vlam *et al.* 2014). Although demographically resembling remnant populations that lack significant constant regeneration (Ogden and Stewart 1995), these species do not actually suffer from either poor seed production or dispersal limitation. These species establish following severe disturbances, and then become common or even dominant vegetation components for long periods of time (even centuries). For these species, rare severe disturbance events such as windstorms or fire create large canopy openings and a disturbed understory promoting recruitment. These long-lived pioneers differ from gap-colonizing short-lived pioneers in having slower adult growth rates, large seed sizes, and much longer life spans (centuries) (Ogden and Stewart 1995). Dependence of these species on disturbance makes their usage in restoration a relatively easy task through the creation of canopy gaps and the planting of seedlings in forest openings.

In fact, restoration trials have started to include threatened species (Adjers *et al.* 1995; Alvarez-Aquino *et al.* 2004; McNamara *et al.* 2006; Avendano-Yanez *et al.* 2016; Subiakto *et al.* 2016). There are also a few case studies that both use threatened plant species in active restoration and also evaluate plant establishment over a reasonable time period (Morgan 1999; Shono *et al.* 2007b; Cordell *et al.* 2008; Millet *et al.* 2013; Subiakto *et al.* 2016). These studies show suitability and similar prospects of the establishment of threatened plant species in comparison with nonthreatened species in restoration projects. Three studies described in Section 6.4, performed in South-East Asia as reforestation trials in degraded grasslands, abandoned farmlands, and cleared forests, showed that many of the tested threatened tree species were among the best performing species (Shono *et al.* 2007; Schneider *et al.* 2014; Subiakto *et al.* 2016). A study of García-Hernández *et al.* (2019) demonstrated that 12 forest tree species, of which nine were threatened, all exhibited high seedling survivorship and good growth over a 2-year period in nine degraded forests of southern Mexico to which they were introduced. In another study performed in Mexico, introduction of the threatened tree *Oreomunnea mexicana* into a secondary forest was successful not only when done through the planting of nursery-grown seedlings but also via seeding (Atondo-Buen *et al.* 2018). In Vietnam, 12 native tree species, of which 11 were threatened, were used to convert a post-logged plantation into secondary forest with high conservation value. Four years after planting, the survival rate of seedlings of all the species but one was above 70% (Millet *et al.* 2013). Although natural regeneration was not assessed in this study, which precludes calling it a complete success, high survival of the majority of the species even in a degraded environment is encouraging.

When talking about the usage of threatened species in restoration, the *inter situ* restoration approach must be mentioned. The supposed meaning of the *"inter situ"* term is "between sites" and therefore to be more grammatically correct the approach should be called *"inter situs"* restoration (Heywood 2014; Volis 2017a). This approach was originally proposed as an off-site collection maintained within the natural habitat (Husband and Campbell 2004). Such extremely general description allowed different interpretations, but current use of this term appears to be limited to the interpretation of Burney and Burney (2007) as establishing imperiled species and associated biotic interactions outside the current range but within the past range of the species (Burney and Burney 2007). According to Burney and Burney, this approach should be applied to a location with some degree of environmental degradation, and the priority targets must be locally extinct and/or still present but highly threatened species. The *inter situ* actions include intensive horticultural and agricultural management, invasive species control, and protection, with gradual withdrawal of care for the reintroduced plants. The goal of this approach is that the reintroduced species complex eventually becomes a natural ecosystem, approaching, as closely as possible, one that existed in that area in the past. In short, the proposed strategy is ecological restoration with explicit conservation goals, utilizing threatened species.

FIGURE 2.6 The Makauwahi Cave restoration project in the Hawaiian
Archipelago. (a) and (b) show an aerial view of the Makauwahi Cave
reserve with the restored area in the center, and a section of a nature
trail providing information to visitors about the restoration techniques
and the plants used, as it winds through the restorations. (c) The
former sugar cane field in front of Mt. Haupu that has been restored
using native Hawaiian threatened and rare plants. Despite use of heavy
equipment for site preparation and planting in rows like crops, mixing
many species and several growth forms in each row allows the restored
site to become indistinguishable from a naturally regenerating forest
in a decade or less. Photos by Ellen Coulombe, Lida Pigott Burney, and
David Burney. *A black and white version of this figure will appear in
some formats. For the color version, please refer to the plate section.*

Examples of the application of this strategy include the Makauwahi
Cave restoration project in the Hawaiian Archipelago (Burney and Burney
2007, 2016) (Figure 2.6), restoration of offshore islets in the Seychelles
Archipelago (Kueffer *et al.* 2013), and conversion of deforested lands
of Rodrigues and Mauritius Islands of the Mascarene Archipelago into
tropical forest (Owen Griffith, personal communication) (Figure 2.7).

FIGURE 2.7 Recreation of the tropical forest using native threatened and rare plant species on Rodrigues (a, b) and Mauritius (c) islands of the Mascarene Archipelago. The upper photos show the landscape view (a) before and (b) during restoration. Photos by Arnaud Meunier and Christine Griffiths. *A black and white version of this figure will appear in some formats. For the color version, please refer to the plate section.*

## 2.9 INTEGRATION OF *EX SITU* AND *IN SITU* STRATEGIES

Active conservation in the majority of cases requires introducing plant material, predominantly in the form of seedlings or saplings. Plant germplasm maintained and propagated *ex situ* can be used for this purpose. The potential of *ex situ* collections for *in situ* actions via storing and propagating plant material has been recognized for a long time (Cugnac 1953; Raven 1981), but obvious space and other limitations of *ex situ* collections for achieving this are also well known (Hamilton 1994; Schoen and Brown 2001; Maunder *et al.* 2004b; Volis and Blecher 2010) and will be discussed in Sections 3.7 and 3.8. These limitations motivated the search for efficient integration of *ex situ* and *in situ* approaches by creating living collections of the required capacity outside botanic gardens and arboreta. For

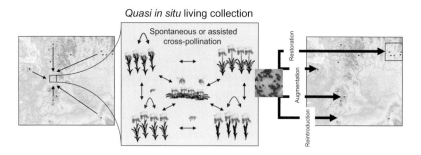

FIGURE 2.8 Role of *quasi in situ* living collections in conservation-oriented restoration. A *quasi in situ* collection (gray rectangle) is established in a protected location having the same environmental conditions with the extant species populations (circles). Seeds that result from spontaneous or assisted cross-pollination within a *quasi in situ* site are used to produce the outplants for *in situ* actions.

example, "establishing botanic garden facilities within protected areas" (Maunder *et al.* 2001b) and "that documented gene banks be established adjacent to protected areas" (Maunder *et al.* 2002) were proposed by Maunder with colleagues but not elaborated on.

More recently, a concept called *quasi in situ* has been introduced. This concept proposes the creation of depositories of genetically variable source material for species of conservation concern as living plants maintained under natural or close to natural conditions (Volis and Blecher 2010; Volis 2016c) (Figure 2.8). The *quasi in situ* site maintains a living collection of individuals from populations sharing the same climatic zone and biotic/abiotic environment, has natural or seminatural conditions, and is legally protected. The detailed guidelines for the choice of material, planting, and management of the *quasi in situ* living collections will be presented in Section 3.9.

Besides preserving species genetic variation, *quasi in situ* living collections can be a reliable source of seeds for *in situ* conservation and restoration projects. Seed banks cannot fulfill this task due to space limitations and problems with storing nonorthodox seeds, while collecting large quantities of seeds in natural populations is undesirable because of the negative impacts of seed harvesting on local population dynamics (Broadhurst *et al.* 2008).

Plants grown in *quasi in situ* living collections can be expected to be locally adapted and properly represent the species genetic variation, i.e., contain adapted and genetically different individuals. As all plants in the collection originate from the same environment, there are no maladapted genes to participate in recombination and segregation, and cross-pollination of these plants should not lead to the breakdown of coadapted gene complexes or dilution of local adaptation. Therefore, the offspring of cross-pollination occurring in the collection suit *in situ* actions well and can be produced and collected in the large quantities needed for translocation or conservation-oriented restoration to be successful.

## 2.10 LEGISLATION ALLOWING ACTIVE INTERVENTIONS

Habitat protection has a crucial role in plant conservation because legal protection prohibits activities that can damage, destroy, or modify natural habitats. However, even if the transition from designation of areas as nature reserves or national parks into actual protection is straightforward and well done (which often is not the case), this does not guarantee a halt to threatened species population decline or further degradation of the habitat (e.g., Tang *et al.* 2011; Bissessur *et al.* 2017; Qian *et al.* 2018). The latter is often impossible without well-organized interventions and clear recovery criteria to follow. But these interventions must be allowed by the protection status of the target site. Thus, a restoration site must have a proper protected area category, allowing management through active interventions but forbidding any unauthorized activities.

Since 1994, IUCN has recognized, based on management objectives, approaches, and regulations, six different protected area categories ranging from sites where human access is banned to landscapes with settled human communities and the permitted extraction of natural resources (Dudley 2008). These categories are used for the purposes of planning, setting regulations, and negotiating land and water use. They reflect a complexity of situations in which decisions about planning and management must be made. Thus far, assignment of a category has depended more on how the management authority intended to resolve an inevitably present conflict between nature and local or visiting human populations rather than on the other criteria. In defining the protection categories, interventions other than those induced by undesirable human activities traditionally were ignored, and as a result are not relevant for most of the existing six categories (I–VI). For example, categories Ia and Ib specify that the sites are strictly protected from human influence, with category Ib being less restrictive for tourist visits. These areas, by definition, do not require "substantial and on-going intervention to achieve the conservation objectives" (Ia), and allow for only "low-impact minimally invasive educational and scientific research activities, when such activities cannot be conducted outside the wilderness area" (Ib) (Dudley 2008). The areas in Ib can include "somewhat disturbed areas that are capable of restoration to a wilderness state, and smaller areas that might be expanded or could play an important role in a larger wilderness protection strategy as part of a system of protected areas that includes wilderness, if the management objectives for those somewhat disturbed or smaller areas are otherwise consistent with the objectives set out above."

Although not stated explicitly, this definition assumes that "restoration to a wilderness state" will occur naturally, as a result of halted human-caused disturbance. Categories II and III emphasize the role of protection against human activity to preserve an ecosystem (II) or some natural features (III), with visitation and recreation usually being encouraged. Categories V and VI assume continuous human interaction with nature by some form of land use. The only category allowing and promoting active conservation management is category IV, which has the aim of protecting particular species or habitats with "management reflecting this priority," and "many category IV protected areas will need regular, active interventions to address the requirements of particular species or to maintain habitats." This is the only category that suits conservation-oriented restoration, although, by definition, it "provides a management approach used in areas that have already undergone substantial modification, necessitating protection of remaining fragments, with, or without intervention." This definition somewhat limits use of category IV to cases when the target areas are either small fragments of natural habitats surrounded by hostile environment or are degraded to some degree. However, active management can be necessary for the last remaining populations of endangered species or threatened habitats located in intact or almost intact natural areas. Moreover, these target sites can be within already strictly protected areas having the status of category Ia or Ib. The latter protected areas are not usually considered suitable for "restoration through time-limited interventions to undo past damage" such as "reintroduction of extirpated species; replanting to hasten forest regeneration; seedling selection; thinning; removal of invasive species" (Dudley 2008).

One solution to allowing management through active interventions in protected areas is their re-categorization, i.e., that a protected area is allowed to be re-categorized not only from category IV to categories Ia or Ib after successful restoration, but also from categories Ia and Ib to category IV to make restoration possible. When a protected area designated for conservation-oriented restoration is embedded in a larger protected area, the latter should be re-designed to incorporate category IV. Such re-categorization, however, requires a simultaneous raising of a level of protection from illegal human activities in category IV to the same level as categories Ia and Ib.

Another solution, which appears to be less controversial and easier to implement, is to permit restoration activities under proper supervision and control in all six categories (Figure 2.9).

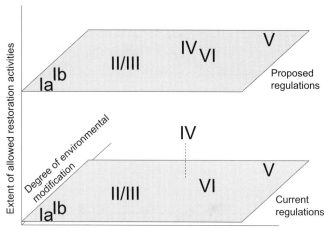

FIGURE 2.9 IUCN protected area categories arranged along three axes of extent of allowed local residents' and visitors' activities, degree of environmental modification, and extent of allowed restoration activities, respectively. To make possible active interventions needed for ecosystem restoration, I propose that restoration activities under proper control are permitted in all six categories.

In 2001, the concept of Plant Micro-Reserves (PMRs) was introduced (Laguna 2001). PMRs are small land plots (up to 20 ha in Spain where the concept has been developed and originally applied, but up to 200 ha in other regions) with legal protection status and the aim of long-term monitoring and conservation of plant species and their habitat. Figure 2.10 shows PMRs established in a variety of habitats of five European and Near East countries. Within PMRs, various forms of active management of plant populations and their habitat are allowed – seed collection, population reinforcement and introduction, herbivore exclusion, weeding, etc. Legislation of PMRs is easier and a less time-consuming procedure than legislation of strictly protected areas having one of the IUCN categories. Besides, in contrast to the latter, PMRs can be established on private or publicly owned land (Laguna 2001; Laguna *et al.* 2004).

PMRs, due to their small size and frequent dependence on the good will of a land owner, cannot be a substitute for large protected

FIGURE 2.10 Six PMRs, each with the main plant species it protects (inset) and the country location. (a) Cap d'Or (*Silene hifacensis*), Spain; (b) Lavajo del Tio Bernardo (*Marsilea strigosa*), Spain; (c) Chryssoskalitissa (*Phoenix theophrasti*), Crete; (d) Lyubash Mounts (*Lathyrus pancici*), Bulgaria; (e) Metropolitan Geawargios Haddad (*Iris bismarckiana*), Lebanon; (f) Mystero (*Ophrys kotschyi*), Cyprus. Photo of the PMR Metropolitan Geawargios Haddad by Magda Bou Dagher Kharrat. Other photos by Emilio Laguna. *A black and white version of this figure will appear in some formats. For the color version, please refer to the plate section.*

areas belonging to the above IUCN protected area categories. The PMRs were proposed as, and proved to be, a valuable complement to an existing network of nature reserves and national parks, often being the only option of protecting the populations of threatened plant species in fragments surrounded by land unavailable for conservation purposes. But, as natural habitat fragmentation is becoming a ubiquitous phenomenon, and unfragmented large natural areas are disappearing all around the globe, the relative ease and promptness of PMR legislation, and especially their amenability for active management can make this form of protection an attractive and widespread alternative to traditional IUCN protected area categories in the near future.

# 3 Restoration of Threatened Species

## 3.1 REGIONAL BASE

Modern plant conservation utilizes systematic conservation planning, a coherent strategy of decision-making and implementation for the sake of biodiversity (Margules and Pressey 2000; Knight *et al.* 2006; Sarkar and Illoldi-Rangel 2010). Since its introduction (Margules and Pressey 2000), systematic conservation planning has become a widely accepted operational approach covering a set of steps from planning to conservation actions. This approach, with its latest modifications (Pressey and Bottrill 2009; Sarkar and Illoldi-Rangel 2010) necessarily includes the following steps (Figure 3.1):

- delimitation of the planning area
- collection of data
- setting of conservation goals
- evaluation of the existing protected area network
- design of expansions
- development of management plan and its implementation, and
- long-term maintenance of biodiversity in the network.

Delimitation of the planning area within a spatial framework of units based on ecological or political criteria is the first step of systematic conservation planning. Global assessments of biodiversity are important for prioritization and focusing attention on broad regions of the world that are of highest conservation concern. However, most decisions about protective management are made at finer spatial scales, and a more detailed assessment within each of these regions for area prioritization is called regional conservation planning. Biologically defined units of area are more efficient for designing networks of conservation areas, as the distribution of

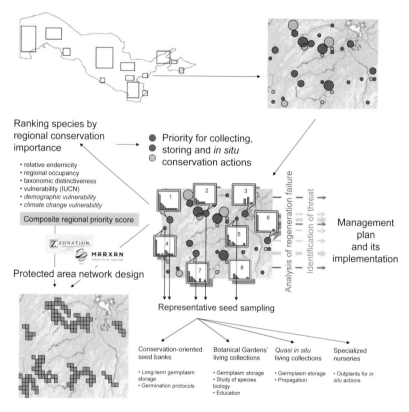

FIGURE 3.1 A scheme of regional conservation planning (from Volis 2018). Each colored circle denotes a population of one of three species with the circle size and color corresponding to a population size and species identity, respectively. All populations of one species (in red) are provided with size class distributions. In size class distribution histograms the *x*- and *y*-axes are size classes and plant density per unit area, respectively. *A black and white version of this figure will appear in some formats. For the color version, please refer to the plate section.*

species and communities rarely coincides with administrative units (Olson *et al.* 2001).

Thus, the first step of systematic conservation planning should be dividing the country into regional conservation units based on geomorphology, climate, and vegetation types. The process of ecological land classification, which fuses the ecological concept of ecosystems with the geographic concept of regions for mapping ecological regions (ecoregions) dates back to Crowly (1967), and is one of the most important concepts in understanding and managing landscape and biodiversity. Ecoregions can be defined at any scale as long as they represent discrete areas that are homogeneous in terms of ecosystem patterns and within which natural communities and species interact with the geographically distinct physical elements of the environment (Omernik 1987, 1995; Omernik and Bailey 1997). Defined from the broadest scale at which the spatial patterns repeat through various hierarchical levels to the local landscape scale from climate, landform, hydrology, vegetation, and soil data, ecoregions provide a consistent spatial framework for ecological modeling, biodiversity conservation policies and management, and systematic conservation planning at the national and subnational level (Cleland et al. 1997; Bryce et al. 1999; Dinerstein 2000; Groves et al. 2000, 2002; Bottrill et al. 2012). Conservation assessments within the framework of ecoregion units are favored by major international conservation organizations and many governmental agencies (e.g., Mittermeier et al. 1998; Ricketts 1999; Groves et al. 2000). The Convention on Biological Diversity (CBD 2009) and the Global and European Strategies for Plant Conservation (GSPC 2010) specifically direct their targets toward the effective conservation of ecoregions.

Conservation strategies that consider biogeographic units at the scale of ecoregions are optimal not only for protecting narrow endemics or localized habitat types, but also for those species and ecological processes that require large areas (Noss et al. 1999b; Soulé and Terborgh 1999; Groves et al. 2000; Margules and Pressey 2000). Only assessments and planning conducted at this scale can identify whether conservation areas are redundant or complementary across administrative borders (Olson et al. 2001). Although conservation planning conducted at a finer level than ecoregion scale can detect some small but important areas overlooked by regional planning, only the latter can provide an explicit regional integration (including corridors for migration, gene flow, and range shifts) (Simberloff *et al.* 1999; Huber *et al.* 2010).

## 3.2 POPULATION DEMOGRAPHIC SURVEY AND IDENTIFICATION OF A THREAT

Within each delimited region, it is not enough to have information about the population locations of threatened species and rough estimates of their sizes. The regeneration status of these species must be evaluated because a species future can only be ensured by its successful regeneration. An example of such evaluation is an assessment of regeneration status of 68 of the most ecologically and economically important Bolivian forest tree species (Mostacedo and Fredericksen 1999). In this study, each species was assigned one of four categories of regeneration status and provided with identified mechanisms of poor regeneration. The information provided is invaluable for any conservation planning involving these species.

In order to compile such an assessment list, every population of a threatened species must be visited for at least a preliminary survey of the population demographic structure (Figure 3.1). The latter is important because even a snapshot of demographic structure can reveal regeneration problems and allow identification of the major threats. For example, a demographic characterization of the seven remaining populations of critically endangered New Caledonian endemic *Araucaria nemorosa* revealed very poor production of seedlings in one population and lack of intermediate size categories in another. These results helped Kettle *et al.* (2012) to propose a conservation plan for these two populations that included, in addition to protection, control of the increasing density of competing angiosperms, stimulation of dispersal, and colonization of the adjoining habitat.

Although in some cases a single year demographic survey may suffice, ideally it should be followed by properly organized long-term observations on population demography and reproductive

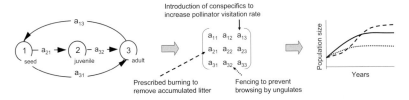

FIGURE 3.2 Use of population demographic data via population viability analysis to choose the optimal intervention. The procedure involves three steps: estimation of the life-cycle stages' transition probabilities, construction of the transition probability matrix, and evaluation of the interventions' effect on population size over time. The three line styles (dotted, dashed, and solid) correspond to three hypothetical interventions affecting different stage transitions.

phenology. The former is needed to work out the most appropriate interventions through PVA (Figure 3.2), and the latter is necessary for making a seed collection calendar. A demographic survey should be accompanied by mapping all the reproducing adults, and a study of species biology, viz. a mode of pollination and main pollinators, seed dispersal, breeding structure, presence of seed dormancy, age at maturity, and pattern of fruit/seed production.

A demographic survey must focus on species regeneration because many extant populations of threatened plant species, including those in protected areas, do not regenerate naturally, and their lack of recruitment has so far received insufficient attention. As a result of the demographic survey, the populations can be classified into two categories: those with naturally occurring regeneration (even if limited) and those with regeneration problems (i.e., with some life-cycle stages missing) (Figure 3.1). These two types of populations require different management. Populations from the first category can be subjected to augmentation to boost population growth. However, augmentation not supplemented by other actions will often make no sense in populations with no recorded regeneration. In such populations, a focus should be on long-term monitoring and a study program to better understand the reasons for lack of regeneration. The observations on flowering and fruiting, and introduction experiments manipulating germination and growth conditions can reveal the causes of the population regeneration failure. Depending on the cause, possible solutions can include thinning of neighboring vegetation, creation of canopy gaps, and introduction of nurse plants or extirpated pollinators/seed dispersers. Most of the actions to stimulate natural regeneration can be done only within a habitat restoration framework and therefore will be discussed in Sections 4.7, 4.8, and 5.1.

## 3.3   PRIORITIZING SPECIES

The basic resource detailing the global conservation status of plants and serving as a reference for many conservation decisions is the IUCN Red List. This uses a set of detailed and generally accepted criteria to evaluate the extinction risk of species and to allocate which category they should be assigned (Figure 3.3). It is also often used for determining the threat status at a national level (although many countries also employ their own criteria for setting conservation priorities for species). An ecoregion inventory of threatened species will result in a list of locally occurring species with each species assigned one of three IUCN categories (CR, EN, and VU). This rough categorization, however, can be misleading in some cases. Species with the same IUCN category can differ dramatically in many important attributes. For example, one of two endangered species can be regionally endemic, whereas the second one has its major distribution outside the region. Or one can represent a monotypic genus and the second one as a genus with several hundred species closely resembling each other. Thus, regional conservation planning must have a more nuanced procedure for ranking species by their conservation priority. A way to do so is to use, instead of a single qualitative variable (IUCN category), a composite variable combining several quantitative estimates of species rarity and vulnerability.

Prioritization of threatened species can utilize a unified system of scoring species in regional conservation planning such as one based on Freitag and Van Jaarsveld (1997) and presented in Table 3.1. Freitag

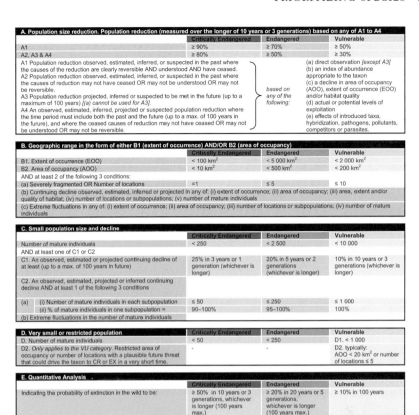

FIGURE 3.3 Five criteria (A–E) used to evaluate if a taxon belongs to a threatened category (Critically Endangered, Endangered, or Vulnerable) (Standards Petitions Working Group 2017). *A black and white version of this figure will appear in some formats. For the color version, please refer to the plate section.*

with colleagues proposed complementing quantitative analogs of the IUCN categories (RV) with three other estimates: relative endemicity (RE), regional occupancy (RO), and taxonomic distinctiveness (RTD) (Freitag and Van Jaarsveld 1997; Freitag *et al.* 1997). These estimates can be calculated in different ways depending on spatial scale and quality of the species occurrence data using either special units (e.g., map grid cells), area occupied, or number of populations. Below are examples of their calculations.

Table 3.1 *Components of a composite regional priority score used for ranking threatened species by their conservation priority*

| Component | Calculation | Definition | Relevance | Source |
|---|---|---|---|---|
| Relative endemicity score (RE) | $$RE = \frac{\text{Ecoregion area covered } (km^2)}{\text{Total area covered } (km^2)} \times 100$$ | A proportion of the species total distribution range or "extent of occurrence" falling within the ecoregion | Species that are increasingly restricted to the ecoregion receive increasingly larger scores. Species with a score of unity have distributions entirely restricted to the ecoregion (true regional endemicity) | Freitag and Van Jaarsveld (1997) |

| Regional occupancy score (RO) | $$RO = \sqrt{\dfrac{1}{\text{No. of populations within ecoregion}}}$$ | A more refined estimate of regional extent of occurrence | Regionally less common species (those with smaller areas of occupancy and smaller number of populations) get higher scores | Freitag and Van Jaarsveld (1997) |
| Relative taxonomic distinctiveness score (RTD) | $$RTD = \sqrt{\dfrac{1}{(f \times g \times s)}}$$ where $f$ is the number of families in the order, $g$ is the number of genera in the family, and $s$ is the number of species in the genus to which a particular species belongs | An estimate of the taxonomic distinctiveness of a species | Taxonomically more distinct taxa get higher scores as contributing proportionately more to regional biodiversity | Freitag and Van Jaarsveld (1997) |

*(continued)*

Table 3.1 (cont.)

| Component | Calculation | Definition | Relevance | Source |
|---|---|---|---|---|
| Relative vulnerability score (RV) | These scores are based on Red Data Book categories, i.e., "critically endangered," "endangered," "vulnerable," and "near threatened" having weightings of 1.0, 0.75, 0.5, and 0.25, respectively | A quantitative analog of the IUCN categories | More endangered species get higher scores | Freitag and Van Jaarsveld (1997) |
| Demographic vulnerability score (DV) | $DV = \dfrac{\text{No. of inviable populations}}{\text{Total no. of populations}}$ | A proportion of the populations having regeneration problems such as lack of seedlings, young plants, or reproducing adults | Species with regeneration problems receive increasingly higher scores | Volis (2017b) |

| Climate change vulnerability score (CV) | $CV = \dfrac{\text{Predicted range by 2080 (km}^2)}{\text{Predicted range under current climate (km}^2)}$ | A proportion of the habitat within the ecoregion that will remain suitable for a species despite climate change | Species with increasingly higher proportion of their range that will have remained within the ecoregion receive increasingly higher scores | Volis (2017b) |
| Composite regional priority score (RPS) | $RPS = \dfrac{RE + RO + RV + RTD + DV + CV}{6}$ | | | Freitag and Van Jaarsveld (1997) |

*Relative endemicity score (RE)* is a proportion of the species total distribution range or "extent of occurrence" falling within the ecoregion. Using distribution area in km², RE will be:

$$RE = \frac{\text{Ecoregion area covered }(\text{km}^2)}{\text{Total area covered }(\text{km}^2)} \times 100$$

According to this equation, species that are increasingly restricted to an ecoregion receive increasingly larger RE scores. Species with RE scores of 1 have distributions entirely restricted to the ecoregion (true regional endemicity).

*Regional occupancy score (RO)* is a more refined estimate of the regional extent of occurrence. Higher scores signify regionally less common species (those with smaller areas of occupancy and smaller number of populations). For example:

$$RO = \frac{1}{\sqrt{\text{No. of populations within ecoregion}}}$$

*Relative taxonomic distinctiveness score (RTD)* estimates the taxonomic distinctiveness of a species. Higher scores signify taxonomically more distinct taxa as contributing proportionately more to regional biodiversity (Vane-Wright *et al.* 1991, 1994). An equation that can be applied to all hierarchical classifications was proposed by Freitag and Van Jaarsveld (1997):

$$RTD = \frac{1}{\sqrt{(f \times g \times s)}}$$

where $f$ is the number of families in the order to which the taxon belongs, $g$ is the number of genera in the family, and $s$ is the number of species in the genus to which a particular species belongs.

The above three estimates complement the quantitative analog of the IUCN categories.

*Relative vulnerability scores (RV)* are based on Red Data Book categories, i.e., "critically endangered," "endangered," "vulnerable," and "near threatened" with weightings of 1.0, 0.75, 0.5, and 0.25, respectively. Species not recorded in the Red Data Book are given a score of zero.

The values for each of these four criteria range between zero and one and are similarly distributed (Freitag and Van Jaarsveld 1997).

In previous sections of this book, I discussed the importance of healthy population demographic structure evident in the presence of natural regeneration, and of possible climate change effects. Therefore, I propose two novel criteria, namely, demographic vulnerability and climate change vulnerability, for ranking threatened species by conservation value.

*Demographic vulnerability score (DV)* is the proportion of the populations having regeneration problems such as lack of seedlings, young plants, or reproducing adults:

$$DV = \frac{\text{No. of inviable populations}}{\text{Total no. of populations}}$$

According to this equation, species with regeneration problems receive increasingly larger scores.

*Climate change vulnerability score (CV)* is the proportion of the habitat within an ecoregion that will remain suitable for a species, despite ongoing climate change:

$$CV = \frac{\text{Predicted range by 2080 (km}^2)}{\text{Predicted range under current climate (km}^2)}$$

The predicted ranges can be routinely obtained from species occurrence data through SDM.

For simplicity, and following Freitag and Van Jaarsveld (1997), these six scores can be given equal weights, summed, and averaged to obtain a total composite score for each regionally occurring species, referred to as regional priority score (RPS):

$$RPS = \frac{RE + RO + RV + RTD + DV + CV}{6}$$

## 3.4   RESERVE DESIGN

### 3.4.1   *Area Selection*

Because in any given region the total amount of land available for conservation is limited, the goal of regional systematic conservation planning is to establish a network of protected areas in which all local species are represented and efficiently preserved (Margules and Pressey 2000). This makes area prioritization necessary for protection.

Traditionally, this prioritization was based on expert opinions, but nowadays area prioritization, as well as the design of the protected area network are worked out using computer-based decision-support tools. The latter are data-driven, inherently quantitative approaches using spatial data on biodiversity and other environmental and socio-economic factors.

Reserve selection (also known as site selection, area selection, reserve design, or reserve network design) can be viewed as an optimization problem – a search for a compromise between resource allocation (land area) and conservation efficiency. The ultimate goal of this optimization is, while minimizing cost, to create a protected area network that will protect all conservation targets (species or habitats), include as many elements of biodiversity as possible, be sufficiently large, and be well connected. Even if the creation of a regional conservation plan starts from the identification of core areas that contain regionally important habitats or charismatic species (e.g., Noss *et al.* 1999a; Shriner *et al.* 2006), it should proceed to formal computer-based reserve selection, as suggested by the findings of Andelman and Fagan (2000). These authors analyzed datasets of diverse taxa on three different geographic scales, and found limited utility of umbrella and flagship species as surrogates for regional biodiversity.

Despite the recent proliferation of algorithms and software packages for reserve selection (reviewed in Sarkar *et al.* 2006; Moilanen *et al.* 2009b), these techniques share several common basic principles. First, the region area is divided into a set of spatial units either based on existing boundaries (e.g., administrative, ecological, watershed, or land ownership boundaries) or grid cells of a required resolution level. Each of the discrete spatial units ("planning units") is assigned a conservation value estimated according to the focus, which can be an individual species, habitat, or ecosystem. Quantification of a conservation value can be based on the species occupancy (presence or absence), species abundance, the area of a species habitat within each planning unit, habitat conditions, etc. The problem formulation can be as simple as the binary decision of whether to include a planning unit in the selected set, or more complicated, such as which actions to implement within a planning unit. In either case, the value of each planning unit for inclusion, or with respect to an action, must be quantified, along with some measure of the cost of implementing the action. The targets for achieving conservation goals can be set for each species of conservation concern but the common target in the reserve design is a habitat area to be protected. The outcome of the area selection procedure depends on the data input, the initial and subsequent selection rules, and the sequence in which selection rules are used in the algorithm.

Selecting sites by species richness is inappropriate because a proper solution is not a set of the richest sites, but a set of sites whose species assemblages complement each other and together encompass the largest species pool (Pressey et al. 1993; Williams et al. 1996). Therefore, most of the existing selection algorithms are based on the concept of complementarity, i.e., a measure of the degree to which a site contributes to the representation of biodiversity features that are not adequately represented in the existing set (Pressey et al. 1993). In addition to complementarity, two other key concepts have guided the reserve selection algorithms: irreplaceability and vulnerability.

Irreplaceability is a measure of how any particular selected site is essential for achieving targets. Sites with lower irreplaceability can be replaced by other unselected sites, if for some reason the former cannot be included in a conservation area network, while sites with high irreplaceability cannot be replaced and therefore negotiated (Pressey et al. 1994). Technically, irreplaceability value is the summed number of times each planning unit was chosen in a set of runs of the algorithm. Complementarity is implicit in irreplaceability. Units with a high value include high numbers of species with low species overlap compared with other units, while those that have a low value do overlap in species composition with many other units, although they can still have high biodiversity.

Vulnerability is about persistence of biodiversity in the planning unit, i.e., the probability of biodiversity loss due to current or imminent threatening processes (Pressey *et al.* 1996; Wilson *et al.* 2005) for which the nonexhaustive list includes urbanization, infrastructure development, mining, agriculture, logging, grazing, and the spread of invasive plants and animals. A comprehensive assessment of vulnerability should consider the threat effects as well as the dynamic responses of threats to conservation actions. The unit vulnerability score for multiple threats can be calculated by differentially weighting threats to reflect their relative impacts.

All the currently utilized reserve selection algorithms produce either a set of sites that represent the maximum number of species in a given number of sites, or all species in the minimum number of sites (known as maximum coverage and minimum set problems), referred to as a "solution." A solution for the former is found via simulated annealing (e.g., implemented in Marxan software, Ball *et al.* 2009) and for the latter via a reverse stepwise search (e.g., implemented in Zonation, Moilanen *et al.* 2009a). Weighting of species by their conservation values has mostly been overlooked by these methods. Several authors have discussed the usefulness of weighting species by their taxonomic or genetic distinctiveness, rarity, endemicity, or economic value in reserve selection (e.g., Freitag *et al.* 1997; Arthur *et al.* 2002; Onal 2004), but very few have actually used species weighting in reserve selection algorithms (Freitag *et al.* 1997; Arponen *et al.* 2005; Fiorella *et al.* 2010).

Most existing algorithms for reserve selection give equal weights to the species, but there are exceptions. For example, in the

Zonation program (Moilanen *et al.* 2009a), the weighting of species by their conservation values is possible. This software optimizes a solution by capitalizing on the natural patterns of species co-occurrences in the landscape to include the maximum possible proportion of each species distribution within a solution. Utilization of the software option of weighting species by their conservation value will change the focus in reserve network design, as was shown by Fiorella *et al.* (2010), from maximizing overall species protection to maximizing protection of the most threatened species. Species can be prioritized not only by their IUCN Red List categories (Fiorella *et al.* 2010) but also by a combination of criteria (Freitag *et al.* 1997; Arponen *et al.* 2005). The regional priority scores presented in the preceding section can be particularly useful for this purpose.

The idea of using weighting of species by their conservation value in reserve selection has some analogy with the idea of using a suite of focal species, each of which is thought to be sensitive to a particular threatening process, for this purpose (Lambeck 1997). The focal species, according to Lambeck, might be the most area-sensitive, dispersal-limited, resource-limited, and ecological process-limited taxa in a landscape. Lambeck (1997) claimed that "because the most demanding species are selected, a landscape designed and managed to meet their needs will encompass the needs of all other species." In other words, a suite of focal species acts as an "umbrella," and the reserve areas identified to serve the needs of focal species are presumed to offer similar protection for other species that fall under their "umbrella" (Lambeck 1997; Caro and O'Doherty 1999).

The search for an efficient reserve design must take into account the issue of climate change (Araujo *et al.* 2004; Hannah *et al.* 2007; Game *et al.* 2011; Brambilla *et al.* 2017). To enable species to adjust their ranges in response to changing climatic conditions, increasing ecological connectivity is often recommended, i.e., the flow of propagules, organisms, and ecological processes across landscapes. This can be achieved by either increasing suitable habitat at range margins or establishing corridors to make possible movements of species (Vos *et al.* 2008; Heller and Zavaleta 2009; Krosby *et al.* 2010). The creation of corridors as a management response dominated the early global change literature (e.g., Peters and Darling 1985) and is still popular today despite a complexity of technical questions about the spatial configuration of corridors, such as the optimal width and edge-to-area ratios (Halpin 1997). In fact, there is little guidance in the literature for corridor implementation beyond commonsense reasoning (Heller and Zavaleta 2009). In many cases, because of uncertainty in quantifying the benefits of connectivity *per se* and the technical challenges of establishing corridors, the connectivity can be coincidentally improved by a mere increase in habitat area and habitat quality (Hodgson *et al.* 2009).

The poleward expansion of a reserve's boundaries or buffer zones is commonly suggested as the attempted first solution (Noss 2001; Hodgson *et al.* 2009). However, one should keep in mind that a species can track a 3°C increase in temperature by moving less than 500 m upward in elevation, while tracing an analogous latitudinal shift would require poleward movement of almost 400 km (Krosby *et al.* 2010). Thus, in situations where an environment sharply changes over short geographic distances, an increase in the reserve size can provide enough opportunities for a species to find a more favorable habitat where it can survive. For example, *Abies religiosa*, the preferred host for overwintering migratory monarch butterfly (*Danaus plexippus*) is currently confined to the Monarch Butterfly Biosphere Reserve (Mexico). The predicted suitable habitat for *A. religiosa* will shift rapidly over the course of the century and by 2090 will move outside the reserve. Thus, what is needed is realignment of *A. religiosa* to new locations upwards (275 m by 2030) (Sáenz-Romero *et al.* 2012), making these new locations a part of the reserve (Figure 3.4).

A more common scenario is when a change in reserve size or configuration will not significantly change the extent of variation in habitat conditions. For many species, the distances to move to adjust their ranges (the temperature isoclines are expected to shift more than 1 km/year in many systems, Loarie *et al.* 2009) will be

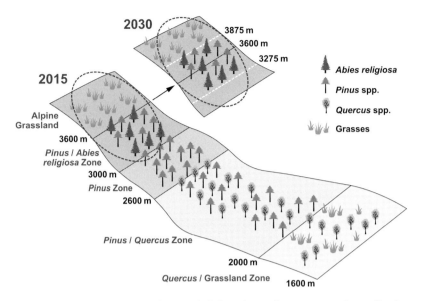

FIGURE 3.4 Predicted range shift for *Abies religiosa* currently confined to the Monarch Butterfly Biosphere Reserve, and the proposed assisted colonization of 275 m upwards by 2030 to mitigate the effect of changing climate. However, the proposed area is currently outside the reserve (from Dumroese *et al.* 2015).

too great to be accommodated by simply expanding reserve boundaries. For these species, the creation of stepping-stone reserves, even of a small size, along climatic gradients can be a better means to enable range shifts. Analysis of the relationship between the species current distribution and climate through SDM, especially if the relevant paleobotanic and paleoclimatic data are available, can make it possible to work out the optimal reserve locations and sizes, as well as to identify areas that will have suitable conditions in the future and therefore a high probability of protecting the target species.

Conservation of Joshua trees (*Yucca brevifolia*) in and around Joshua Tree National Park, California, illustrates the relevance of the above considerations. Today, the Joshua tree population within the park shows regeneration problems due to increased aridity, and 90% of the habitat suitable for the species in the park is predicted to be lost by 2100. The fossil data suggest that the species cannot move >2 m/year.

Under such a slow migration rate, and because Joshua tree's major dispersal agent, Shasta ground sloths, are extinct, tracking climatically suitable areas by shifting the distribution northward is unlikely. Instead of expanding the park area into adjacent areas north, management of the species should focus on more distant land within the predicted suitable range. The options include the acquisition of new lands for protection, and the assisted colonization of already protected areas (Cole *et al.* 2011).

### 3.4.2   *Minimum Reserve Size*

A reserve must always be sufficiently large to ensure the persistence of the target species and the whole ecosystem. For biotopes that are harboring populations of threatened species and are naturally small or have clear boundaries (e.g., rare soil types, ponds, gorges, or cliffs), reserve design is usually straightforward. Whenever possible, these habitats should be surrounded by buffer zones or integrated into larger protected areas (Bucking 2003), but a minimum area to protect is not an issue here. However, in the majority of situations encountered by conservation planning, it is important to set minimum reserve size thresholds, as protected areas that are too small may not sustain viable populations of their flora and fauna, are more vulnerable to anthropogenic disturbances, and have higher management costs. The importance of such thresholds motivated the creation of MinPatch, a software package that manipulates outputs from Marxan reserve design software, so that every area planned for protection meets the user-defined size threshold (Smith *et al.* 2010). However, how do we define such a threshold? For a single target species, it will be the area required to maintain a viable population. But in reserve design, we deal with many species, so for a particular area there will be a set of range sizes. In plant species, the pollen/seed dispersal distance, natural density, and abundance determine the local range sizes. For the majority of the widespread tree species, even 1 ha can host a viable population, whereas individuals of rare tree species are usually scattered over great distances.

For three such low-abundance neotropical tree species, Stacy *et al.* (1996) estimated, based on the mating patterns observed, the smallest area required for a natural breeding unit, defined as the minimum area within which 95% of the pollen received by a centrally located adult originates. From the results obtained, the authors generalized that for low-abundance species with evenly distributed reproductive adults, the latter unit is 60 ha, while for species with a clumped distribution of reproductive trees it is 40 ha. The authors suggested that these results should apply to a broad range of tropical forest species. Indeed, in the later two studies performed on predominantly outcrossing tropical tree species, the estimated breeding units were close to these predictions, being 68 ha for *Neobalanocarpus heimii* with a population density of 0.7 reproductive trees/ha (Konuma *et al.* 2000), and 12 ha for *Shorea lumutensis* with a population density of 4.3 trees/ha (Lee *et al.* 2006).

In animals, the individual home range matters the most. It was evaluated that 20–50 ha will generally satisfy the territorial needs of micro- and mesofauna inhabiting Middle European forest reserves (Table 3.2). But to sustain populations of large herbivores and their predators, much larger areas are needed. For example, one roe deer (*Capreolus capreolus*) lives in an area of about 25 ha, and therefore a deer population will require a minimum of several hundred hectares (Strandgaard 1972). Gurd *et al.* (2001) estimated the minimum reserve area in eastern North America, allowing the persistence of terrestrial mammals, including the rarest and most vulnerable species, to be 5037 km$^2$ (95% CI: 2700–13 296 km$^2$). Based on these considerations minimum reserve sizes for some forest types are shown in Table 3.2.

So, a seemingly straightforward but too simplistic solution for defining a minimum reserve size threshold would be to arrange the range sizes of the local species and to use the largest one as a threshold. However, this species-based approach ignores the processes that occur at the community and ecosystem levels. Conservation planning should take into consideration and have procedures for determining the area required to maintain ecological processes (Poiani *et al.* 2000;

Table 3.2 *Minimum areas of European strict forest reserves (from Bucking 2003)*

| | Minimum area (ha) |
|---|---|
| **Based on forest structure research** | |
| Extreme sites | 5–20 |
| Mixed forests | 10 |
| Beech, beech-oak, and beech-fir forests | 50 |
| Mixed alpine and mountain forests | 70–100 |
| **Based on faunistic and side studies** | |
| Micro- and mesofauna | 50–100 |
| Large mammals/birds | >>> 100 |
| Typical site mosaic and landscape fraction | 100 |

Leroux *et al.* 2007). The conceptual foundation for such considerations is set by a minimum dynamic area (MDA) as proposed by Pickett and Thompson (1978) and defined as "the smallest area with a natural disturbance regime, which maintains internal recolonization sources, and hence minimizes extinction." In the MDA concept, patch dynamics induced by a habitat-specific disturbance regime can be used to identify the size of reserve "islands" that will support a quasi-equilibrium landscape state. Landscapes in a quasi-equilibrium state are expected to exhibit relatively stable community biomass, composition and age structure (Sprugel and Bormann 1981; Sprugel 1991), and thus to support stable population sizes of target species. This conceptual linkage between habitat patch dynamics and population persistence is of direct relevance to reserve design (e.g., Poiani *et al.* 2000; Groves *et al.* 2002; Leroux *et al.* 2007).

Currently available reserve design computational methods incorporate habitat requirements of focal species, but not MDA, to inform reserve size. The MDA concept has a large potential for conservation planning, but it is currently underutilized due to lack of estimated minimum dynamic areas for different habitat types. Nevertheless, combining habitat-specific MDA and the target species

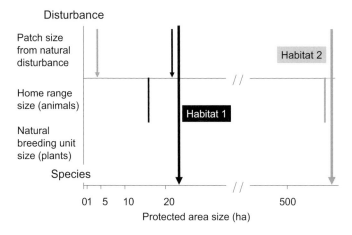

FIGURE 3.5 Determination of minimum reserve size simultaneously considering two factors: patch sizes from habitat-specific natural disturbances and either the smallest area required for a natural breeding unit of sparsely distributed plant species or the home range needed to accommodate a viable population of wide-ranging animal species (based on Groves *et al.* 2002). The large arrows show the minimum reserve size for two hypothetical habitats worked out from the relevant patch size and home range/breeding unit estimates.

area requirements for determination of the minimum reserve area (Figure 3.5) has been recognized as a way forward in conservation planning (Groves *et al.* 2002), and this approach is already in use. Using this approach, the minimum size for forested conservation areas in the Northern Appalachians Ecoregion was set at approximately 12 000 ha (Groves *et al.* 2002).

## 3.5 GENETIC CONSIDERATIONS: ADAPTIVE VERSUS NEUTRAL INTRASPECIFIC DIVERGENCE

Evolution proceeds through changing genetic composition of populations, and therefore availability of genetic variation is a prerequisite for a species evolutionary change. What should be considered for how best to conserve the intraspecific variation present across the species geographic range? Giving priority to particular populations is usually based on their uniqueness and level of divergence from other populations. This divergence can arise from two different evolutionary processes: local selection and isolation, as discussed below.

Consideration of intraspecific genetic variation in an evolutionary context, as relevant for conservation, goes back to the concept of evolutionary significant unit (ESU) first formulated by Ryder (1986). Ryder defined ESUs as "subsets of the more inclusive entity species, which possess genetic attributes significant for the present and future generations of the species in question" and suggested that delineation of the subsets utilizes different information types (e.g., ecological, genetic, physiological) and seeks their congruence. However, because genetic markers and phenotypic traits are influenced by two different evolutionary forces (gene flow and natural selection, respectively), such congruence is rarely observed. Moreover, because genetic markers are predominantly neutral while the majority of phenotypic traits are adaptive, use of DNA markers for inferences about intraspecific adaptive divergence or potential is discouraged (Reed and Frankham 2001; Mittell et al. 2015; Volis et al. 2016b).

For the aforementioned reason, definitions of ESU after Ryder have recognized intraspecific ecological and genetic variation as two distinct species properties important for species viability and evolutionary potential. Both distinct lineages that resulted from isolation and ecotypes adapted to different environmental conditions must be conserved because each is functionally important for the species (Vogler and Desalle 1994; Moritz 1999; Fraser and Bernatchez 2001; Thorpe and Stanley 2011; Rodriguez-Quilon et al. 2016). Especially important for conservation is intraspecific adaptive diversification producing differently adapted geographic variants (Dizon et al. 1992; Crandall et al. 2000). Ideally, the ESU judgment should be made from ecological and genetic data, but when these data are lacking (at least at the initial stage of a conservation program), knowledge of the species distribution in many cases can be a good proxy to presumed intraspecific adaptive divergence or historical isolation.

Analysis of the species distribution and intraspecific variation must be based on recognition of the importance of environmental (e.g., soil, altitude, climate) and ecological (e.g., vegetation type) conditions that vary within the species range. Utilization of the concept of ecoregions described in Section 3.1 can be a useful first step in dividing the species range into geographical units that are relatively homogeneous inside but differ from each other in environmental/ecological conditions. However, because this concept is ecosystem and not species based, and because determination of the importance of the environmental characteristics in the utilized method is subjective (Omernik 2004), the delineation of the relevance of the target species ecoregions will require not only detailed information on its distribution but also a good knowledge of the species biology. If the ecoregions are defined on the subjectively decided importance of such characteristics as geology, physiography, vegetation, land use, climate, soils, etc., the relative importance of these characteristics will vary from one species to another, and for the same species for different parts of the species range. For example, an endangered plant *Iris atrofusca* is represented in Israel by populations occupying distinctly different ecoregions in terms of climate, soil, and vegetation, but not land use (Volis *et al.* 2016c). However, within one of the ecoregions, the Northern Negev Desert, land use becomes the major factor for delineation of the two subregions, the Goral Hills and the Arad Valley areas. In these two areas, differential land use that existed for centuries appears to result in two eco-morphs, one of which (in the intensively grazed Goral Hills area) resembles an unpalatable *Asphodelus ramosus* (Shimshi 1979/80).

## 3.6 SEED SOURCING AND COLLECTING

### 3.6.1 *Sampling Design for* Ex Situ *Conservation*

In discussing appropriate strategies for seed collecting, we need to distinguish general seed collecting, for which the purpose is to properly cover the genetic diversity of the species and then preserve it for a long time in a seed bank, and specifically targeted collecting, which has a narrower goal. For general collecting, the basic sampling questions were outlined by the Center for Plant Conservation (1991):

1. How many and which populations to sample?
2. Within each population, how many, and which individuals to sample?
3. For each individual, how many, and what type of propagules to collect without threatening the population? Later guidelines for sampling wild populations of plants (Guerrant 1992, 1996; Guerrant *et al.* 2004; IUCN/ SSC 2013) added a temporal aspect to these questions.
4. To collect propagules all at once, or over a period of years?

Although overall very useful, these questions nevertheless ignore a spatial and environmental context in which the populations exist.

The importance of environmental and ecological conditions that vary within the species range for sampling design has been recognized (Volis and Blecher 2010; Guerrant *et al.* 2014; Hoban and Schlarbaum 2014). The concept of ecoregions can be used not only for understanding the species spatial pattern of distribution and identification of ESUs, but also for working out the sampling design. The ecoregions should be the highest level in the stratified design, subdivided further if needed. In this spatially stratified design, the lower level of neutral variation is nested within a higher one of adaptive variation, i.e., sampling in each (sub)ecoregion should include several geographically isolated populations (Volis and Blecher 2010).

Once regions for collecting are defined, targeting populations for collecting within each region can use different approaches. For example, for populations distributed along some environmental gradient, such as altitude, some arbitrarily defined elevational increment can be used as strata to embrace the whole local elevational range occupied by the species (Guerrant *et al.* 2014).

Collection from 5 to 50 populations is generally recommended (Brown and Marshall 1995), but this recommendation does not account for the population spatial structure and environmental/ ecological variation across the species range. With a hierarchical sampling design, when the species range is subdivided into distinct environment or habitat regions and subregions, five populations per (sub)region seems to be a reasonable minimum population number. However, if there are fewer than five populations per (sub)region, ideally all the populations in the area should be sampled.

Recently, guidelines on efficient sampling explicitly accounting for regional spatial structure were proposed by Hoban and Schlarbaum (2014). These authors used simulation modeling to quantify the expected probability of capturing alleles (especially rare alleles) for sampling designs that differed in sample sizes and spatial distribution of sampled populations. Using different classes of genetic markers in their simulations, they found that for the same number of populations (four) the "spatially dispersed strategy" (one population in each of four regions) consistently outperformed the "spatially constrained strategy" (four populations in only one region), for all sampling intensities, and all allelic categories. The spatial effect was strongest for poorly connected (i.e., with low gene flow) species. On the other hand, they found that sampling more than 25 individuals per population under the "spatially dispersed" sampling strategy results in diminishing returns for all allelic categories.

Although these results are not surprising, they are important, as they provide strong quantitative support to the idea that sampling must take into account not just spatial configuration of the populations, but their nestedness within regions. One important comment must be made. As the authors admit, their modeling considered hierarchical population structure only in terms of population connectivity via migration. The possible role of environmental conditions inducing spatially varying local selection in their model is ignored. However, if local adaptation is important and differs among regions, then the authors' recommendation that "moderate sampling (25–30 individuals per population) from few but widely-spaced populations performs optimally" would not be a good strategy. Not sampling "few but widely-spaced populations," but sampling at least one (and ideally more!) population per habitat, environmental type, or vegetation community would be optimal. Of course, usage of the collected germplasm for *in situ* actions should also be based on population nestedness within ecoregions or habitat types. The conclusion about sampling 25–30 individuals per population also needs some refinement and upward adjustment to account for anticipated attrition during storage and unavoidable losses due to mortality in *in situ* actions (Guerrant *et al.* 2015), as well as considerations about the effective population size necessary to prevent inbreeding depression and maintain evolutionary potential (Frankham *et al.* 2014).

The number of plants to be sampled in each population will depend on the extent of genetic variation that exists within a species. For example, for vegetatively reproducing plants occupying spatially homogeneous environments, such as many aquatic plants, a population sample does not need to be large. But for outcrossing and especially self-incompatible plants, as well as for predominantly self-pollinated plants with localized seed dispersal, large population samples are necessary. The number of sampled individuals ideally should be kept approximately the same across populations. However, some weighting criteria can be used, such as population size or plant density.

### 3.6.2   *Seed Sources for* In Situ *Introduction*

The guidelines for general seed collecting are straightforward and easy to follow. It is more difficult to provide general recommendations for collecting seeds to be used *in situ*. Success in either restoration of the plant community or translocation of an endangered species will depend on matching the plant material to the site, i.e., that the introduced plants will survive and reproduce. Therefore, one of the most important issues in planning conservation or restoration introductions is a location of source populations. A seemingly obvious recommendation to use local seed sources is not as obvious as it appears. On the one hand, local or nearby seed sources are likely to be adapted to conditions of the habitats targeted for introduction (McKay *et al.* 2005). On the other hand, the local sources can have limited genetic diversity or be too scarce to provide enough material even after propagation, while introduction of nonlocal sources can cause such negative consequences as: genetic swamping, when the nonlocals replace locals; and outbreeding depression, when hybridization with locals leads to fitness reduction in the offspring due to the dilution of adapted genes or disruption of coadapted gene complexes (Hufford and Mazer 2003). Thus, it is important to find source material that is not necessarily local but adapted to the local conditions and genetically variable. This requires understanding the factors determining local fitness and limiting a geographic range within which source material will be adapted to the same conditions. When local conditions differ little among the locations (e.g., aquatic plants' habitats) genetic analysis can be helpful. Genetic analysis revealed that the most appropriate male plants to reestablish reproduction in the last three Italian female populations of the dioecious aquatic *Stratiotes aloides* were not from the most geographically proximal German population, but were from the more distant Dutch and Romanian populations (Orsenigo *et al.* 2017).

As stated above, although use of local seed is generally recommended, local population(s) can be too small, genetically depauperate, or inbred to be used as a source material. Poor performance, low reproductive success, and increased susceptibility to pests and pathogens are the commonly documented effects associated with genetic erosion and inbreeding (Lienert 2004). Over time, reduced fitness and productivity will expose small populations to decline through poor or no recruitment. Therefore, using seed from small, inbred populations is likely to create more small, inbred populations with poor prospects of developing into a self-sustaining population.

If we are to use more than one seed source (i.e., population), we need guidelines on how local the collected plant material should be to balance genetic diversity with the risk of maladaptation, or, in other words, how far the collected material can be moved from its native environment. Such guidelines provide seed transfer zones, the delineated areas within which seed can be transferred with little risk of maladaptation. The concept of transfer seed zones was originally developed in forestry (Ying and Yanchuk 2006) and involved common garden studies evaluating a large number of populations sourced across a species range in a set of sites that were representative of the variety of environmental conditions experienced by a species. However, these kinds of studies are expensive and logistically challenging. As a result, seed transfer zones have been derived for only a limited number of important timber species and for the majority of native plant species, data on adaptive genetic variation do not exist. Thus, there is still a need for seed zone guidelines that would prevent maladaptation, could be applied across species, and are operationally manageable. In an effort to provide such general guidelines on seed sourcing for species lacking experimentally determined seed transfer zones, some authors have suggested employing provisional seed transfer zones defined by relevant climate variables (e.g., winter minimum temperature, maximum mean monthly temperature, and annual precipitation) in combination with the ecoregion classification system used by the US Environmental Protection Agency (Johnson *et al.* 2010). In a study by Bower *et al.* (2014), when Omernik's level III ecoregions (Omernik 1987) were superimposed over 64 provisional seed zones representing areas of high climatic similarity, the combined model resulted in more variation partitioned among zones than either approach alone (Bower *et al.* 2014). The areas distinguished by a combination of the two approaches are climatically similar yet ecologically different, and therefore account for adaptation not only to climate but also to the other environmental features that vary spatially. Despite the higher proportion of variation explained, finer subdivision of provisional seed zones (i.e., usage of level IV ecoregions) should be discouraged as it produces too many zones to be practical from a land management perspective (Bower *et al.* 2014).

Because genetically diverse sources are preferred to those having low genetic variation, the mixing of genotypes can be generally recommended (1) when they share a seed transfer zone or an ecoregion (i.e., are "regional mixtures"; Falk *et al.* 2001), or (2) when the remaining populations of a species are extremely scarce, small, and fragmented. If the aim of seed sourcing is the augmentation of the remnant populations, it is important that the introduced genotypes will not swamp the local genotypes, i.e., the nonlocal gene flow should be less than 20% (Hedrick 1995). However, mixing too broadly (from different seed zones or ecoregions) should in general be discouraged.

The above considerations assume that the environmental conditions are stable and do not change over time. For this reason, the focus of sourcing germplasm, until recently, was on maintaining local genetic–environmental relationships, i.e., local adaptation (Keller *et al.* 2000; McKay *et al.* 2005), as discussed above. However, in the new realities of global climate change, sourcing germplasm for conservation and restoration must include the goal of promoting adaptation to a changing climate because the latter may erode local genetic adaptations, making the established genetic–environmental relationships no longer relevant. Thus, a focus on local adaptations must account for climate-caused range shifts, and address the need for a sufficiently high population adaptive potential (Weeks *et al.* 2011; Aitken and Whitlock 2013; Thomas *et al.* 2014). This has been recognized as the need to move away from a strict focus on local provenancing, toward both addressing risks of inbreeding in small and fragmented populations and considering adaptation potential in relation to changing environments (Broadhurst *et al.* 2008; Breed *et al.* 2013). "Predictive" provenancing (Crowe and Parker 2008; Wang *et al.* 2010; Sgro *et al.* 2011), explicitly targets adaptation to climate and proposes the use of genotypes that are experimentally determined to be adapted to projected conditions. The latter requirement and the requirement for precise knowledge of the expected climate changes therefore strongly limit the utility of this concept.

For a situation when changes in climate conditions are expected but likely responses are not known, two other approaches were proposed. Broadhurst *et al.* (2008) proposed "composite" provenancing that mimics natural patterns of gene flow by mixing seed from local provenances with progressively smaller amounts of seed from more distant sites, while Breed *et al.* (2013) proposed "admixture" provenancing as a way to increase adaptive potential by mixing a wide variety of provenances from sources across a species range without regard to the location of the planting site. Prober *et al.* (2015) combined the above approaches in his "climate-adjusted" provenancing in which the germplasm pool is composed of genotypes from a climatic gradient, biased toward environments more likely to be encountered in the future, and also includes local genotypes. The aim of this strategy is to maximize the potential for selection of pre-adapted genotypes and new gene combinations under predicted – with some certainty – new climate conditions, while maintaining genetic diversity for adaptation to other factors (Isaac-Renton *et al.* 2014). However, as the authors admit, the application of climate-adjusted provenancing needs to be tested within the context of potential risks of outbreeding depression and the disruption of local adaptation to nonclimatic factors. To avoid such a risk, I propose that seed sourcing (1) utilizes the ecoregion concept; (2) simultaneously considers the strength of local selection (evident in gene–environment interactions between local and nonlocal environments), scale of gene flow, and the extent and certainty of predicted climate change effects on the species range; and (3) takes into account the risk of inbreeding depression and loss of genetic variation that occur as a result of habitat degradation and fragmentation.

According to this concept, under no climate change effect, strong gene–environment interactions, and low gene flow, local provenance sourcing (i.e., sampling within an ecoregion defined at the appropriate scale) should be optimal (Figure 3.6). If an effect of climate change on the species range is expected, composite provenancing (i.e., a mix of local and nonlocal provenances) (Broadhurst *et al.* 2008) should be favored, and a proportion of nonlocal material in the mix should increase with the predicted impact of climate change on the species, the importance of climatic variables in defining ecoregions, and the intensity of gene flow (Figure 3.6). Even if the rate and severity of climate change cannot be predicted with certainty but

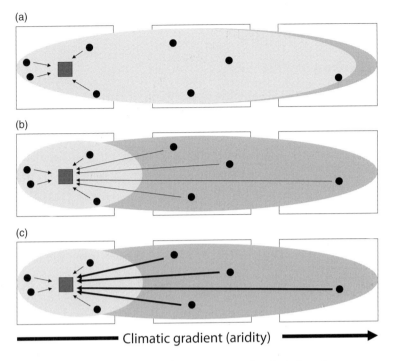

FIGURE 3.6 Seed sourcing for restoration. The shaded quadrat represents a restoration site within one of the three ecoregions located along the aridity gradient. Small circles are the existing populations of a target species. Two large shaded areas are the current and predicted species ranges, respectively. Multiple population sourcing is recommended in both scenarios. (a) Ecoregion-limited provenancing when no range change is predicted. (b and c) Ecoregion-mixed provenancing when range change is predicted. Proportion of nonlocal material in the mix (indicated by line width) increases with importance of climate in defining ecoregions.

directionality of change is clear, this strategy can be appropriate under both weak and strong gene–environment interactions if the latter are predominantly due to climate. Combining the idea of composite provenancing (Broadhurst *et al.* 2008) with the recommendation that nonlocal gene flow should not exceed 20% (Hedrick 1995), I suggest as a rule of thumb 80% and 20% for local (within the ecoregion) and nonlocal (outside it) material, respectively, when a climate-induced shift in species range is expected but additional information is unavailable.

### 3.6.3   Sampling Protocol

Information on the geographic distribution of rare species is usually scarce (Pulliam and Babbitt 1997), and getting precise distribution maps for endangered and rare species is difficult and often requires intensive surveys. A method of choice for detecting new populations and individual plants is adaptive cluster sampling (Thompson and Seber 1996; Philippi 2005). In this approach, when the species is found at one of the plots included in the original design, the neighboring areas within a neighborhood window of specified shape are searched more intensively. The procedure is repeated every time the species is found within the window until no additional occurrence is found. This method is shown schematically in Figure 3.7. Although adaptive cluster sampling is not universally optimal (Acharya *et al.* 2000; Bried 2013), it works well when the clusters are few and far between (Smith *et al.* 2004; Abrahamson *et al.* 2011), i.e., for species with limited dispersal abilities, low abundance, and narrow niches. As these attributes are characteristic for the majority of rare and threatened species, adaptive cluster sampling should be a preferred method for locating their individuals.

Because rare species are usually difficult to detect, a search for new occurrences of the target endangered species can greatly benefit from utilizing niche-based modeling, especially when knowledge and data on the abundance and distribution of the species are limited (Edwards *et al.* 2005; Guisan *et al.* 2006; Aitken *et al.* 2007; Williams *et al.* 2009; Le Lay *et al.* 2010). Model-based sampling that uses predictions from species distribution models can enhance the efficiency of fieldwork by limiting the searched area to areas of high habitat suitability. The protocol developed by Guisan with colleagues for sampling rare species (Guisan *et al.* 2006) deserves a detailed description. At first the model, based on topographic and climatic predictors, is fitted with available species occurrence data.

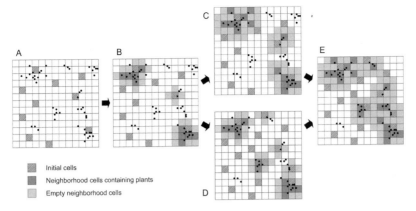

Initial cells

Neighborhood cells containing plants

Empty neighborhood cells

FIGURE 3.7 Adaptive cluster sampling (based on Philippi 2005). After a cell grid is superimposed on the target area where plants occur as eight spatially segregated clumps, sampling is initially done in a specified number of cells randomly located within the area (a). If a plant is found, the search continues in the neighborhood defined as a specified number of adjacent cells. If a neighboring cell also includes plants, it is added to the network and its neighbors are recursively sampled as well. In the set of networks produced, each network comprises initial and neighboring cells with plants and is surrounded by a ring of empty cells (b). The efficiency of the search can be increased by increasing: (c) neighborhood size (from 4 adjacent cells to 8); (d) intensity of initial sampling (from 10 initial cells to 15); or (e) both.

Then the spatial predictions derived from this initial model are used to stratify the field sampling. For this purpose, the predicted area map obtained is superimposed on a geographic map to identify sub areas with similar environmental conditions. Data obtained during sampling these subareas serve to update the dataset and fit improved models to inform the next sampling stage, and so on iteratively over several field seasons. Such model-based sampling of rare species, involving reiterative alternation of modeling and field sampling, was recognized by conservation biologists as very promising and cost-effective (e.g., Le Lay *et al.* 2010; Maschinski *et al.* 2012).

Decisions about sampling a population once or multiple times should be based on several considerations; in general, relatively small but more frequent samples are less harmful to a population than larger samples taken more infrequently (Menges *et al.* 2004). If a population undergoes strong demographic fluctuations (either stochastic or due to fluctuating environmental conditions), or mass flowering and fruiting happen only after disturbances (e.g., fire), adequate sampling during a single season may not be possible.

Beside limited seed production or variable reproductive output (e.g., masting), a reason for sampling a population over multiple seasons can be its small size. Within a season, several visits can be desirable if the flowering time is highly variable among the individuals or flowering conditions are unpredictable. For species with very limited seed production, the Center for Plant Conservation (1991) recommends that the number or propagules per sampled plant are 1–20, but for highly fecund species these numbers can and should be greatly increased to account for inevitably occurring attrition at each *ex situ* conservation stage (seed processing, storage, and propagation) and at the following *in situ* stage (outplanting). However, no more than 10% of annual production (Guerrant and Raven 2003) and less than 20% of the mature fruits per individual should be collected in order not to affect the natural regeneration of the sampled population. Seeds collected from individual plants should ideally be kept separately, but if they are to be bulked, approximately equal amounts of seed should be obtained from each sampled plant.

Whenever possible, at least 50 individual plants per population should be sampled; preferably, samples should be greatly in excess of 50 individuals (Way 2003; Guerrant *et al.* 2004, 2014) with the sampled plants separated spatially to increase the chance of sampling genetically unrelated individuals. The inter-sample distance will depend, on the one hand, on the species life form, breeding structure, and mode of seed dispersal, but also on the population size and a spatial pattern of plant distribution. Populations of species with predominantly clonal reproduction (e.g., many Iridaceae) often represent many ramets of very few or even one genet. In contrast, species with sexual reproduction are usually genetically heterogeneous. For annuals, the sampled individuals should be at least 1 m apart and for woody perennials the sampled plants should be separated by several crown diameters.

The right time for seed collection is also critical, as there is usually a short window of only a few days or weeks between seed maturation and dispersal, after which they are no longer available for collection. For example, in some *Lepidosperma* (Cyperaceae) viable seeds are shed quickly while nonviable seeds (which are indistinguishable from the viable ones) are retained for several months, meaning that late collection will yield only nonviable seeds (Kodym *et al.* 2010). Therefore, it is important to have a collection calendar with information about onset and duration of fruiting for each species of interest. This calendar will be used for repeated sampling in natural populations (Figure 3.8).

## 3.7 *EX SITU*: SEED BANKS

Seed availability is often a major constraint in reintroduction and restoration projects (Cochrane *et al.* 2007; Merritt and Dixon 2011). Seed banks have an important advantage over living collections in that for the same amount of genetic diversity they require much smaller space. For example, a typical cold room in Kew's Millennium Seed Bank stores 1 billion seeds representing 20 000 seed collections in just 30 m$^3$ (Smith 2014). Seeds in the seed banks are more secure than seedlings in a nursery, the latter being more susceptible to pests and diseases. Except for the species with seeds that have a very slow rate of germination, or, alternatively, are recalcitrant, or germinate immediately and cannot be stored, it makes sense to store plant diversity as seed right up to the time when it is needed (Smith 2014). By minimizing the time from germinating a seed to obtaining a seedling ready to be outplanted, the risk of losing a plant in a nursery due to infection or poor care is decreased.

However, despite many advantages over living collections, seed banks have well-known limitations in their utility for conservation *in situ*. Because the population samples in seed banks are typically small (obtained from fewer than 100 individuals in the

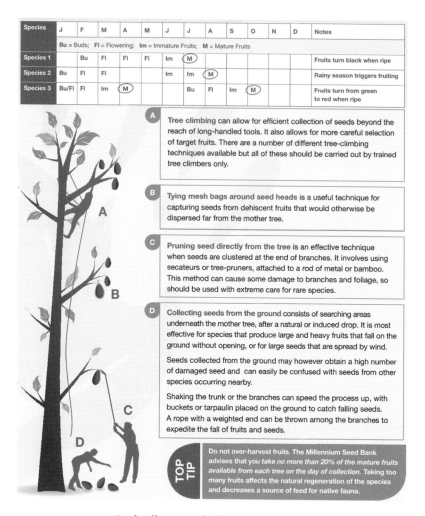

| Species | J | F | M | A | M | J | J | A | S | O | N | D | Notes |
|---|---|---|---|---|---|---|---|---|---|---|---|---|---|
| Bu = Buds;  Fl = Flowering;  Im = Immature Fruits;  M = Mature Fruits | | | | | | | | | | | | | |
| Species 1 | | Bu | Fl | Fl | Fl | Im | M | | | | | | Fruits turn black when ripe |
| Species 2 | Bu | Fl | Fl | | | Im | Im | M | | | | | Rainy season triggers fruiting |
| Species 3 | Bu/Fl | Fl | Im | M | | | Bu | Fl | Im | M | | | Fruits turn from green to red when ripe |

**A** Tree climbing can allow for efficient collection of seeds beyond the reach of long-handled tools. It also allows for more careful selection of target fruits. There are a number of different tree-climbing techniques available but all of these should be carried out by trained tree climbers only.

**B** Tying mesh bags around seed heads is a useful technique for capturing seeds from dehiscent fruits that would otherwise be dispersed far from the mother tree.

**C** Pruning seed directly from the tree is an effective technique when seeds are clustered at the end of branches. It involves using secateurs or tree-pruners, attached to a rod of metal or bamboo. This method can cause some damage to branches and foliage, so should be used with extreme care for rare species.

**D** Collecting seeds from the ground consists of searching areas underneath the mother tree, after a natural or induced drop. It is most effective for species that produce large and heavy fruits that fall on the ground without opening, or for large seeds that are spread by wind.

Seeds collected from the ground may however obtain a high number of damaged seed and can easily be confused with seeds from other species occurring nearby.

Shaking the trunk or the branches can speed the process up, with buckets or tarpaulin placed on the ground to catch falling seeds. A rope with a weighted end can be thrown among the branches to expedite the fall of fruits and seeds.

**TOP TIP** Do not over-harvest fruits. The Millennium Seed Bank advises that you *take no more than 20% of the mature fruits available from each tree on the day of collection*. Taking too many fruits affects the natural regeneration of the species and decreases a source of feed for native fauna.

FIGURE 3.8 Seed collection calendar and a brief introduction to four possible methods of seed collecting (from Hoffmann and Velazco 2014).

wild), the general principles of genetic drift in small populations also apply to seed banks. Seeds cannot be stored indefinitely and must be regenerated regularly due to a decrease of seed viability over time (Schoen and Brown 2001; Godefroid *et al.* 2010). The need for repeated propagation and small initial sample sizes of *ex situ* collections make them prone to such inevitable consequences of genetic drift as loss

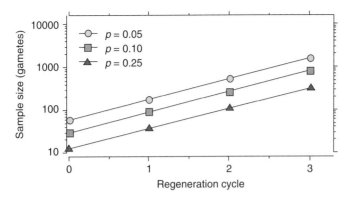

FIGURE 3.9 Maintaining the genetic variation in a seed collection through regeneration (from Schoen and Brown 2001): the relationship between sample size required for allele retention (with 95% probability), frequency of alleles to be retained (p), and number of regeneration cycles. The y-intercept shows the required size of the regeneration sample.

of genetic variation, increase in the level of inbreeding, and accumulation of deleterious mutations in regenerated collections. The effect of genetic drift on variation at a single locus depends on population size and frequency distribution of alleles. For example, to retain in a sample, with a probability of 95%, a particular allele that has a frequency of $p$ or more in a natural population, the number of gametes collected must equal approximately $3/p$. For $p = 0.05$, this is equivalent to sampling 30 diploid individuals in a population of an outcrossing species, and 60 individuals in a population of a self-fertilizing one (Schoen and Brown 2001). With each regeneration of a sample, to conserve all alleles captured in the initial sample with a 95% probability, the sample size must be three times as large as the previous one (Brown *et al.* 1997). This relationship between the sample size required for allele retention (with 95% probability), frequency of alleles to be retained ($p$), and number of regeneration cycles is shown in Figure 3.9, reproduced from Schoen and Brown (2001). As with loss of genetic variation, the level of inbreeding and accumulation of deleterious mutations will increase with the number

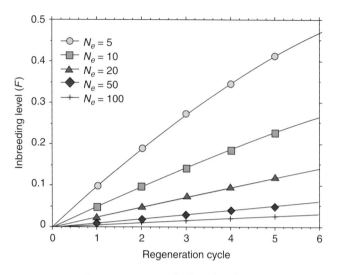

FIGURE 3.10 Changes in expected inbreeding level ($F$) with an increase in the number of regeneration cycles for different effective population sizes ($N_e$) (from Schoen and Brown 2001).

of regeneration cycles but decrease with larger initial sample sizes (Figures 3.10 and 3.11). For these reasons, repeated propagation should be done with as small a number of regeneration cycles as possible and fresh collections should be made at regular intervals.

Obtaining sufficiently large collections of threatened species is often extremely difficult or impossible, e.g., when populations are small, seed production is limited, and the proportion of reproducing individuals in a population is small. In these situations, the number of source plants and quantities of seeds per plant can be well below target levels. Of 332 critically endangered collections held in storage by the Western Australian Threatened Flora Seed Center, only five had >10 000 seeds, while 50% had <1000 seeds (Cochrane *et al.* 2007). For comparison, Guerrant and Fiedler (2004) estimated, using modeling, that the required initial collection size, taking into account expected losses through the reintroduction process for a population with 1000 mature plants for *Panax quinquenifolium*, would be 67 000 seedlings. Similarly, Cohrane *et al.* (2007), using data from 20 reintroductions,

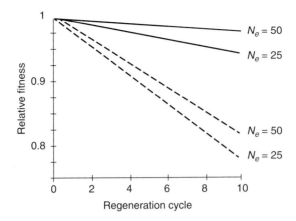

FIGURE 3.11 Fitness reduction due to the accumulation of deleterious mutations with an increase in the number of regeneration cycles for two different effective population sizes ($N_e$) (from Schoen and Brown 2001). Solid lines show the reduction in the case where selection acts to reduce mutation buildup. Dashed lines show the reduction in fitness when selection is minimized by equalization of family sizes.

estimated the required initial collection size for *Dryandra ionthocarpa* subsp. *ionthocarpa* to be 12 500 seeds. In restoration projects, the required seed quantities are even higher. For example, in the tropical forests of Borneo, even at a minimal planting density of 500 seedlings/ ha, over 7 billion seedlings would be required to restore the estimated 14.3 million ha of degraded forest (Kettle *et al.* 2011).

   Thus the quantities required for successful reintroduction or restoration can hardly ever be collected in natural populations even with multiple visits. They will not only require repeated collecting, but also an intermediate step of propagation. In these situations, a good approach is to distinguish short- and long-term seed collections that will have different purposes (Volis 2016c). The long-term collection will create a source of germplasm that can be used only for propagation or renewal but not for *in situ* actions. The short-term collection, on the other hand, will be oriented exclusively toward *in situ* actions with the collected material used for this purpose either directly or after propagation (e.g., Forte Gil *et al.* 2017).

Besides distinguishing two types of seed collections, several other improvements to the seed bank concept and management are needed. Although integrating seed bank collections into *in situ* conservation efforts, as sources for population reintroduction and restoration has been cited as an important justification for seed banks for many years (e.g., Hurka 1994; Holsinger 1995), threatened species are still a minority in major germplasm banks. For example, the China Germplasm Bank of Wild Species based at Kunming Institute of Botany, which has the largest seed bank collection in China and the whole of Asia, had in its collection in 2016 a total of 8855 species. However, only 47 of them (less than 0.6%) were species from critically endangered, endangered, and vulnerable categories. Moreover, this gene bank has never supplied seed to any reintroduction or restoration project. In North America, only around 21% of 9496 extinct in the wild, endangered, or vulnerable plant species native to North America (the United States, Mexico, and Canada) are maintained in *ex situ* seed bank collections in these countries (Kramer *et al.* 2011).

The only way to make seed collections useful for *in situ* conservation and their managers willing to invest efforts in collecting seeds of species, which pose a collecting challenge, is to establish seed banks that specialize in conservation and restoration, and make the collected seeds available for *in situ* actions. These seed banks should be designed to store, handle, and use a large number of seeds per species rather than just a large number of species, and they must place an emphasis on the collecting and storage of as much of the species genetic variation as possible. The idea of collecting and storing only those species to be used for *in situ* actions is utilized in restoration seed banks, facilities created to collect, store, and propagate germplasm to be used in restoration projects (Merritt and Dixon 2011). Experience of existing restoration seed banks should be learned and adopted for the specific needs of conservation seed banks.

Although ideally all threatened and rare species should be represented in seed bank collections, seed collecting is a time- and money-consuming process, and therefore species prioritization is necessary. The species most threatened globally are a priority, but prioritization should also take into account regional and local concerns (Guerrant and Raven 2003). As Guerrant and Raven (2003) pointed out, even though one particular species might not be as globally rare or threatened as another, there might be particular threats known to the local botanists that would warrant extra protection through seed collection. Other criteria listed in Section 3.3 can also be used for species prioritization.

Seeds from each maternal plant should, whenever possible, be maintained separately. Although maintaining seeds from each maternal plant separately is a bit more labor intensive than combining the seeds from all the plants into a single bulk collection, the potential rewards of keeping accessions separately greatly outweigh other considerations because this practice allows for the maximum control and management of genetic diversity. Knowledge of genetic identity of the seeds is important for a number of reasons. First, many rare and endangered species have only very small populations with regeneration problems and specific genotypes may be needed to reinforce sexual reproduction in these populations. Second, the creation of reintroduction gene pools may need an intermediate step of targeted breeding prior to introduction, and, in this case, only collections organized by family can provide the information necessary for breeding designs. Third, knowledge of the seed identity is necessary for maximizing family number and equalizing family size in reintroduction gene pools (Guerrant 1996; Guerrant and Pavlik 1997; Havens *et al.* 2004). From 2500 to 10 000 seeds per accession is the amount that should be sufficient for the diverse uses of seed collections, including long-term preservation, germination testing, research, and restoration trials (Meyer *et al.* 2014).

The international standard for long-term seed storage temperature is −18°C or lower, with a seed moisture content between 3 and 7% depending on the species (FAO/IPGRI 1994). The moisture content can be achieved under drying conditions of 10–15%

relative humidity at 10–15°C. Efficient management of the seed bank collection also requires control of the quality and viability of collected seeds (Godefroid *et al.* 2010; Ferrando-Pardo *et al.* 2016). Before successive germination tests to monitor the changes in viability, a reliable germination protocol for each target species must be developed. In the absence of any knowledge of a species germination requirements, a small number of seeds (usually five seeds for each treatment) is typically subject to the following four standard germination conditions (Guerrant and Raven 2003):

- direct placement into a 20°C chamber with an 8h light and 16h dark cycle;
- direct placement into a chamber with alternating temperatures: 20°C during the light portion of the cycle (8h) and 10°C during the dark portion (16h);
- 8 weeks of cold stratification followed by placement into the 20°C chamber (with the same light/dark cycle);
- 8 weeks of cold stratification followed by placement into the chamber with alternating 10/20°C temperatures.

## 3.8    *EX SITU*: LIVING COLLECTIONS IN BOTANIC GARDENS

### 3.8.1    *Conservation Potential of Botanic Gardens and Their Limitations*

A botanic garden (including an arboretum) can be defined as "a place with an orderly, documented, labeled, collection of living plants, that is open to the general public, with collections used principally for research and education" (Watson *et al.* 1993) or as an institution "holding documented collections of living plants for the purposes of scientific research, conservation, display and education" (Wyse Jackson 1999). Currently, there are more than 2700 botanic gardens around the world. The fundamental conservation contribution of botanic gardens is in preserving, through living collections, imperiled or extinct in the wild populations and species (Raven 1981; Griffith *et al.* 2011). The CBD recognizes the value of *ex situ* conservation in botanic gardens (Glowka *et al.* 1994; Wyse Jackson 1997) with these activities being undertaken "preferably in the country of origin" and as a support to the "recovery and rehabilitation of threatened species and for their reintroduction into their natural habitats." The IUCN SSC policy on *ex situ* conservation recognized the primary goal of *ex situ* activities as being "to help support the conservation of a threatened taxon, its genetic diversity, and its habitat" (IUCN 2002), and later stated that "for a growing number of taxa *ex situ* management may play a critical role in preventing extinction as habitats continue to decline or alter and become increasingly unsuitable" (IUCN/SSC 2014). The Global Strategy for Plant Conservation highlighted that role by setting the requirement of a minimum of 75% of threatened plant species to be preserved within *ex situ* collections, with at least 20% available for recovery and restoration (Wyse Jackson and Kennedy 2009). Conservationists and botanic gardens have recognized the potential of botanic gardens for preserving threatened plant species as well as producing outplants for reintroduction and restoration programs (Hardwick *et al.* 2011; BGCI 2012; Cibrian-Jaramillo *et al.* 2013).

For species that are extinct in the wild, botanic gardens can often be the last chance of survival and reintroduction. For example, *Hyophorbe amaricaulis* is represented by a single individual in Curepipe Botanic Gardens (Ludwig *et al.* 2010). *Corypha taliera* has no more than 20 living individuals in four botanic garden living collections (Dhar 1996; Griffith *et al.* 2011). *Encephalartos woodii* is widespread in collections as a vegetatively propagated single male individual. All the individuals of *Sophora toromiro* grown in botanic gardens and private collections originated from a few seeds collected in the wild before species extirpation (Maunder *et al.* 2000). Greuter (1994) lists 37 extinctions from the Mediterranean, of which four survived in cultivation, namely *Coincya pseuderucastrum* subsp. *puberula*; *Diplotaxis siettiana*; *Lysimachia minoricensis*, all from Spain, and *Tulipa sprengeri*, from Turkey. Other examples include *Bromus interruptus*, UK; *Myosotis ruscinonensis*, France; *Cochlearia polonica*, Poland (Maunder *et al.* 2001b), and *Tecophilaea cyanocrocus*, Chile (Maunder *et al.* 2001a). Unfortunately, attempts to reintroduce such species into the wild are very rare and mostly unsuccessful (e.g., *Sophora toromiro*; Maunder *et al.* 2000). *Lysimachia minoricensis*, which was successfully reintroduced into the wild (Fraga *et al.* 1997), is a lucky exception.

Maintenance in botanic garden living collections is also vital for many critically endangered species. For example, only two individuals of *Pritchardia aylmerrobinsonii* remain in the wild, but at least 30 botanic gardens maintain this species in their living collections (Chapin *et al.* 2004). *Hemithrinax ekmaniana*, growing on just two hilltops in central Cuba, is now preserved and is being reproduced at Montgomery Botanical Center (Griffith *et al.* 2011).

On the other hand, botanic gardens provide only a limited time opportunity for an imperiled species to survive. That threatened species and especially those that are extinct in the wild cannot be kept in cultivation indefinitely is a sad fact. For example, the last known specimen of *Vicia dennesiana* endemic to the Azores died in cultivation (Maunder *et al.* 2001b). Similarly, the last individuals of Hawaiian endemic *Cyanea kuhihewa* grown in the National Tropical Botanical Garden were killed by an unknown disease (Griffin-Noyes 2012). *Bromus eburonensis*, the only Belgian endemic, was cultivated for many decades at the Botanical Garden of the University of Liège but went extinct when the area was transformed into a public park (Govaerts 2010). Of 844 plant taxa listed as extinct in the wild in 2010, 5% had been in collections but were subsequently lost (Govaerts 2010). Thus, the opportunity of reintroducing these species is lost forever. This happens, at least in part, due to insufficient cooperation between botanic gardens and conservation organizations and a lack of *ex situ* conservation plans.

However, this does not mean that cooperation between botanic gardens and conservation agencies does not exist. There are examples of the utilization of botanic garden living collections in various reintroduction and restoration programs. Over 600 rescued individuals from six extirpated wild populations of *Amorpha herbacea* var. *crenulata* maintained at Fairchild Tropical Botanic Garden were used in a translocation program (Wendelberger *et al.* 2008). Similarly, more than 200 seedlings and juvenile plants of *Pseudophoenix sargentii* were produced from a living collection maintained at Fairchild Tropical Botanic Garden and used for reintroduction (Fotinos *et al.* 2015). The Berry Botanic Garden has been directly involved in experimental reintroductions of three endangered taxa. It supplied 1000 seedlings to reintroduce *Stephanomeria malheurensis*, was directly involved in designing and executing the reintroduction of *Lilium occidentale*, and in augmenting with seeds and young plants an existing population of *Arabis koehleri* var. *koehleri* (Guerrant and Raven 2003). Bok Tower Gardens actively participated in producing outplants for the augmentation and experimental introduction of *Ziziphus celata* (Menges *et al.* 2016). Outplants for the translocation of *Dianthus morisianus* were produced at the Botanic Gardens of Cagliari University (Fenu *et al.* 2016). Among the 25 rare or extinct in the wild species introduced at restored wetland sites in Switzerland, outplants for seven species were from populations maintained at the Botanical Garden of Bern (Noël *et al.* 2011). The Australian Botanic Garden Mount Annan, using plants maintained in its living collection, produced more than 100 outplants of critically endangered *Wollemia nobilis* for experimental translocation (Zimmer *et al.* 2016). Brackenhurst Botanic Garden was the main driving force in a forest restoration project in Kenya throughout its planning and implementation; this involved collecting seeds, producing outplants, and planting (Shaw *et al.* 2015).

Nevertheless, despite the above examples, cooperation between botanic gardens and conservation organizations is still very poor. Partly, this is because of numerous limitations undermining the utility of botanic gardens for conservation. A need for the regeneration of planted material leads to genetic erosion (Schoen and Brown 2001) and the divergence of an *ex situ* collection from the wild source population over time (Ensslin *et al.* 2011; Rucinska and Puchalski 2011; Lauterbach *et al.* 2012). Due to space limitations, and because the collecting and regenerating of seed samples is costly, garden living collections rarely accommodate sufficiently large population sizes. Usually the population samples in seed banks are small and often from fewer than 50 individuals in the wild. The small sample sizes of *ex situ* collections and the need for regeneration inevitably lead to genetic drift and an increase in the level of inbreeding in regenerated collections (Schoen and Brown 2001). The latter can result in fitness decline due to inbreeding depression (Havens *et al.* 2004; Vitt and Havens 2004). Genetic drift in garden collections increases with the duration of cultivation and number of generation cycles in the same way as in the seed bank collections (Figure 3.9). Plants with short generation times such as annuals, biennials, or short-lived monocarpic perennials, are especially vulnerable to drift, compared to long-lived perennials. Maintaining a viable *ex situ* collection for even a small fraction of the overall genetic diversity demands significant resources, in terms of land size and budget.

Another negative consequence of cultivating plants *ex situ* is the potential risk of adaptation to the *ex situ* environment with a loss of adaptations to the original natural environment (Havens *et al.* 2004; Ensslin *et al.* 2011, 2015). Very few studies have so far addressed trait change and adaptation as a result of *ex situ* cultivation and for this reason the study of Ensslin *et al.* (2011) requires close attention. Ensslin *et al.* (2011) compared, using a common garden experiment, *ex situ* populations from 12 botanic gardens with five natural populations of the perennial herb *Cynoglossum officinale*. Garden populations exhibited strikingly lower seed dormancy than natural populations, and garden plants had larger inflorescences but fewer flowering stems than wild plants. These changes are consistent with domestication syndrome, which typically includes loss of seed dormancy and the production of larger inflorescences (Zohary *et al.* 2012; Iriondo *et al.* 2018). In gardens, the staff act in the same way as the early farmers did, imposing unconscious selection: they usually plant out only the early germinants and collect seeds mainly from tall plants with a long main inflorescence. The resulting trait changes can be maladaptive in nature (e.g., too early germination).

Among other drawbacks of living collections is their poor representation of genetic variation in wild populations (Christe *et al.* 2014; Griffith *et al.* 2015), and commonly observed lack of information on accession sampling locality and mislabeling (Hurka 1994). Physical proximity of plants leads to a high risk of infestation by pathogens and, if they have different origin, may result in spontaneous hybridization. A risk of hybridization seriously limits the utilization of botanic garden *ex situ* collections for conservation purposes because the hybrids may lack genetic integrity and harbor maladaptive gene combinations (Maunder *et al.* 2004b). To prevent these risks, sampled individuals must be maintained separately or through controlled breeding and pedigree design, which is problematic because of gardens' space limitations and high cost of maintenance.

### 3.8.2   *How to Improve the Conservation Utility of Botanic Garden Living Collections?*

The limitations of botanic gardens were recognized a long time ago, leading to an understanding that preservation by cultivating plants in botanic gardens is unlikely to succeed in the long run (Raven 1981; Maunder *et al.* 2001b) and that historical botanic garden accessions should be used for *in situ* actions with caution (Maunder *et al.* 2001b, 2004b; Havens *et al.* 2006). On the other hand, there always will be a need for botanic gardens to directly support *in situ* conservation, and, despite their limitations, botanic gardens are an important and currently underutilized resource for the management and conservation of threatened species. Maunder *et al.* (2001b) considered botanic gardens central to an integrated *in situ* and *ex situ* conservation strategy. However, there are logistical limitations and most importantly, an "insufficient understanding of the potential contribution of *ex situ* conservation hampers the implementation of truly integrated plant conservation strategies" (Havens *et al.* 2006). The effective *ex situ* conservation of threatened species requires a new concept. In this new concept, the current focus of the living collections on maintaining limited numbers of species representatives must be changed to maintaining species genetic diversity and an explicit orientation toward supporting reintroduction *in situ*.

Recently, Cibrian-Jaramillo *et al.* (2013) proposed an approach for the management and use of botanic garden living collections that is aligned with *in situ* conservation goals. In this approach a particular species living collection is assigned a conservation value based on a species risk assessment, the genetic representation of the collection in the context of the total species genetic variation, and the operational cost of maintaining a collection, with this information shared via online databases (e.g., BGCI's PlantSearch database). An approach like this one can greatly assist coordination among botanical garden collections, on the one hand, and between collections and interested organizations pursuing *in situ* conservation on the other.

Following an idea of Cibrian-Jaramillo *et al.* (2013) to create a unified management strategy of botanical gardens collections, I propose my version of a strategy for the management of threatened plants in living collections (Figure 3.12). This strategy includes setting regional conservation priorities for the species, the creation of genetically representative collections for high priority species, and the use of these collections in *in situ* actions. The value of the existing and future species living collections for conservation will be a function of the species conservation status and how well the collection represents its natural genetic variation. The strategy includes the following components:

1. Prioritization: Many gardens grow threatened plants in their collections, but there are often many more threatened species than the country's conservation infrastructure including botanic gardens can recover (Maunder *et al.* 2004a; Havens *et al.* 2006). To address this, the existing collections can be optimized by the gardens focusing on those species most threatened, endemic, and adapted to the local climate. Selection

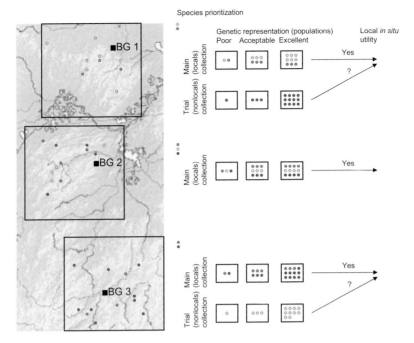

FIGURE 3.12  A proposed unified management strategy of botanical gardens' living collections exemplified by collections in three hypothetical botanic gardens (BG1-3) (from Volis 2017b). Colored circles denote populations of three different species in three ecoregions denoted by rectangles. Species in each ecoregion are prioritized based on a set of criteria. Representative collections are created for species with the highest regional priority. Only collections representing all known populations in the region (excellent representation) or more than two populations (acceptable representation) can be used for *in situ* actions. In addition to the main collections of regionally local species, trial collections of nonlocal threatened species can be used for regional *in situ* actions based on plants' performance and SDM predictions. *A black and white version of this figure will appear in some formats. For the color version, please refer to the plate section.*

of threatened species to be maintained in a garden can utilize the regional conservation planning species scoring described in Section 3.3. Utilization of this scoring system will give the highest priority to the most vulnerable local endemics representing distinct taxa.

2. Genetic diversity: Botanic garden collections must better conserve original genetic diversity (i.e., that present in natural populations). Unfortunately, the majority of the botanic garden stocks of threatened species are genetically depauperate relative to the wild populations. For example, *Lysimachia minoricensis*, while numerically secure in cultivation, is thought to derive from a single founder, and the cultivated stocks of *Lotus berthelotti* are self-incompatible and probably represent only one clone (Maunder *et al.* 2001b).

Thus, the collecting strategy must be based on principles described in Section 3.6, and each population in the living collection should be represented by multiple accessions which are managed separately (Griffith *et al.* 2015). In the living collections, minimization of genetic threats can be achieved by maintaining large population sizes, providing close to natural growing conditions, decreasing the number of generations in captivity, and periodic immigration from wild populations (Havens *et al.* 2004, 2006).

In order to become useful for *in situ* actions, existing collections of threatened species that do not properly represent the species genetic diversity must be enriched by either the exchange of accessions among gardens or by collecting more from the wild.

3. Redundancy: For every critically endangered species, there must be more than one living collection to prevent accidental loss due to an extreme climatic event or disease, or as inevitably happens in small collections, genetic drift. For example, *Attalea crassispatha* is imperiled by habitat reduction and seed consumption in its natural environment in southwest Haiti, where it has fewer than 30 individuals (Timyan and Reep 1994). Three large *ex situ* living collections ensure that this critically endangered species can survive in cultivation (Griffith *et al.* 2011). Another example is *Brighamia insignis*, an endemic Hawaiian succulent species which is functionally extinct in the wild as it is represented in nature by only one remaining extant individual. Fortunately, it is cultivated *ex situ* in more than 50 botanical collections around the world (Fant *et al.* 2016).

If done in a range of environmental conditions, duplicating collections through sharing plant material can, in addition to increased likelihood of long-term survival, also provide vital information about species climatic tolerance. Botanical gardens have largely unutilized utility for climate change research. Because one of the major goals of botanical gardens traditionally was (and still is) the creation of collections of taxonomically and ecologically diverse flora, and because plants in these collections are maintained under the most optimal conditions as possible for these species (e.g., mulching, weeding, fertilization, pest control), the effects of climate on plants in these collections are not confounded with other effects, allowing inferences of origin by climate interactions, as in common garden experiments (Primack and Miller-Rushing 2009). Observations on key phenological events (leaf bud burst, flowering, fruiting, leaf color changes, and leaf senescence), besides mortality and reproduction, across botanical gardens representing different climatic zones, are invaluable for understanding the effects of changing climate.

Often disadvantageous from a conservation biologist's point of view, that botanic gardens give an opportunity to plant and monitor both local and nonlocal species, can be turned into an advantage. If done properly, this can allow reliable inferences about the impacts of climate change on a target species. For example, there are 25 gardens located in Mexico and the southwestern United States of which eight are located within the boundaries of the Mojave, Sonoran, or Chihuahuan Deserts. A reciprocal garden network using these eight desert botanic gardens alone would span a 6°C mean annual maximum and 10°C minimum temperature gradient. This network could be used for the reciprocal planting of a number of threatened Cactaceae species in a common garden framework and provide precious information about the anticipated responses of the planted species to climate changes (Hultine *et al.* 2016).

4. Integration with *in situ* conservation: For every extinct in the wild and critically endangered species maintained in a botanic garden, there must be a program explicitly oriented toward its reintroduction. This requires close coordination with conservation agencies. The potential of garden collections for reintroduction and even restoration of habitats has often been suggested (Pavlik 1997; Maunder *et al.* 2001b; Hardwick *et al.* 2011; Cibrian-Jaramillo *et al.* 2013; Griffith *et al.* 2015), but practical implications of this idea are very modest and limited to a handful of cases of supplying a small number of outplants for experimental translocations or recommendations based on the analysis of the genetic variation preserved in species collections (e.g., Wendelberger *et al.* 2008; Da Silva *et al.* 2012; Yang *et al.* 2015; Menges *et al.* 2016; Zimmer *et al.* 2016).

Without close coordination with conservation agencies and participation in the development of regional conservation and habitat restoration plans, botanic gardens will never become an integrated part of threatened species conservation.

### 3.9   *QUASI IN SITU* LIVING COLLECTIONS

The *quasi in situ* approach (Volis and Blecher 2010) directly addresses the key *ex situ* issue of the long-term storage of species genetic diversity. To overcome the space and logistic limitations of garden living collections, in this approach, the living collections of required capacity are created in sites that have legal protection status and natural or seminatural conditions. These can include even the least valuable and degraded, to some degree, parts of archeological, memorial, cultural, and other, not only natural, protected areas, as well as buffer zones in nature reserves. The choice of a site explicitly takes into account the issue of local adaptation, i.e., it assumes that the structure of species genetic variation is related to the site's environmental conditions and, therefore, requires a stratified design with a lower level (i.e., population) nested within a higher one (i.e., ecoregion or habitat) for both sampling and establishing collection sites. This means that for a species whose range covers different habitats (e.g., soil types, regions of different aridity, vegetation communities), several geographically isolated populations are sampled in each habitat and there is a close environmental match of *ex situ* location and locations of the sampled natural populations. Different genotypes are planted separately at a distance from each other allowing subsequent identification of those planted genotypes. This can be important for the identification of superior, in terms of survival and seed production, genotypes, maintaining collection genetic variation and allowing controlled pollination, if necessary.

Plants, grown in this collection, can be expected to contain adapted and genetically different individuals. As all plants at a particular *ex situ* site originate from the same environment, there will be no maladapted genes to participate in recombination and segregation, and the cross-pollination of these plants should not lead to a break-down of coadapted gene complexes or the dilution of local adaptation.

The features described above make *quasi in situ* well suited to the purpose of the long-term preservation of species genetic variation. On the other hand, this strategy can provide a real bridge between *ex situ* and *in situ* conservation because the offspring of cross-pollination occurring in the collection can be used for *in situ* actions such as reinforcement, reintroduction, or habitat restoration (see Figure 2.7). It is often a problem for the nurseries producing seedlings of rare and threatened species to plan the numbers of seedlings that might be produced with certainty because of the low predictability of seed production in many such species. For species which have low or very variable seed output, or have very few, and isolated reproducing adults in natural populations, *quasi in situ* collections can be a solution.

The *quasi in situ* approach has been successfully used for conservation of a threatened rhizomatous perennial *Iris atrofusca*, which is endemic to Israel and Jordan (Volis *et al.* 2010). Rescue collecting of adult plants has been done in several locations of Israel where the future of the plants was either highly uncertain due to anthropogenic threats (plow, grazing) or where infrastructure construction was planned. The species has been planted in two national parks which have relatively intact soil and are located within the Negev Desert, the area of the species natural distribution. One of these living collections is shown in Figure 3.13. As *I. atrofusca* occupies two distinct environments in the Negev Desert that differ, first of all, in the soil type, two national parks were chosen and the planting achieved exact soil-type matching.

The *quasi in situ* approach has a close analogy in seed orchards used in forestry (Figure 3.14). The purpose of seed orchards is also to preserve intraspecific genetic diversity and produce seeds. A good example of the utilization of seed orchards in a restoration project with conservation goals is a program of the production of autochthonous tree-planting stock in Flanders, Belgium (Vander

FIGURE 3.13 A *quasi in situ* living collection of *Iris atrofusca* created in Tel Beer Sheva National Park, Israel, 5 years after planting. Note the large number of fruits produced by the plants. Photo by the author. *A black and white version of this figure will appear in some formats. For the color version, please refer to the plate section.*

Mijnsbrugge 2014). In order to preserve the remaining autochthonous populations of woody species of Flanders, a survey was conducted to locate these remaining autochthonous populations, followed by the collection of seed from at least 30 seed-bearing plants per species within each of four regions of provenance, and the creation of seed orchards. Seed orchards have been established for all woody species that are regularly or occasionally planted in Flanders. As there are four main regions of provenance in Flanders, four seed orchards for every woody species, one for each region of provenance, have been set up. The purpose of the seed orchards is to preserve the gene pool of the populations from which the plants in the orchard originated, with the planted orchard individuals serving as parent material for breeding in the future. Every seed orchard contains a minimum of 50 genotypes per species, collected from at least five different sites, and up to four ramets per genotype. In addition, it is planned to duplicate each seed orchard at another location within the region of provenance. Once established, the autochthonous seed orchards

FIGURE 3.14 The forestry seed orchards preserving genetic diversity of native woody species of Flanders, Belgium: (a) *Crataegus monogyna*, (b) *Malus sylvestris*, (c) *Prunus padus*. Photos by Kristine Vander Mijnsbrugge. *A black and white version of this figure will appear in some formats. For the color version, please refer to the plate section.*

are officially approved as seed sources (category "source identified") and the seeds from them can be certified. The majority of orchards are situated on land owned by the Agency for Nature and Forest, under the Flemish Forest Administration, while some have been established on municipal land and land owned by nature conservation organizations (Vander Mijnsbrugge 2014).

An example of using seed orchards for the conservation of critically endangered tree species is provided by Ducci (2014). *Abies nebrodensis*, or Sicilian fir, is represented by a single relict population of approximately 30 adult trees spread over an area of 150 ha in Sicily. The population suffers from a high level of selfing, poor production of cones, low seed viability, and infrequent germination. To preserve the species genetic diversity, two seed orchards were created through massive clonal replication, with grafts from all adult trees of the original population. Four-year-old individuals of closely related *Abies alba* served as the rootstock. As a result, besides preserving the original species genetic diversity, the seed orchards produced several hundred seedlings of diverse genetic origin (Ducci 2014).

## 3.10 OBTAINING OUTPLANTS: MANAGED BREEDING

When the number of plants acting as parents for seed production is small, and fruit/seed set is low, managed breeding is necessary. Robichaux *et al.* (2017) provide two excellent examples of successful managed breeding that made it possible to produce large amounts of outplants of two critically endangered species – Hawaiian endemics *Argyroxiphium kauense* and *Clermontia peleana*.

At the start of the breeding program, silversword (*A. kauense*) was represented by two remnant populations, having 381 and six individuals, respectively, and these two populations served as source populations for reintroduction efforts. Flowering wild plants and those cultivated at the Hawaiian Volcano Rare Plant Facility were pollinated with the pollen collected from all other plants flowering at the same time, and thoroughly mixed (Figure 3.15). The bulk pollen was used in order to increase the genetic diversity of offspring for each maternal founder. From the produced offspring, all maternal founders were represented in the reintroduced population, but no maternal founder accounted for >2.5% of the total number of seedlings to avoid overrepresentation. The number of outplants produced was >21 000 seedlings derived from 169 founders.

Lobeliad (*C. peleana*) had six remnant plants in the wild, all growing epiphytically in heavily degraded montane wet forest habitat, and one individual growing in cultivation at the Rare Plant Facility, when the managed breeding program started. Because the six remnant plants grew in an area deprived of honeycreepers, their natural pollinators, their hand pollination was highly problematic (the plants grew epiphytically high in the forest canopy, and the pollen could be washed away by frequent rains), an air-layering technique was used. Five of the six remnant plants produced air layers which were harvested and grown as rooted cuttings at the Rare Plant

FIGURE 3.15 Managed breeding program for critically endangered Hawaiian endemic *Argyroxiphium kauense*. The photos show (a) the silversword plants cultivated at the Volcano Rare Plant Facility and (b) in the remnant population that acted as both mother plants and pollen donors; (c) hand pollination of a mother plant; (d) seedlings produced for outplanting at the Kahuku site in Hawaii Volcanoes National Park; and (e) the reintroduced plants. Photos by Robert Robichaux and David Boyle. *A black and white version of this figure will appear in some formats. For the color version, please refer to the plate section.*

Facility. Each of these cultivated founders flowered in one or more subsequent years, and was successfully hand pollinated. Multiple mature fruits produced via either selfing or outcrossing from each of the six founders provided >1000 seedlings used as outplants.

In the two examples above, breeding has been done either in the wild or in a facility. However, when a wild population is not within easy reach and the adult plants are large, living collections suit managed breeding better. For example, seed orchards helped to produce several hundred seedlings of diverse genetic origin and high heterozygosity of critically endangered Sicilian fir *Abies nebrodensis* (Ducci 2014), as described in the preceding section. This was achieved by replicating the remaining 30 wild adult trees through clonal propagation and planting the mother trees in a complete single-tree random design to enable efficient pollen exchange among genotypes. In the natural population, pollen flow is severely constrained by the very low density of adult trees.

Managed breeding is an option for plant species whose existence is threatened by emerging invasive diseases. The most striking examples are American chestnut (*Castanea dentata*), European ash (*Fraxinus excelsior*), and Dutch elm (*Ulmus glabra*). For these species, the creation of the outplanting stock must be done in cooperation with plant pathologists and involve the identification of resistant individuals or their genetic modification followed by a breeding program. The goal of such a program is to develop resistance while ideally preserving as much as possible of the extent and structure of the remaining species genetic diversity in the wild (Jacobs *et al.* 2013; Pautasso *et al.* 2013; McKinney *et al.* 2014; Budde *et al.* 2016; Steiner *et al.* 2017).

### 3.11 OBTAINING OUTPLANTS: TYPES OF SEEDS AND THEIR PROCESSING

Knowledge of seed longevity in storage and of seed handling, processing, and germinating procedures is essential to raise sufficiently large quantities of seedlings and to meet a specified planting date. This information is lacking for the majority of threatened species. Therefore, some general guidelines for different kinds of seeds, such as presented in Table 3.3, can serve as a first step in developing species-specific protocols. With growing utilization of rare and threatened

Table 3.3 *Some general germination methods recommended for a range of rainforest seed types (from Lott* et al. *2005 with changes)*

| Type of seeds | Seed processing |
| --- | --- |
| Fleshy fruits with hard stones | Slow germination over years unless the correct treatment to break dormancy is applied. Both fresh and old fruits can be collected and sown densely together into the same tray. |
| Capsuled fruits with arils and firm seeds | Rapid germination; viability is usually very short. The fruits should be collected from trees to prevent insect infestation, and seeds cleaned and sown as soon as possible after collection. |
| Soft, fleshy fruits with several to many small seeds | Viability times vary greatly. If the viability time is not known, early processing and sowing is preferred. If possible, the flesh should be removed. If the seeds are too tiny to sieve the whole fruits can be mixed with fine sand to a crumbly texture before sowing. |
| Firm, fleshy fruits with one to a few large seeds | Germination is generally rapid; seed may have short viability. The seeds should be peeled immediately after collection, soaked to drown insect larvae, and sown as soon as possible. |
| Very large seeds | Prompt sowing and enough space to accommodate vigorous early root systems are recommended. |
| Winged seeds | Long-lived and generally storable if kept dry. Seed should be collected just as the first capsules open. When sown, seeds need only a light cover by the germination medium. |
| Very small seeds, usually in dry capsules | Long-lived and generally storable if kept dry. When sown, seeds should not be covered by a germination medium. The latter must be a fibrous, airy but moisture-holding medium, sprinkled sparsely on top. |
| Hard-coated (i.e., with water-impermeable coats) seeds | Special treatment is often needed such as scarification, brief boiling and soaking, or alternating the temperature, and light conditions. |

species in restoration projects, region-specific species lists with detailed information on seed storage and germination requirements will become available. An example of such a list can be found in Sautu *et al.* (2006), which presents 100 tree species native to the seasonal moist tropical forest in the Panama Canal Watershed, selected as candidate species for use in reforestation projects in Panama. Another example is a list of 21 palm species of Tirimbina rainforest, Costa Rica, assessed for suitability in forest restoration (Ley-López and Avalos 2017).

Seeds of many species have dormancy, and will not germinate without first breaking this dormancy. Naturally, dormancy break usually happens after exposure to season-specific temperatures, and the prescribed dormancy-breaking treatment should simulate the latter. For many taxa, however, the information about a sequence of summer → fall → winter temperatures to break dormancy does not exist or can be difficult to obtain. A good solution is the "move-along experiment" or double-germination phenology technique proposed by Baskin and Baskin (2003) to determine the temperature, or temperature sequence, for seed dormancy break in seeds that are permeable to water. Before initiating this experiment, the intact seeds/germination units must be checked to see whether they are dormant and their seed coat is water permeable. The former is done by incubating them for 2 to 4 weeks under several temperature regimes and checking germination. The latter is done by weighing seeds (or peeled fruits) on an analytical balance before and after they have been incubated at room temperature on a moist substrate for one to several days. If the seed/fruit weight does not increase, then the germination unit is impermeable to water.

Water-permeable dormant seeds are sown in Petri dishes (50 or 25 for seed-limited species per dish) on moist soil, sand, or filter paper. The temperature regimes to which the seeds will be subjected normally should include 5°C to simulate winter stratifying temperatures, 25/15°C for summer, 20/10°C for early fall and late spring, and 15/6°C for late fall and early spring; these are 12h day temperatures alternated with 12h night temperatures. These

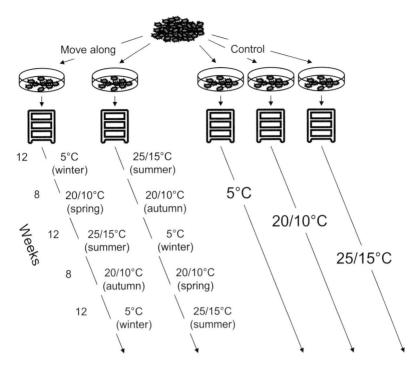

FIGURE 3.16 The move-along experimental procedure of Baskin and Baskin (2003) for determining the temperature or temperature sequence needed to break seed dormancy. Two "move-along" treatments show progression of temperatures for seeds starting in "winter" and "summer." The 20/10°C treatment is omitted.

temperatures may need to be adjusted to those occurring in the target region. If the number of seeds and/or incubators is limited, 15/6°C can be dropped from the design and 20/10°C can be used to simulate the entire fall and spring. The move-along portion of the experiment consists of two treatments: one treatment is started at 5°C and the other at 25/15°C. Seeds in these two treatments are moved through the next three seasons (Figure 3.16) in contrast to the control portion of the experiment in which four temperature regimes (5, 15/6, 20/10, and 25/15°C) are maintained permanently.

 If seeds require only summer temperatures for dormancy break, the move-along seeds that start at 25/15°C will start germinating when moved to fall temperatures, but not at any of the four control temperatures, or when started at 5° and moved to spring temperatures. Conversely, if seeds require only winter temperatures for dormancy break, the move-along seeds that start at 5°C will germinate when moved to spring temperatures, but not if kept continuously at any of the four temperatures, or in the seeds started at 25/15°C and moved to fall temperatures.

For seeds that require the summer through winter temperature sequence (or warm followed by cold) to break dormancy, the move-along seeds that start at 5°C will not germinate when moved to spring temperatures (neither control seeds will), but those that start at 25/15°C will germinate after being moved to spring temperatures. Further, seeds started at 5°C and subsequently moved from spring → summer → fall → winter → spring will germinate in the second spring.

There are two general ways of sowing seeds in which dormancy is no longer a problem. Large seeds can be sown directly in single- or multi-plant plastic containers (Figure 3.17). However, for small-seeded species, it is necessary to grow seedlings in germination trays (Figure 3.17) to a manageable size before they can be transferred to containers (a procedure called pricking-out).

Germination trays should be placed on wire grid benches. Using benches increases air circulation and causes the medium surfaces to dry more rapidly after watering. The latter decreases plants' susceptibility to diseases. In the germination trays, seeds should be sown on the surface of the growing medium and then covered with a thin layer of the medium approximately 2–3 times the seeds' size. The sown seeds should be at least 1–2 cm apart to prevent overcrowding. Seedlings should not be pricked out before their first pair of true leaves is fully expanded. The pricking-out procedure is shown in Figure 3.17. The medium surface should be 1–2 cm below the container's rim and the seedling's root collar (the junction between the root and shoot) be at the medium's surface.

For the majority of tropical tree species a mixture of forest topsoil with coconut husk or sand will create a germination medium

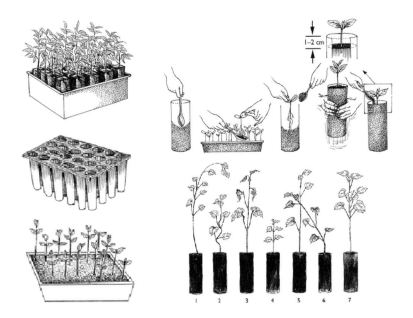

FIGURE 3.17 Sowing seeds, pricking-out seedlings, and seedling quality control (adapted from Elliot *et al.* 2013). The shown seedlings represent: 1. A plant with unbalanced root and shoot growth: the shoot is too long and thin and may well break during handling. 2. A plant with a malformed stem which compromises future growth. 3. A plant that has been attacked by insects. 4. A plant with stunted growth when compared with other plants of the same age. 5. A disease-infected plant. 6. A container that was knocked over and spent some time lying on its side, resulting in a non vertical stem. 7. A properly developed, well-balanced, and disease-free plant; all of the plants in the nursery should look like this.

with, on the one hand, mycorrhizal fungi required by most tropical forest tree species, and, on the other hand, good aeration and drainage.

## 3.12   OBTAINING OUTPLANTS: RAISING SEEDLINGS

The use of seeds for *in situ* conservation actions is known to be ineffective, as compared with seedlings, because of high mortality and low establishment of the former (Guerrant and Kaye 2007; Menges 2008; Godefroid *et al.* 2011). Germinating and establishing seedlings from seed is by far the most commonly used method to obtain material for *in situ* actions to yield large numbers of individuals, but raising a large number of seedlings from seeds in botanical gardens is rarely possible for logistical reasons (high cost and space limitations). Although not without exceptions (e.g., De Motta 2010), specialized nurseries established in proximity to the introduction site are a better option (Figure 3.15). Some detailed and simple-to-follow guidelines on designing and setting up a nursery, as well as planning and managing nursery operations, are readily available (Longman 2003; Gosling 2007; Elliott *et al.* 2013; Hoffmann and Velazco 2014; Stott and Gill 2014). The necessary information for the efficient running of the nursery includes knowledge on:

- how much seed is required to grow a given number of outplants;
- how long each seedling takes to grow to planting size;
- germination and seedling growth protocols;
- planting time.

This information should be tracked and summarized for each species on a nursery production calendar and in plant development records (Figure 3.18). However, besides the technical challenges of obtaining genetically diverse seedlots that are of a sufficient quantity, the cost of producing the seedlings from seed is higher than for the other methods.

The use of vegetative propagation can be a less labor-intensive and cheaper alternative to producing a large number of seedlings. The two common forms of vegetative propagation are cuttings, which are typically 20–40 cm long, taken from young branches or shoots of trees and shrubs; and stakes, 2.0–2.5 m long parts of branches that are pollarded from trees. Itoh *et al.* (2002) analyzed the rooting ability of 100 tropical trees in Malaysia and found that fast-growing species of a smaller mature stature typically root more readily. The ability to establish is also related to the type of cutting used; mature branches (harvested further down a stem) establish more readily than apical cuttings, and leafy cuttings appear more successful at rooting than leafless cuttings (Zahawi and Holl 2014). Many threatened

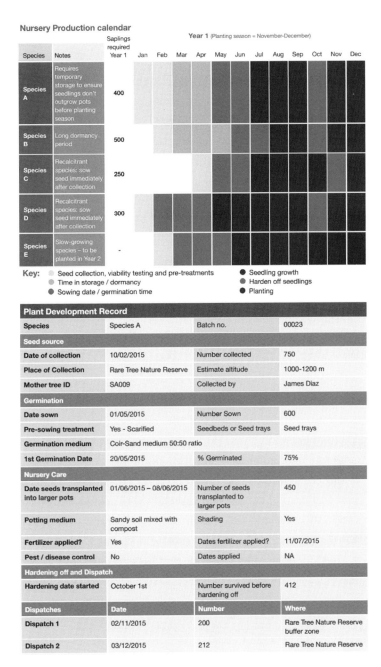

FIGURE 3.18 Nursery production calendar and plant development records (from Stott and Gill 2014). *A black and white version of this figure will appear in some formats. For the color version, please refer to the plate section.*

and extremely endangered species, e.g., conifers such as *Wollemia nobilis*, can be relatively easily propagated through cuttings. Usage of woody species cuttings is not uncommon in restoration projects, e.g., enrichment planting of dipterocarp forests in Indonesia (Kettle 2010), but is rare in threatened species reintroduction (Dalrymple *et al.* 2012), apparently because rooted cuttings are inferior to seedlings in their establishment success (Ray and Brown 1995).

Whether raising plants either directly from seeds or via vegetative propagation, both require use of dedicated conservation nurseries with their staff having necessary horticultural expertise. There are, however, two low-cost approaches that do not involve this stage to obtain large quantities of seedlings. "*In situ* seedling banks" (Pritchard *et al.* 2014) is the method of sowing seeds and maintaining seedlings in the forest understory. This method is easy to apply, because the seedling banks can be established in a wide range of forests: natural, degraded, or planted, including monoculture tree plantations. However, it can only be used for species that produce numerous seeds with low requirements for successful germination. The other limitations of this approach are that seedlings cannot withstand a long time in the shady understory and advance to saplings, and a need to prevent seedling herbivory.

Another approach is to transplant seedlings that have established in natural settings (often referred to as "wildlings") (e.g., Palmiotto 1993; Adjers and Otsamo 1996; Parrotta and Knowles 2001; Kettle 2010). Clearly, this method will also apply only to those species that do not suffer from germination failure, but this is a method of choice for species in which seed collecting is a challenge, such as dipterocarps (Kettle 2010). Beside overcoming technical difficulties in seed collecting, advantages of wildlings are their high genetic diversity because they are likely to represent several cohorts rather than a single one, a higher chance of subsequent establishment in comparison with the seeds collected, and because they have already succeeded in germination and initial establishment on the forest floor (Adjers and Otsamo 1996; Kettle *et al.* 2008). Two examples of wildlings' utility for conserving Chinese threatened tree species are given in Volis (2016b). Seedlings of *Liriodendron chinense* are rarely observed in natural settings but planting this species in Gaowangjie nature reserve along the local road resulted in dense stands of seedlings and saplings in proximity to the mother plants. These wildlings can be directly used for the creation of new populations of *L. chinense* in this area. Another example is *Phoebe bournei*, planted and maintained by farmers in the village of Baojing County. The numerous fruits produced by the trees are dispersed by birds into the surrounding village forest, where they readily germinate and develop into seedlings. However, strong anthropogenic disturbance (grazing and cutting for firewood) have made forest environmental conditions unsuitable for this species, and seedlings do not develop into saplings. Thus, wildlings of *P. bournei* that have no future in this environment can be used for the creation of new populations in a more suitable natural location. Survival of wildlings is usually high and field performance is good. Two years after the planting of 10 dipterocarp species wildlings, survival exceeded 40% for eight of the species (Adjers and Otsamo 1996).

Containers are used to grow seedlings to the required size and then they are transported to the introduction site. Smaller containers have reduced handling and transportation costs, while larger containers allow better plant development and thus a higher chance of establishment. Native topsoil should constitute a substantial part of the soil mix used as growing medium in such containers.

The use of low-quality seedlings is one of the reasons for the low survival rate of seedlings in reintroduction programs. Therefore, control of seedling quality is important (Figure 3.17). Table 3.4 presents the criteria for assessing the seedlings' quality, which gives a cumulative score ranging from 0 to 15. The assessment is based on the sampling in the nursery of 50 seedlings of plantable size (i.e., at least 20 cm tall) and recording five morphological criteria: seedling health, stem form, root form, sturdiness, and shoot-to-root ratio (Gregorio *et al.* 2017).

Table 3.4 *Criteria for assessing the quality of seedlings and description of scales for each criterion (Gregorio* et al. *2017)*

| Criterion and weight | Description |
|---|---|
| Health | Absence of pest and diseases |
| 0 | Poor – more than 15 samples affected by pests and diseases |
| 1 | Moderate – 10–15 samples affected by pests and diseases |
| 2 | Good – only 5–9 samples affected by pests and diseases |
| 3 | Excellent – fewer than 5 samples affected by pests and diseases |
| Stem form | Straightness of the stem |
| 0 | Poor – more than 15 samples with two or more stem leaders and bent shoots more than 30° from stem axis |
| 1 | Moderate – 10–15 samples with two or more stem leaders and bent shoots more than 30° from stem axis |
| 2 | Good – only 5–9 samples with two or more stem leaders and bent shoots more than 30° from stem axis |
| 3 | Excellent – fewer than 5 samples with two or more stem leaders and bent shoots more than 30° from stem axis |
| Root form | Evidence of root deformation (e.g., J-roots, pot-bound roots, and root curling) and roots growing out from the container |
| 0 | Poor – more than 10 samples have deformed root system |
| 1 | Moderate – 5–10 samples have deformed root system |
| 2 | Good – only 1–4 samples have deformed root system |
| 3 | Excellent – only 1 sample has deformed root system |

Table 3.4 (*cont.*)

| Criterion and weight | Description |
|---|---|
| Sturdiness | Robustness of the stem; assessed using sturdiness quotient (SQ), the ratio of stem caliper to stem length; ideal value is less than 6 |
| 0 | Poor – more than 15 samples have SQ of more than 6 |
| 1 | Moderate – 10–15 samples have SQ of more than 6 |
| 2 | Good – only 5–9 samples have SQ of more than 6 |
| 3 | Excellent – fewer than 5 samples have SQ of more than 6 |
| Shoot–root ratio | Balance of shoot biomass to root biomass (ideal value is 1 but 2 is acceptable) |
| 0 | Poor – more than 15 samples have S:R value greater than 2 |
| 1 | Moderate – 0–15 samples have S:R value greater than 2 |
| 2 | Good – only 5–9 samples have S:R value greater than 2 |
| 3 | Excellent – fewer than 5 samples have S:R value greater than 2 |

Seedlings left too long in the container suffer root deformation which negatively affects plant survival and growth. Before planting, seedlings must be "hardened" to reduce their susceptibility to heat, cold, or water stress in the field. This is usually done by reducing watering and increasing exposure to sun and ambient temperature. Hardening starts about 2 months before planting, when outplants are moved to a separate area in the nursery where their shade and frequency of watering is gradually reduced.

Seedling transportation is an operation that requires considerable attention because seedlings must arrive in a good condition for planting. Care must be taken to avoid damage from exposure to wind, loss of soil from the roots, or excessive transpiration. Heavy watering of plants is generally recommended before they leave the nursery, with a minimum delay between delivery and planting.

## 3.13   *IN SITU*: TRANSLOCATION

### 3.13.1   *Types of Translocation*

Conservation translocation is defined by the IUCN as "the intentional movement and release of a living organism where the primary objective is a conservation benefit: this will usually comprise improving the conservation status of the focal species locally or globally, and/or restoring natural ecosystem functions or processes" (IUCN/SSC 2013). Different types of conservation translocation are recognized based on whether individuals are released within or outside the species indigenous range, the latter either inferred from historical records or physical evidence of the species occurrence. Reinforcement (also augmentation or enhancement) is any effort to enhance population viability (e.g., through increasing population size or improving demographic structure) by movement of individuals into an existing population of conspecifics. Reintroduction is movement of individuals into a part of the species indigenous range from which it has disappeared (IUCN/SSC 2013). The latter term is often used in the broader sense as any kind of controlled placement of plant material into a natural location with the aim of increasing a species persistence (Akeroyd and Wyse Jackson 1995; Jusaitis 2005; Godefroid *et al.* 2011). According to IUCN/SSC (2013) reinforcement and reintroduction, applied within the species known range, are population restoration approaches, and differ from conservation introductions performed outside the species indigenous range. Conservation introductions involve transfer of organisms for the creation of new populations in a variety of new locations; these locations can be very far from the current range or just small range extensions into contiguous areas. These *in situ* actions are extremely important for rescuing the imperiled species: analysis of 181 Recovery Plans for endangered plant species (Hoekstra *et al.* 2002) revealed that 72% of species require some form of translocation.

If a species is still represented in nature by at least one or more natural populations, reinforcement usually is a preferred option because "existing individuals indicate that the survival of the species at a specific location is principally possible, at least when the factors contributing to a population decline have been identified and eliminated" (Betz *et al.* 2013). However, for reinforcement to be effective, it is necessary that the remaining populations of a threatened species are located in nondegraded and protected areas. If the locations are unprotected and have a low chance of being protected in the future, the populations are almost inevitably doomed to disappear by either direct effect (e.g., harvesting, logging) or destruction/ alteration of their habitat. When the population habitat is degraded and cannot sustain a viable population, supplementation of this population will not compensate for local mortality (Seddon 2010). Unless the whole habitat is restored such population reinforcement would just be a waste of money and valuable plant material.

The erosion or loss of genetic diversity and greatly reduced reproductive potential or sterility are the common problems of remnant populations as a result of fragmentation and degradation of their habitat. For these populations to achieve an increase in population growth rate and a decrease in the probability of extinction, which is the aim of reinforcement, the necessary measures can include the introduction of specific genotypes when the presence of different flower morphs or S-alleles are necessary for successful reproduction, plants of particular gender to correct the sex ratio in a population, adults needed to increase the pollinators' visitation rate, or young plants to rejuvenate the degraded populations.

Ideally, the material for reinforcement must originate from the same locations but be genetically diverse. However, as the majority of populations of threatened species have usually already undergone dramatic reduction in size and genetic variation, propagation and introduction of local genotypes will often not improve the population's genetic make-up and demography. Several studies convincingly demonstrate the importance of genetic rescue (positive population-level demographic response to the introduction of new beneficial alleles or genotypes) for small imperiled populations (DeMauro 1993; Pavlik *et al.* 1993; Weekley *et al.* 2002; Young and Pickup 2010). Extinction of these populations can be prevented only by recreating genetically diverse and sexually reproductive populations. What can be recommended for this purpose is to use a combination of plants of local origin and plants from the other populations within the same habitat (ecoregion) (Volis 2016c). Introduction of plants from multiple populations will increase (1) the chance of sampling and introducing different S-alleles needed to restore the outcrossing potential of populations of self-incompatible species; and (2) the fitness of offspring due to mating among genetically divergent individuals (heterosis). However, in reinforcement projects the level of nonlocal gene flow into recipient populations should not exceed 20% to prevent losing uniquely adapted alleles in the recipient population (Hedrick 1995).

Reintroduction is traditionally considered the first choice strategy for threatened species (and especially those that are extinct in the wild or are on the brink of extinction) whose natural populations are unprotected and under threat if there are some protected areas within the species indigenous range that can be used to create new populations (Maunder 1992; Albrecht *et al.* 2011; Godefroid *et al.* 2011; IUCN/SSC 2013). However, for species whose natural habitat is destroyed or degraded and which cannot be preserved in protected areas within their known historical range, the recommended conservation strategy is assisted colonization, i.e., introduction of the species into seemingly suitable protected areas with no past history of its existence. If there is no remaining area within a species historic range able to sustain viable population(s), there is no alternative to assisted colonization. Other reasons, besides destruction of the habitat, are escape of diseases and climate change effects. For many species, suitable habitats with all or most of the species fundamental

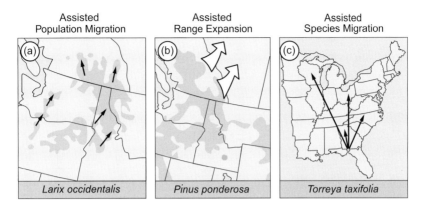

FIGURE 3.19 Different types of translocation and their application to particular tree species: (a) within the current distribution, (b) beyond but near to the current range, and (c) far away from the current range (from Dumroese *et al.* 2015).

niche attributes may exist outside the species recorded distribution, with dispersal limitations being responsible for the difference between fundamental and realized niches. At least for some species these limitations could result from fragmentation of previously continuous habitat due to human activity. Some authors consider it useful to distinguish two kinds of assisted colonization – movement of populations either long distances beyond a current species range (assisted species migration) or proximate to it (assisted range expansion) (Figure 3.19) (Ste-Marie *et al.* 2011; Winder *et al.* 2011; Williams and Dumroese 2013).

Reintroduction, including climate change motivated movement of populations within the current species range, and assisted colonization (either near or a long distance away from the current range) (Figure 3.19) share many challenges and potential solutions. In both, individuals are moved from *ex situ* or vulnerable *in situ* locations to protected *in situ* sites for long-term conservation. The immediate goals are successful establishment and the control of immediate ecological threats (e.g., predation, desiccation, or weeds), and the long-term goals are successful reproduction and resilience to stochastic events (either environmental or demographic) (Akeroyd and Wyse Jackson 1995; Milton *et al.* 1999; Vallee *et al.* 2004). For both, prior knowledge of the species biology, including reproduction, demography, environmental requirements, and ecological interactions, is absolutely vital. For both, not the geographic proximity, but environmental and ecological similarity between the source and recipient locations is important (Bowman *et al.* 2008; Lawrence and Kaye 2011; Noël *et al.* 2011). Ecological similarity to locations of extant populations and protection are the two major criteria in evaluating the suitability of a site for introduction, although the assumption that similarity to extant populations is the best criterion for successful establishment of a created population is not universal (Maschinski *et al.* 2012). This assumption may be misleading when (1) climate changes, anthropogenic disturbances, and any disruptions of natural ecosystem processes cause spatial shifts of an environment that is suitable for the species; and (2) when the species extant populations are located in a fragmented and degraded environment that does not support their long-term viability (Maschinski *et al.* 2012). In such cases, the occupied habitat can be less reliable as a reference for introduction planning than one inferred from knowledge of historic species range and the species ecological requirements. In any case, detailed information on the ecological requirements of the target species and of the habitat conditions at potential reintroduction sites are essential. Valuable information for estimating the suitability of a potential site can come from a comparative study of ecological factors in locations still occupied by the species and locations where the species became extinct.

Elaborating reliable criteria for choosing potential introduction sites for a threatened species is a challenging task due to the complexity of the species environmental and biotic interactions, often poorly documented historical range, and usually very limited number of extant populations (Maunder 1992; Maschinski *et al.* 2012; Rünk *et al.* 2014). In a study by Noël and colleagues, sites that had all appeared suitable based on expert opinion at the time of introduction turned out to differ in ecological similarity and, as a result, in introduction success (Noël *et al.* 2011). Identification of the most suitable habitats for the species using quantitative tools is based on establishing a relationship between ecological and environmental features in the landscape and either (1) species occurrence in space prior to introduction or (2) population establishment after introduction. The former utilizes SDM. However, usage of SDM is limited to large geographic scales and species with documented and relatively wide distributions. Species that occupy only a limited number of locations including a patchy distribution at the fine geographic scale, or have little information on prior distribution, require an experimental approach. Such experimental introductions across multiple (micro)sites can be performed either within the current or documented historic species range (reintroduction) or within the presumed ecological niche of the species (assisted colonization) (Collins and Good 1987; Jusaitis *et al.* 2004; Bottin *et al.* 2007; Guerrant and Kaye 2007; Volis *et al.* 2010; Albrecht *et al.* 2011; Lawrence and Kaye 2011; Maschinski *et al.* 2012; Rünk *et al.* 2014; Menges *et al.* 2016). Moreover, even for species that have detailed occurrence records, modeling can delimit the potential habitat, but only actual introduction will allow identification of the species realized niche (Maschinski *et al.* 2012).

Because environmental/ecological requirements and biology of rare and endangered species are usually poorly known, their translocation inevitably involves substantial uncertainty about the space, time, planting protocol, etc. This uncertainty makes trial and error experimentation a highly appropriate approach allowing hypotheses to be tested and methods to be compared, thus maximizing the likelihood that at least one method will have high success (Griffith *et al.* 1989; Falk and Olwell 1992; Guerrant and Kaye 2007; Menges *et al.* 2016).

### 3.13.2 *Translocation: General Recommendations*

In general, translocations using seedlings or saplings rather than sown seeds (Guerrant 1996; Jusaitis *et al.* 2004; Guerrant and Kaye

2007; Menges 2008; Godefroid *et al.* 2011; Albrecht and Maschinski 2012), introducing growing rather than dormant plants (Batty *et al.* 2006; Smith *et al.* 2009), conducted over multiple years (van Andel 1998; Kirchner *et al.* 2006; Kaye 2008), and with carefully chosen favorable (micro)sites that match the species niche (Jusaitis 2005; Menges 2008; Dalrymple *et al.* 2012; Knight 2012; Wendelberger and Maschinski 2016) have a higher chance of success. It is also crucial to control pre- and posttransplantation pests and pathogens to avoid their introduction into the translocation site (Sainsbury and Vaughan-Higgins 2012; Langlois and Pellerin 2016).

The founder population should be large enough (substantially larger than 50 individuals; Albrecht and Maschinski 2012) to overcome the demographic and genetic constraints associated with small population size (Guerrant 1996; Frankham *et al.* 2014), and planting density must be sufficiently high for successful wind or animal pollination (Abeli *et al.* 2016).

Although in some cases the size of outplants matters little for growth or survival (Alley and Affolter 2004), in the majority of situations larger plants perform better during reintroduction (Wendelberger *et al.* 2008; Albrecht and Maschinski 2012). In rhizomatous and tuberous plants, larger outplants with more buds have a higher chance of survival and better growth (Smith *et al.* 2009; Brzosko *et al.* 2018). For slow-growing species (e.g., cycads and palms), introduction of young reproductive adults should be preferred over introduction of seedlings or immatures, despite the long development time to maturation in these species. Reaching the reproductive stage for plants raised in a nursery can be achieved much faster than in natural settings due to more favorable conditions, such as lack of competition and an optimal nutrient, light, and watering regime.

For example, 2-, 4-, and 7-year-old outplants of a cycad, *Dioon edule*, were monitored after reintroduction and compared with cohort plants grown in the nursery. Mortality of the reintroduced plants was about 20% during the first year of monitoring with no mortality during subsequent years. However, the reintroduced plants exhibited no measurable growth increment 10 years after planting in 1997. In contrast to this stasis observed *in situ*, in the nursery the same cohorts exhibited much faster development. The reintroduced 7-year-old plants maintained the same stem diameter of c. 4 cm with three leaves per crown, whereas plants of the same age in the nursery reached a stem diameter of 12 cm with 15 leaves per crown by 2005 and a stem diameter of 25 cm with 43 leaves by 2009 (Vovides *et al.* 2010). Therefore, although it took 15 years for male plants and 17 years for female plants to reach reproductive age in the nursery (Vovides *et al.* 2010), maturation of individuals introduced as seedlings/saplings may be an indefinitely long process. Another example is provided by Maschinski and Albrecht (2017). The palm *Pseudophoenix sargentii* can achieve reproductive maturity in cultivation within 14 years, however, an individual grown during its first 10 years under favorable garden conditions after reintroduction to the wild required an additional 25 years to produce flowers and fruit. Thus, introduction of reproductive adults or large plants that are able to gain reproductive maturity more rapidly than the majority of the plants in a population can improve the age structure of the population and boost its regeneration.

Using outplants of different ages/life stages should be considered for translocations of long-lived species (Albrecht and Maschinski 2012). This can be advantageous if the environmental conditions at the recipient site vary spatially or temporally, and this variation differentially affects plants of different age or life stages. For example, the light conditions supporting the translocated adults of *Tephrosia angustissima* var. *corallicola* were found to differ distinctly from the light conditions favoring the recruiting seedlings (Wendelberger and Maschinski 2016).

Pre- or postplanting intensive management interventions can be necessary for establishing outplants, e.g., caging or fencing (Monks and Coates 2002; Maschinski *et al.* 2004; Daws and Koch 2015; Fenu *et al.* 2016), weeding and burning (Pavlik *et al.* 1993; Bowles *et al.* 1998), soil preparation and mulching (Sinclair and Catling 2003;

Monks *et al.* 2012), adding fertilizer (Kaye and Brandt 2005), or supplemental watering (DeMauro 1993; Dillon *et al.* 2018). The success of a translocation to a large extent depends on whether the recipient site has legal protection status (Godefroid *et al.* 2011).

### 3.13.3   Translocation: Within-Population Genetic Diversity

Because the first generation of the newly established population determines the subsequent natural regeneration at a site, the founder population must be sufficiently large and genetically variable. It is very common in translocation projects that a large proportion of outplanted individuals fails to survive beyond 2–3 years (Brown and Briggs 1991). This implies that translocations are subject both to the founder effect (the result of the initially limited gene pool) and subsequent genetic bottlenecks due to mortality (Falk *et al.* 2001). If the founder population originated from a small number of related plants, the initially low genetic diversity, subsequent inbreeding, and genetic erosion can lead to fitness decline in future generations (Stacy 2001; Reed and Frankham 2003), especially if the original planting material was derived through vegetative propagation. Comparison of selfed and outcrossed offspring of clonal *Pseudotsuga menziesii* 33 years after establishment revealed reduced survival and average diameter at breast height of selfed trees (39% and 59% of that of the outcrossed trees, respectively) (White *et al.* 2007). Also Kephart (2004) found significantly greater survival and reproduction over 3 years in reintroduced populations of *Silene douglasii* var. *oraria* that originated from outbred versus inbred progeny. The combined effects of initially low genetic variation and subsequent inbreeding may not be immediately evident, but will appear later as reduced growth, survival, and reproduction, and lower persistence during periods of more extreme climatic fluctuations. The traits conferring tolerance of disturbance or climatic extremes often have a strong genetic basis, and therefore a genetically narrow population may be able to survive only in a narrow range of conditions.

The latest update to the recommended number of introduced families and individuals is provided by Frankham *et al.* (2014). The authors advise doubling the effective population sizes, $N_e$ = 50 and 500, traditionally recommended to prevent inbreeding depression and retain population evolutionary potential, respectively (Franklin 1980; Soulé 1980; Lande and Barrowclough 1987). For conversion of effective into census population size the recommended $N_e$ /$N$ is 0.1–0.2 as the first approximation (Frankham *et al.* 2014) (Table 3.5).

Creating a genetically diverse population from a single population source is rarely possible as the remnant populations of threatened species are usually genetically depauperated. There is growing evidence that mixed population sources lead to greater success in translocations than single population sources from both short-term studies (Vergeer *et al.* 2005; Godefroid *et al.* 2011) and long-term observations accounting for extreme climatic events (Maschinski *et al.* 2013). A decision about source material to be used for reintroduction or assisted colonization must weigh two potential dangers: of inbreeding and outbreeding depression (Fenster and Dudash 1994; Frankham 2010; Frankham *et al.* 2011; Neale 2012). The negative consequences of inbreeding in reintroductions appear to overcome potential risks of outbreeding depression, but only if the local conditions of source populations do not differ much from the conditions at the recipient site (Frankham *et al.* 2011). A decision tree for determining the probability of outbreeding depression between two populations (Frankham *et al.* 2011) (Figure 3.20) can be used for decision-making about whether to mix population sources in an introduction. Considerations of the species life history and breeding structure are also important. For example, self-pollinated and gravity-dispersed species have a lower geographic scale of local adaptation and therefore are less suitable for population mixing than outcrossing and wind-dispersed species.

The importance of diversifying genetic sources in translocation is perfectly illustrated by the story of the eastern daisy (*Hymenoxys acaulis* var. *glabra*), a federally threatened plant of the Great Lakes

Table 3.5 *Genetics in conservation management: revised recommendations for the 50/500 rule, Red List criteria, and population viability analyses (Frankham et al. 2014)*

| Topic | Revised recommendation |
| --- | --- |
| 1. Avoid inbreeding depression | $N_e \geq 100$ |
| 2. Maintain evolutionary potential | $N_e \geq 1000$ |
| 3. Extrapolating from $N_e$ to $N$ | Use different $N_e/N$ according to life history of species, but current default is 0.1–0.2 |
| 4. Fragmented populations and connectivity | Evaluate on case basis, but distinguish current and historical gene flow |
| 5. Genetic factors in PVA | Routinely include inbreeding depression |
| (i) Inbreeding depression | Routinely apply realistic levels (~12 lethal equivalents on total fitness) |
| (ii) Evolutionary potential | Include in long-term and environmental change contexts |
| 6. MVPs | Apply common standard and specify (suggest 99%) |
| | Standardize and specify in generations (suggest 40) |
| | Routinely include all systematic and stochastic factors (including genetic) |
| 7. IUCN Red List criterion Population size | |
| Critically Endangered | <500 |
| Endangered | <5000 |
| Vulnerable | <20 000 |

region of the United States and Canada. Reproductive failure in one remnant Illinois population turned out to be the result of a lack of compatible mating types among the remnant individuals, and this genetic knowledge served as the basis for the species recovery program (DeMauro 1993). The reintroductions utilized individuals

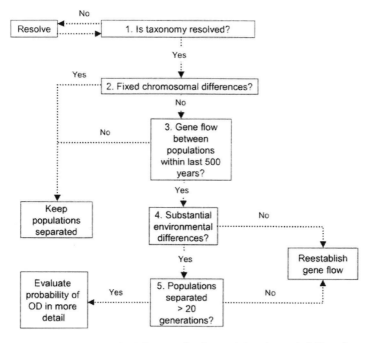

FIGURE 3.20 Decision tree for determining the probability of outbreeding depression between two populations (from Frankham *et al.* 2011).

from multiple populations to increase the likelihood of capturing different mating types and the transplant locations were chosen to maximize the potential for outcrossing (DeMauro 1994). As a result, production of viable seeds was evident in all reintroduced populations in the years immediately after restoration.

The importance of using multiple seed sources in translocation was shown in several other studies. Assessment of genetic diversity in remnant and reintroduced populations of the purple prairie clover (*Dalea purpurea*) revealed that reintroduced sites established from multiple local (within 80 km) seed sources retained as much diversity as the remnant populations (Gustafson *et al.* 2002). Mixing individuals of multiple-source populations significantly increased the genetic diversity of the reintroduced compared to the local-source population of *Arenaria grandiflora* and this increased diversity has been maintained for over 10 years (Zavodna *et al.* 2015).

However, while mixing population sources in translocation reinstates the needed level of population genetic diversity, for species with high probability of outbreeding depression creation and introduction of inter-population mixes should be avoided. In the experimental introduction of *Asclepias meadii*, an endangered self-incompatible perennial herb of tallgrass prairie, a heterosis effect was observed for seed viability, with greater germination among seeds derived from inter-population crosses. Nevertheless, cumulative growth of planted juveniles as well as population growth in locations with suboptimal conditions tended to be lower for propagules derived from inter-population crosses, demonstrating outbreeding depression (Bowles *et al.* 2015).

### 3.13.4 Translocation: Among-Population Genetic Diversity and Gene Flow

An important question in translocations is whether to reinstate gene flow between fragmented populations or preserve locally adapted populations. The answer to this question requires knowledge of the extent and structure of genetic variation, potential for dispersal via seeds and pollen, and scale and magnitude of local adaptation (Munzbergova *et al.* 2005; Volis *et al.* 2005; Frankham 2010; Maschinski *et al.* 2013). This knowledge is essential for the evaluation of the probability of reduction in fitness of the offspring, i.e., outbreeding depression.

A decision tree for determining the probability of outbreeding depression between two populations (Frankham *et al.* 2011) (Figure 3.20) can be used for decision-making not only about whether to mix population sources in introduction, but also about whether to establish gene flow between introduced and existing populations. If there is a low probability of outbreeding depression, the introduced populations should be preferably located within a seed/pollen dispersal distance from the extant populations. Otherwise, they should be sufficiently distant from extant populations to prevent gene flow. When the scale and importance of local adaptation or spatial genetic structure are not known, considerations of the species mating system and dispersal mode can be used (but with considerable caution). In general, establishing gene flow between introduced and extant populations of outcrossing and wind-dispersed species is desirable, while for self-pollinated and gravity-dispersed species less so. However, there are cases when life-history traits can be misleading. In contrast to expectations, an insect-pollinated herb *Geranium pratense* displayed low genetic diversity, high among-population and pronounced within-population genetic differentiation (Michalski and Durka 2012), and *Alexgeorgea nitens*, a dioecious, clonal perennial with limited seed dispersal, displayed high levels of genetic diversity within populations (Sinclair *et al.* 2010).

### 3.13.5   Assessment of Translocation Success

By far the most commonly used measures of reintroduction success are survival of outplants over a specified time period or first-generation establishment evident in reaching adulthood and reproduction. These, and more specific short-term objectives (e.g., obtaining a particular age or size distribution, in a population reaching a desired level of genetic diversity, seed production or seed bank density), however, are insufficient to conclude that a new location is suitable (Bell *et al.* 2003). A particular species can exhibit good performance in survival or growth but a worse performance in recruitment. In a restoration trial performed in degraded rupestrian grassland of Brazil, assessment of the translocation success of 10 native shrub species revealed a tradeoff between survival and reproduction in outplants. Species with high long-term survival had very low or zero recruitment and vice versa (Gomes *et al.* 2018). But only recruitment of subsequent generations is a crucial parameter for long-term population persistence, and the extent of recruitment must attain and maintain an MVP size as defined by a 95% probability of survival after 100 years. According to Pavlik (1996), criteria for introduction success are the population's ability to reproduce, persist, and adapt to changing environmental conditions, measured as abundance (establishment, vegetative growth, fecundity, and population size), self-sustainability, and genetic variation. Ideally, the established population should show signs of producing satellite populations through colonization of adjacent areas. In judging the success of a particular conservation program the following definition can be useful: "Species are no longer considered threatened if the population number is over 10 and mature plants exceed 1000 per population, provided these populations are neither declining nor under threat" (IUCN 2001). This definition is implicitly based on analysis of population dynamics, central to conservation and management of rare and endangered species (Schemske *et al.* 1994).

As discussed in Section 3.2, detailed characterization in terms of population size and demographic structure across the distribution range of a threatened species is needed to provide an accurate picture of its current conservation status, and studying the dynamics of populations prior to restoration actions is necessary to assess the threats and devise an appropriate management strategy (Figure 3.2). Evaluation of translocation success also requires long-term demographic monitoring of the introduced populations and the assessment of the likelihood of their persistence through PVA. Comparison of the introduced and natural populations in their demography can help to optimize the management of introduced populations or future translocations (Colas *et al.* 2008; Laguna *et al.* 2016; Maschinski and Quintana-Asciencio 2016; Menges *et al.* 2016). For example, demographic comparisons of the rare *Centaurea corymbosa* showed that reintroduced populations had higher survival but lower fecundity than natural populations, with a management recommendation to increase plant density to improve mate availability for self-incompatible flowering individuals (Colas *et al.* 2008).

Unfortunately, very few studies, besides the abovementioned study of Colas *et al.* (2008), have used PVA to compare demographic characteristics of reintroduced and natural populations (Bell *et al.* 2003; Maschinski and Duquesnel 2006; Knight 2012; Halsey *et al.* 2015; Bladow *et al.* 2017) or to evaluate introduction strategies (Bell *et al.* 2003; Satterthwaite *et al.* 2007; Halsey *et al.* 2015). This scarcity of reintroduction PVAs is due to many reasons, as discussed in Monks *et al.* (2012). Translocated compared to natural populations, are likely to have only a subset of the life stages in the early years after reintroduction, such parameters as fecundity and survival of adults can be difficult to obtain or they may not be accurate, and even transition probabilities for younger stages or smaller size classes can be unreliable due to small population sizes and because the latter may differ from those of naturally recruited individuals. Thus, reliable transition matrices cannot be constructed until transplants complete their life cycle, and all stages are present among naturally recruited individuals. For these reasons, in contrast to a natural population for which 2-year monitoring is often sufficient to construct a transition matrix, many more monitoring years are necessary for translocated populations. Long-term monitoring is also important because an increase in monitoring duration increases the likelihood that rare events having an impact on the population will be accounted for. Often true population trends can be detected only many years after translocation. A good example is translocation of *Grevillea calliantha*, an endangered long-lived woody perennial of Western Australia. From the outplants, 55–95% survived the first year, then survival decreased to 68% after 4 years, and dropped to 18% after 10 years, apparently as a result of drier-than-average years (Monks *et al.* 2012).

All the above inherent limitations of PVA preclude its wide applicability in species conservation planning. In ESA recovery planning, of the 258 final recovery plans for the 642 listed plant species, only 15% mentioned or recommended PVA (Zeigler *et al.* 2013).

In critically endangered species, the remaining natural populations may already be too small and genetically depauperate, have no recruitment, and grow in highly degraded habitats. In these cases, the introduced populations can only be compared with sister taxa or other species with similar functional traits (Monks *et al.* 2012).

Translocation projects tend to be most effective if the process allows for iterative feedbacks and updates (i.e., adaptive management), and when translocation actions are performed as designed experiments involving clear hypothesis formulation and testing (Kaye 2008). As noted by Kaye (2008), when results of translocation experiments are evaluated through monitoring and hypothesis testing, poorly performing methods can be discarded, while more effective methods can be deployed more widely; adaptive management will allow testing further hypotheses and continued protocol improvements.

# 4 Restoration of Threatened Species Habitat

4.1 CONSERVATION-ORIENTED RESTORATION
VERSUS SPECIES-TARGETED CONSERVATION AND
TRADITIONAL RESTORATION

If we look at the secondary communities that developed after anthropogenic impact, be it logging, tillage, overgrazing, or any other activity, we will notice an almost inevitable, to some degree, difference in community function and composition from the pre-disturbance communities. Restoration of these habitats usually requires enrichment with late-successional species and absent functional groups, e.g., large-fruited or insect-/bird-pollinated trees (Martinez-Garza and Howe 2003; Garcia *et al.* 2015). Of course, it is easier for many reasons for restoration practitioners to use common and widespread species than functionally equivalent rare and endangered species. However, I provided reasons in Sections 2.7 and 2.8 for the use of the latter in restoration. On the one hand, threatened species will have a future only in the restored habitats. On the other hand, if they belong to a functionally important plant category, a category needed to restore the ecosystem integrity, why not use them for ecological restoration? If we accept the idea that the threatened species in principle can be used in restoration, the immediate questions are under what circumstances, how, and which of them can be used? These topics will be addressed below within the context of habitat restoration driven by conservation objectives.

Habitat restoration has already become a part of recovery efforts for many threatened and rare animal species, as is the case, for example, under the US Endangered Species Act (examples can be found in Wiens and Hobbs 2015). Unfortunately, habitat restoration as a means of the recovery of threatened plant species is less popular and is still in its infancy. To date, habitat restoration has been mostly envisioned as a prerequisite of a threatened species recovery (Birkinshaw *et al.* 2013), but not as a part of the recovery program.

The idea of using threatened plant species in habitat restoration, and making ecological restoration an integral part of conservation planning and implementation, if proven realistic, and producing good results, may become a new conservation paradigm (see Table 1.1). I propose to call this concept "conservation-oriented restoration" to distinguish it from traditional restoration, which usually has more utilitarian goals, e.g., improvement of air quality, erosion control, or soil replenishment. Conservation-oriented restoration should not be seen as an alternative to conservation focusing on particular species (species-targeted conservation) discussed in the preceding part of this book, but as its complement (Figure 4.1). The abovementioned approaches, i.e., conservation-oriented restoration, traditional restoration, and species-targeted conservation, differ not only in broadly defined goals and attributes of their targets, but also in the types of ecosystems they are applicable to.

Hobbs *et al.* (2013) distinguished historical ecosystems, which remain within their historical range of variability, hybrid ecosystems, where the changes are reversible, and novel ecosystems, where the changes are irreversible (Figure 4.2). Despite some overlap (Figure 4.1), traditional restoration predominantly focuses on novel ecosystems, species-targeted conservation on historical ecosystems, and conservation-oriented restoration on hybrid ones.

A step-by-step description of the stages of conservation-oriented restoration, with their corresponding major issues and methodology suitable for solving them (Table 4.1) are provided in the following sections.

## 4.2   PRIORITIZING LOCATIONS

Many readers of this book will agree with the note of Young that "restoration is always a poor second to the preservation of original habitats"

Table 4.1 *Stages of conservation-oriented restoration with corresponding major issues and methodology suitable for solving them*

| Stages | Issues | Methodology |
|---|---|---|
| 1. Restoration site | Prioritization by conservation value | Ecological threshold |
| | Present and future suitability for threatened species | SDM |
| 2. Reference conditions | Considering past, present, and future conditions, search for alternative states | Historical records, fossil data, SDM |
| 3. Choice of species | Number of species and their identity | Species-abundance distribution, regional and local species pool, characteristic and derived diversity, plant functional groups |
| 4. Material for planting | Seed collecting and storage | Restoration seed bank |
| | Obtaining outplants | *Quasi in situ*, seed orchards |
| 5. Site preparation | Considering plant–animal interactions, availability of safe sites | Introduction of animals and facilitating plant species |
| 6. Planting design | Importance of spatial scale heterogeneity and local extinction | Small-scale restoration, replicated multi-species experiments |
| | Number of plants introduced and planting density | Consideration of Janzen–Connell effect, Allee effect, autotoxicity |
| 7. Management and observations | Monitoring, interventions | Population viability analysis, adaptive management |

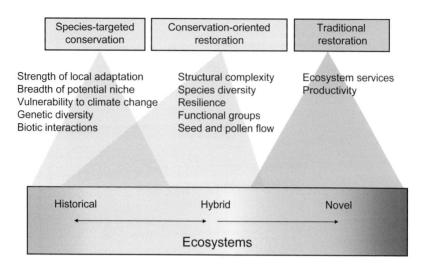

FIGURE 4.1 Three approaches to conservation of species and ecosystems, attributes of their targets on which the approach actions are focused, and the range of ecosystem degradation they can be successfully applied to (terms are *sensu* Hobbs *et al.* 2013, for explanation, see Figure 4.2). *A black and white version of this figure will appear in some formats. For the color version, please refer to the plate section.*

(Young 2000). Clearly, restoration should not be viewed as a substitute for protecting the remaining intact ecosystems (Gibson *et al.* 2011; Shoo *et al.* 2016). Unfortunately, virtually no ecosystems remain intact and, therefore, the number of potentially suitable locations for threatened species could be close to zero if only pristine ecosystems are considered suitable. A solution is to use partly degraded habitats, but ecological restoration of such habitats has obvious limitations. For example, many of the ecological attributes of primary forests cannot be restored in degraded forests even with the most intensive and expensive restoration methods currently available, and the allowance of long periods for succession and growth (Shoo *et al.* 2016).

This has clear implications for prioritizing areas for conservation-oriented restoration, viz. that setting site priorities should be based on identification of the state at which natural regeneration fails and interventions become necessary along the gradient of forest degradation (McIntyre and Hobbs 1999; Chazdon 2008).

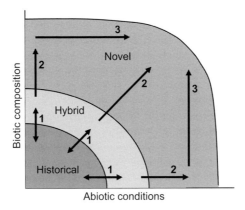

FIGURE 4.2 Types of ecosystems under increasing levels of change in abiotic conditions (x-axis) and biotic composition (y-axis) (after Hobbs *et al.* 2009). A historical ecosystem remains within its historical range of variability; a hybrid ecosystem is biotically and/or abiotically dissimilar to its historical ecosystem but is capable of returning to the historical state; novel ecosystems are biotically and/or abiotically dissimilar to the historical state and have passed a threshold such that they cannot return to the historical state. Pathways represent possible directions of change: (1) shifts from historical to hybrid ecosystems that are reversible; (2) nonreversible shifts from historical or hybrid ecosystems to novel ecosystems; and (3) further possible biotic and abiotic shifts within novel ecosystems. *A black and white version of this figure will appear in some formats. For the color version, please refer to the plate section.*

However, the application of this knowledge is not so obvious. For example, Charron and Hermanutz (2016) suggested that "the highest priority should be assigned to the most heavily altered forest stands that have undergone significant community-level change, while allowing natural regeneration (passive recovery) to occur in the least degraded stands."

For restoration that has conservation goals, this view cannot be accepted as absolutely utopian, given the current scale and rate of environmental degradation. Just the opposite, a focus must be on the least altered ecosystems, which with some restoration work have a reasonable chance of approaching the undisturbed habitats that once existed in the area (see also Yoshioka *et al.* 2014; Tobon *et al.* 2017).

Among the three types of ecosystems distinguished by Hobbs *et al.* (2013) (Figure 4.2) only hybrid ecosystems (beside historical ones) that did not approach a threshold beyond which return to the original state was impossible could be a home for the threatened species. This ecological threshold is the point at which a small change in environmental conditions causes an abrupt change in the ecosystem state (Groffman *et al.* 2006; Suding and Hobbs 2009) (the red line in Figure 1.1). Most of the heavily degraded areas represent novel ecosystems, where the ecological thresholds are crossed. However, less degraded areas in which the ecological thresholds are not crossed, especially those already protected, give an opportunity to reestablish communities that existed in the past or at least to approach them with considerably less investment (Wiens and Hobbs 2015). For the latter, they must undergo restoration driven by conservation goals. In these areas, spontaneous succession and recovery might occur naturally when threats are identified and removed (e.g., prohibition of grazing, logging and cutting for firewood), but, as a rule, a successional change will be too slow or not to a desired state without crucial interventions (e.g., burning, thinning, or plant/animal introductions) (Rey Benayas *et al.* 2008; Larios *et al.* 2017). As soon as ecological constraints that prevent the spontaneous return of a degraded system to the reference state are identified (e.g., disrupted biotic interactions, propagule limitation, or altered disturbance regime), they should be addressed simultaneously or, if this is not possible, prioritized by importance (Suding *et al.* 2004).

Although hybrid ecosystems, in contrast to novel ecosystems, can be targets for restoration, not all hybrid ecosystems will have the same conservation value. As I stated above, priority should be given to the least altered habitats, which still need interventions to restore altered structure and missing ecological functions, but have a reasonable chance of approaching a once existing habitat. But other features, both intrinsic and extrinsic, are also important for ranking the candidate restoration sites by conservation value. The former are

Table 4.2 *Priority areas for restoration*

| Priority rank | System description |
| --- | --- |
| 1 | Systems in which highly endangered plant species still have populations and these populations exhibit natural regeneration |
| 2 | Systems in which highly endangered plant species still have populations but natural regeneration in these populations has not been observed or is depressed |
| 3 | Systems which are least degraded among other similar systems, and which can potentially support the establishment of endangered species currently not growing there |
| 4 | Systems of varying degree of disturbance that are located within protected areas or are important for their connectivity |
| 5 | Systems of varying degree of disturbance that have a low probability of supporting the establishment of endangered species but have a good chance of approaching (after restoration) historical habitats by species structure and composition |
| 6 | Systems of varying degree of disturbance that are important for establishing connectivity between patches of fragmented habitat and maintaining the regional species pool |

the presence of threatened species and their regeneration, and the latter are the total forest cover remaining in the landscape, richness of the regional species pool, and whether the location is protected or is going to become such. Based on these key attributes ranking is presented in Table 4.2.

Restoration interventions should boost naturally occurring succession and change the species composition toward higher species richness and an increased proportion of threatened and rare species. Lack of convergence of the species composition in the restored

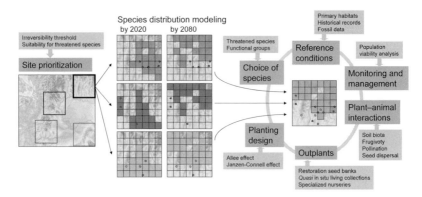

FIGURE 4.3 Stages of conservation-oriented restoration with major issues and appropriate methodology to solve them (from Volis 2018). At the site prioritization stage, four quadrats denote candidate sites with border line width corresponding to the priority rank. As a result of ecological niche modeling for the three species for which extant populations are denoted by colored circles, the small-scale planting sites are defined (grid cells in gray). *A black and white version of this figure will appear in some formats. For the color version, please refer to the plate section.*

habitat with that of the nearest undisturbed one should not be seen as a failure of the restoration project, as will be shown below. Choice of restoration sites should take into account that some locations are more buffered against climate changes than others (e.g., those that acted as refugia during previous climate fluctuations) (Millar *et al.* 2007). Figure 4.3 shows area prioritization for conservation-oriented restoration, followed by the steps that will be discussed in the next sections.

## 4.3   REFERENCE CONDITIONS

Restoration requires a target or model in mind, i.e., a reference, the desired state that the restored ecosystem is expected to attain or approximate. For several decades, a reference based on historical knowledge served as a keystone concept in the science of restoration ecology and the practice of ecological restoration (Clewell and Aronson 2013; Aronson *et al.* 2017). In a broad sense, reference conditions can be equated to a set of conditions under which a restored ecosystem is likely to be functioning and self-sustaining in the long run. A more detailed view refers to the historical range of variability in ecosystem composition, structure, and function that is used for (1) making a comparison with a contemporary ecosystem to evaluate the changes and to design management actions; and (2) measuring the success of ecological restoration (Kaufmann *et al.* 1994; Morgan *et al.* 1994; Swanson *et al.* 1994; Christensen *et al.* 1996). Thus, a reference is essential to understand the structural and functional changes that occurred in the target ecosystem as a result of human activity, to identify the exact causes of these changes, and to work out the appropriate measures to restore the ecosystem.

What serves as a reference most often is a still present historic ecosystem or its remnant(s). However, when the latter no longer, exists it can also be an ecosystem reconstructed and synthesized from the documented ecological descriptions, photographs, and herbarium/museum specimens, i.e., from available sources which may collectively provide a good approximation of the historic conditions (Egan and Howell 2001).

Although there is a general consensus that reference conditions are useful for deciding on ecologically justifiable goals in restoration (Fule *et al.* 1997; White and Walker 1997; Moore *et al.* 1999; McDonald *et al.* 2016), there has always been a debate about how, where, and to what extent historical knowledge of natural variation should be applied (Landres *et al.* 1999; Millar and Woolfenden 1999; Aronson *et al.* 2017). This debate intensified in the last 10 years after the emergence of the concept of "novel" ecosystems that undermined the value of historical references in setting restoration objectives (Harris *et al.* 2006; Hobbs *et al.* 2009, 2013; Suding 2011). As the goal of this concept is the creation and management of ecosystems with no analogs in the past, past conditions are largely irrelevant and historic reference is not needed. But even more moderate opponents of the utility of historic reference list numerous reasons why it should be abandoned. The latter include continuous changes in species assemblages over time and the lack of a clear picture of particular historic conditions, even in those cases when detailed historical records exist (e.g., Chambers *et al.* 1999), the dynamic nature of ecosystems rarely having a single equilibrium state and trajectory, widespread introduction of exotic species, disruption of natural processes, and many anticipated climate change impacts (Harris *et al.* 2006; Blois *et al.* 2013). Responding to these critical points, the following arguments should be relevant.

Indeed, many plant communities are dynamic and the current remnants can be quite different from the communities that existed in the recent and more distant past. However, it was recognized a long time ago that "a single reference ecosystem generally is inadequate and inappropriate for evaluating the degree and effectiveness of restoration" with a recommendation to use "multiple reference sites that account for patch dynamics and physical site heterogeneity" (Clewell and Rieger 1997). The concept of reference has shifted over time, from a former reliance on fixed reference points and species composition to a current focus on a range of variation, multiple potential trajectories, and function rather than composition (Landres *et al.* 1999; Suding *et al.* 2004; Falk *et al.* 2006). Recognizing that a single solution does not exist or is inappropriate (White and Walker 1997; Holl and Cairns 2002; Balaguer *et al.* 2014), we should try to identify and target the alternative states as reference conditions (Hobbs and Norton 1996; Temperton *et al.* 2004) and compare the outcomes. Searching for alternative reference states should be based on the assembly rules theory of theoretical community ecology, in which community assembly is deterministic in the composition of trait-based functional groups, but stochastic in terms of species composition. The functional groups fill the available niches created by the particular environmental conditions while species composition within functional groups is determined by the order of species arrival (also known as "priority effects"; Connell and Slatyer 1977). As a result, local biotic communities can enter alternative stable states, even when they share the same species pool and the same environmental conditions (Connell and Slatyer 1977; Beisner *et al.* 2003; Chase 2003; Schroder *et al.* 2005). That assembly is a historically contingent process in which, under given environmental conditions, assemblages converge in species traits but diverge in species identities, was demonstrated experimentally by Fukami *et al.* (2005).

It is also widely accepted that climatic and other environmental conditions fluctuated significantly from the early Holocene to the present, as did species ranges. Quaternary paleoecological records provide convincing evidence that ecological systems have always been dynamic, characterized by variability at a variety of temporal scales. Because of this variability, knowledge of historical species ranges and responses to climate fluctuations in the past can be vital for determining the limits of their ecological tolerances and search of potential new habitats. The latter can be those areas that supported species in the past under conditions similar to those predicted in the future, and which are outside the current species range (Millar 1998; Millar *et al.* 2007). A good example is the endangered narrow endemic Monterey pine (*Pinus radiata*), which naturalized along the north coast of California distant from its present native distribution in California and Mexico (600 km from the closest current native population). Much of this area was the paleohistorical range for *P. radiata* during climate conditions similar to those expected in the future in California. Based on this paleoecological knowledge, the proposed conservation strategy for *P. radiata* was, instead of focusing on extant populations, to realign the areas for protection according to the species responses to climate effects by protecting the sites where the species has naturalized (Millar 1998).

A realignment approach (Millar 1998) states that because historic ranges of species were constantly shifting, contracting, and expanding in response to changing regional and global conditions, the restored ecosystems should be realigned with current and expected future conditions rather than with pre-disturbance past conditions. The realignment approach and the similar recently proposed restoration concept (Butterfield *et al.* 2017) suggest that if a species prehistory and biological requirements are known, a suitable range of conditions in the present and future can be reliably predicted. This approach is the preferred choice in many situations (Halpin 1997; Harris *et al.* 2006; Millar and Brubaker 2006; Millar *et al.* 2007; Cole *et al.* 2011).

Thus, as was noted by Higgs *et al.* (2014), discounting the importance of historical reference in restoration is both premature and unwarranted, and the role of history becomes more, not less, important under rapidly changing ecological conditions. The motivation to seek historical references in restoration is based on the strongly supported assertion that the integrity of virtually all ecosystems was greater before modern human disturbance than it is now. As noted by Higgs *et al.* (2014), "to ignore the legacy of an ecosystem, even in cases where specific historical information is scanty, is to practice something other than restoration." Therefore, instead of abandoning the search for and use of the references, we need to reinforce the latter (Swetnam *et al.* 1999; Egan and Howell 2001; Hall 2010). Reinforcement can come from recognizing not just a single but multiple alternative stable states for an ecosystem, intensive utilization of paleoclimatic and fossil data, environmental niche modeling, and species ranges realignment (Birks 1996; Millar 1998; Davies and Watson 2007; Brewer and Menzel 2009; Keane *et al.* 2009; Dietl *et al.* 2015; Barak *et al.* 2016; Butterfield *et al.* 2016; Natlandsmyr and Hjelle 2016). The reasons for considering alternative potential reference states in restoration include the following. As described above, the existence of such states is predicted by theoretical community ecology. Second, extant populations of threatened species are usually located in remnants of a natural habitat that are small in size representing only a subset of the habitat's original variation in terms of environment, community composition, and successional stage. Third, paleobotanic data suggest the dynamic nature of many vegetation communities over centuries and millennia. Finally, rapid climate change can make even primary species habitats unsuitable in the near future. The search for these states as reference ecosystems should be done through adaptive learning, as a replicated over space experiment where outcomes are critically evaluated and used to inform the next experiment. As a result, optimal, i.e., leading to the highest restoration success, reference conditions can be determined.

In cases where there are no extant habitats that could serve as reference ecosystems, a proxy to once existing ecosystems may still be found. Long-term data can help to define appropriate reference ecosystems. For example, the natural forests on black marls at an altitudinal belt between 600 and 1200 m do not currently exist in the entire Haute Province, France. However, a regional ecological survey and historical records allowed Vallauri *et al.* (2002) to identify the oak *Quercus pubescens* as the dominant tree species in the preexisting forests and propose an appropriate intervention (thinning of introduced Austrian black pine) to promote establishment of native broad-leaved vegetation. Similarly, analysis of fossil pollen, and microscopic and macroscopic charcoal helped to determine the pre-anthropogenic vegetation and the disturbance regime by humans in Apuseni Natural Park, Romania, over the last 5700 years, and to reveal the anthropogenic cause of decline in abundance of regionally endangered *Abies alba* (Feurdean and Willis 2008). Another example is the rare and endangered plants of the group of "arctic-alpine" species of the Scottish Highlands, northern England, and northern Wales that currently grow only on inaccessible cliff-ledges. Pollen data provided evidence that during the Holocene these taxa were growing in an open herb-rich vegetation at or above the tree limit and, as a result of anthropogenic impact in the last two centuries, were replaced by grazing-tolerant vegetation and forced into locations inaccessible to sheep herbivory (Birks 1996). These and other examples (Muller *et al.* 2012; McCarroll *et al.* 2016; Natlandsmyr and Hjelle 2016) show the high utility of paleoecological data for guiding conservation management, through their ability to disentangle the roles of climate and human activities in long-term community changes.

Because the historical records are fragmentary and often unavailable for particular places or time periods, and because there is always large uncertainty about causation and the potential role of unmeasured variables, extrapolation of reference conditions may be the only resort. But the validity of extrapolating reference conditions will almost inevitably decrease with increasing spatial and temporal distance from the current ecosystem to which those reference conditions are to be applied – a "distance decay" problem (White and Walker 1997). A way to improve confidence in historical interpretations is to use converging lines of evidence by combining comparative analyses and tests with multiple, independent data sources and methods (White and Walker 1997; Swetnam *et al.* 1999; Motzkin and Foster 2002). Examples are the joint use of historical reconstructions and field experiments (e.g., Fastie 1995; Lloyd and Graumlich 1997) or comparisons of model simulations with historical data (e.g., Anderson 1995; Miller and Urban 1999; Cole *et al.* 2011).

A reference is needed not only for choosing the restoration site and designing the interventions, but also for assessing restoration success. Criteria used in judging whether a restoration is successful are numerous (Hobbs and Norton 1996; SER 2004; Ruiz-Jaen and Aide 2005a, b; McDonald *et al.* 2016), but in a restoration project that has conservation goals, the population biology perspective should be a priority, i.e., that the populations of target species are restored to a level allowing their long-term persistence (Montalvo *et al.* 1997; Guerrant and Kaye 2007; Menges 2008). This means that the restored population must possess attributes necessary for successful regeneration and adaptive evolutionary changes, i.e., viable demographic structure evident in the presence of new generations in addition to the founders, and sufficient genetic variation. As well as evaluating these attributes in the restored site, it is necessary to compare them with values from the reference sites. The variation among reference sites in these attributes can be important for understanding restoration failures; thus, more than one reference site (when available) should be used for estimating restoration success (Hobbs and Norton 1996; Ruiz-Jaen and Aide 2005a).

## 4.4   MAKING SPECIES LISTS

### 4.4.1   How Many Species?

One of the next steps of conservation-oriented restoration, after the area for restoration is chosen, is a decision about the number and identity of the species to be introduced as a part of the habitat restoration. There is a relationship between species diversity and the variety of ecosystem functions (Schwartz *et al.* 2000; Hooper *et al.* 2005a; Vila *et al.* 2007; Paquette and Messier 2011; Chisholm *et al.* 2013; Gamfeldt *et al.* 2013), which has a very practical application in restoration ecology, namely "how many species are needed to make a functioning ecosystem?" A complete or nearly complete ecosystem function in most cases can be achieved with a limited number of species, given that they represent all the needed functional groups (Walker 1992; Hooper *et al.* 2005a). However, conservation-oriented restoration has the goal of reinstalling maximum diversity communities, not just functioning ones, and, desirably, providing a home for endangered species.

Based on these considerations, a small number of introduced plant species can hardly be justified in conservation-oriented restoration projects. The only exception would be a situation where the restored habitat still has high species diversity of late-successional species and the introduction list can be limited to only a number of threatened species. This, however, is rarely the case. Although a few studies showed that it was feasible to use as many as 20–30 species (Lamb 2011), and even >50 species (Rodrigues *et al.* 2009; Garcia *et al.* 2014) (Figure 4.4), restorations traditionally use a limited number of species that are easy to collect and propagate, and have high survival and growth rates after planting. Clearly, threatened species usually do not possess these attributes and therefore currently are not the species of choice in restoration projects.

FIGURE 4.4 (a) The nursery workers at Camara Nursery in São Paulo State, Brazil, are assembling mixtures of native tree species from the Atlantic Forest. (b) Boxes, each with a mixture of 50 native tree species, are ready to be transported to restoration sites. Photos by Robin Chazdon. *A black and white version of this figure will appear in some formats. For the color version, please refer to the plate section.*

With lack of seed flow from neighboring locations, local impoverishment within small patches of restored habitat is highly likely due to the negative demographic and genetic effects of small population size, competition, and predation (Shaffer 1987; Montalvo *et al.* 1997; Kramer *et al.* 2008) as experimentally demonstrated by Gibson *et al.* (2013). When restoration starts with a limited subset of the species pool characteristic for the habitat, the above process can only

accelerate species loss. To mitigate this loss, restoration must maximize the number of species introduced, and include in the list species
with a narrow regeneration niche and limited dispersal ability that
are of high conservation value. Introduction into multiple locations
will still result in extinction of these species in some but hopefully
not all locations (Drayton and Primack 2000).

Species in a given plant community can be ranked by their
abundance reflecting their importance in the community. The shape
of the community species-abundance distribution will depend on
system productivity (Whittaker 1965; Hubbell 1979), succession
stage (Bazzaz 1975), and relative importance of filtering and random
assembly (Jabot *et al.* 2008; Qiao *et al.* 2015). The information
provided by species-abundance distributions such as dominance
order, species richness and evenness can have important conservation implications (Maina and Howe 2000; Hubbell 2013). Two ways
to plot species-abundance distribution in a way useful for decisions
about choice of species for restoration are rank abundance diagrams
and empirical cumulative distribution functions (McGill *et al.* 2007).
A challenge in successful experimental restoration is the choice and
number of species and a sufficiently large number of introduced
plants to prevent losses due to their small population sizes. Rank
abundance diagrams and functions can be compared among the
potential restoration and reference sites to determine the number
and identity of species to introduce. Each habitat patch has a small
number of dominant and abundant species, and a large to a very large
number of infrequent or rare species. In habitats with relatively short
"tails" of rare species, such as boreal forests, a list of species even
for large-scale restoration will be short. In contrast, habitats with
very long tails of rare species, such as tropical forests, will require
a much longer list of species for reintroduction to be made, and the
particular combination of species that are reintroduced can vary substantially from patch to patch. The reason is that in communities
with very skewed species-abundance distributions, the occurrence
of rare species varies across patches, often unpredictably, due to

patch-colonization dynamics, and/or dispersal limitation. Decisions based on species-abundance distributions must take into account that some species can be rare but viable under the given conditions while other species can be accidental or otherwise inviable under the same conditions (Maina and Howe 2000).

### 4.4.2 Which Species?

Once the approximate number of species needed is worked out, the next goal is making an operative list of the species to be introduced as a part of the habitat restoration. This list will be a subset of a larger species pool, "a set of species which are potentially capable of coexisting in a certain community" (Eriksson 1993). The species pool concept has been developed as a theoretical framework for explaining species richness at different spatial scales, and states that species richness on a smaller scale is primarily determined by the availability of "appropriate" species at the next larger scale (Partel *et al.* 1996, 2011; Zobel 1997). As a result of environmental filtering and dispersal limitations, the local community species pool (the set of species present in the target community) is a subset of the higher scale (regional) species pool (Partel *et al.* 1996, 2011; Zobel 1997; Zobel *et al.* 1998) (Figure 4.5).

The importance of both local and regional processes in structuring natural communities has been recognized by restoration ecology, stressing the role of the regional species pool as a source of the species that might successfully colonize a restored site (Zobel *et al.* 1998; Brudvig and Mabry 2008). Thus, the regional species pool provides the upper boundary for species richness at restored sites and limits the choice of potentially suitable species. These candidate species must be evaluated for their actual chances to establish in the target community. Zobel *et al.* (1998) proposed a set of ecological, functional, and phytosociological approaches, as well as an experimental approach based on the probability of germination and survival of the introduced seeds, for evaluating species not observed at the restoration site but present in the regional species pool.

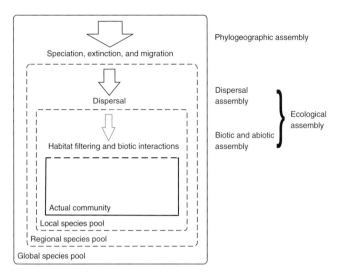

FIGURE 4.5 The different processes involved in community assembly and the relative scales at which they are most influential (from Götzenberger *et al.* 2012). A global species pool defines a regional species pool through the speciation, extinction, and migration of species (phylogeographic assembly). At a given local site the species pool constitutes species from the regional species pool that are able to disperse there (dispersal assembly), but habitat filtering and biotic interactions define the actual assemblage of plant species (ecological assembly).

A study of Brudvig and Mabry (2008) is an example of how to apply the above concept and produce manageable lists of key species for habitat restoration. To make a species list for restoration of degraded Midwestern oak savannas in Iowa, United States, first, the authors assembled a list of species for the regional pool supposed to represent the historic savanna communities in the region. They used three reference lists: (1) a list of species with habitat requirements that match the mixed-light conditions found in Midwestern oak savannas; (2) a list determined by species survival in the experimental introductions; and (3) a list of species found at a pristine reference site. Then, they used life-history traits to target species from the regional species pool. After removing the exotic species, species from non-savanna habitats, and species already present at the restoration sites, they applied a set of filters to select the most appropriate for restoration species. These filters included distance-limiting seed dispersal mode, affinity for intact native habitat, and affinity for a high-light environment. As a result, they were able to narrow down the regional species pool (900 species) to a manageable species list (111 species).

Designing community assemblages in traditional restoration, the goal of which is to restore ecosystem functionality, is based, either explicitly or implicitly, on functional traits (i.e., traits having similar effects on ecosystem functions or similarly responding to particular environmental factors) and functional diversity metrics (Funk *et al.* 2008; Suding *et al.* 2008; Laughlin 2014). In traditional restoration, the species role in an ecosystem (i.e., in primary production or nutrient cycling) is important but not its identity. However, in conservation-oriented restoration the goal is to restore not only all the ecosystem functions, but also the original biodiversity. Therefore, compiling a species list in conservation-oriented restoration should utilize both species-centered and functional approaches. The species-centered approach applies to the endangered and rare species which should be top-listed. Their functional type is secondary. In contrast, for the nonthreatened candidate species, their functional roles in the ecosystem (i.e., in canopy structure, food web, seed dispersal, pollination service) are very important. Functional groups are groups of species sharing not necessarily the same niche but functional traits (Gitay and Noble 1997; Lavorel and Garnier 2002; Rusch *et al.* 2003; Franks *et al.* 2009). The core list of traits considered to be of general importance in the identification of plant functional types can be found in Weiher *et al.* (1999), with more detailed lists and classifications elsewhere (e.g., Box 1996; Diaz and Cabido 1997; Herault *et al.* 2005). In conservation-oriented restoration, the functionally important species must represent functional groups lost during ecosystem degradation and which are necessary for ecosystem functioning. The ideal choice for nonthreatened candidate species are functionally important and co-occurring with the introduced threatened and/or keystone species, because the co-occurring species are known to have a similar ecological niche as the focal species (Halme *et al.* 2009).

The choice of traits for defining the functional groups will depend on the ecological context. In frequently disturbed ecosystems the traits should be related to colonization success, e.g., seed bank persistence, germination cues, and longevity (Tozer *et al.* 2012), or light tolerance, leaf type and succulence, dispersal, and regeneration mode (Gondard *et al.* 2003). In areas with frequent wildfires, the important traits are those conferring fire-resistance, e.g., bark thickness and resprouting ability (Clarke *et al.* 2013; Laughlin *et al.* 2017), while in drought-prone environments such traits are those providing resistance to drought-induced cavitation, e.g., wood density (Hacke *et al.* 2001). In species-rich communities, such as subtropical and tropical forests, the focus should be on the ecosystem's functional diversity, i.e., diverse flower and fruit types that support diverse pollinator and frugivorous fauna, which in turn enhances the functionality of a plant community (Tucker and Murphy 1997; Martinez-Garza and Howe 2003; McConkey *et al.* 2012; Garcia *et al.* 2015). Thus, lists of species for restoration in these environments must include flower types typical for bird, bat, butterfly, moth, bumblebee, and bee syndromes, as well as plants producing edible fruits of different size, which may require inclusion of not only trees in the list, but also shrubs, subshrubs, and herbs (Garcia *et al.* 2015).

Compiling species lists for conservation-oriented restoration can be based on the concept of characteristic and derived diversity (Helm *et al.* 2015). According to Helm *et al.* (2015), the observed community diversity represents two species pools of different historical backgrounds. The first one, called characteristic diversity, consists of species belonging to a habitat-specific regional species pool, and the second one, called derived diversity, represents species not typical to a given community (region) and whose presence is due to intended or unintended human impact. Although the majority of the candidate species will be from the first group, some threatened and functionally important species can be from the second group (Figure 4.6). Of

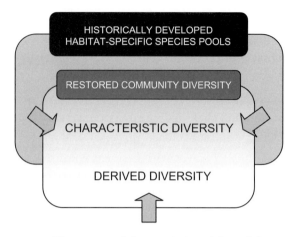

FIGURE 4.6 The concept of characteristic and derived diversity applied to conservation-oriented restoration (modified from Helm *et al.* 2015). A restored community diversity (smaller square) will consist of characteristic and derived diversity, the former being species that belong to a habitat-specific species pool (larger square), and the latter representing species from other species pools. The appropriate candidates from the second group must be threatened and functionally important species able to establish and reproduce under environmental conditions that are typical of the particular habitat.

course, potentially invasive species or those having other negative impacts on the local biota species cannot be among the second group of candidates. All the species from the regional species pool absent in the characteristic diversity pool can be considered representatives of "dark diversity" group *sensu* Partel *et al.* (2011), that is, the set of species in a region that currently do not inhabit a site due to dispersal or establishment limitations. The latter can be a result of human-caused alteration of abiotic and biotic conditions. Introduction and supported establishment can convert dark diversity into characteristic diversity.

The design of a restoration map can start from large spatial units of hundreds of hectares with the same plant community, but then should proceed to delimitation of relatively small working units of a few hectares accommodating small-scale patchy spatial distribution for some species and spatial heterogeneity of assemblages. For high-diversity forest ecosystems, a good strategy appears to be the creation of a mosaic of the compiled species list randomly assembled from local plant communities because, in these ecosystems, species assembly processes during succession at the fine spatial scale are largely random (Hubbell *et al.* 1999; Bruelheide *et al.* 2011).

With knowledge of the species environmental requirements, suitable locations for each listed species within the target area can be identified and mapped using SDM and, after maps are superimposed, the species assemblages can be defined for any spatial scale (see Figures 2.5 and 4.3). This is illustrated with the three following examples. In a study by Garcia del Barrio *et al.* (2013), the distribution maps for 40 native tree species of Spain were used to identify the species pools for each region within the country at different scales down to the grid of 1 × 1 km. In a study of Siles *et al.* (2010a), also performed in Spain, an operative restoration target map was worked out based on predicted species assemblages for 23 target species, for a grid of 100 × 100 m (Figure 4.7). And Rovzar with colleagues identified, using areas of overlap predicted by SDM for eight Hawaiian rare and endangered species, 1292 10 × 10 m multi-species reintroduction plots (Rovzar *et al.* 2016).

## 4.5 COLLECTING SEEDS AND OBTAINING OUTPLANTS

Everything described in Sections 3.6 and 3.10–3.12 about seed sampling and raising outplants for species-targeted *in situ* conservation applies to conservation-oriented restoration too. The difference is that seed sourcing and collecting, and obtaining outplants in the latter, must be done for many species simultaneously. A study by Hoffmann *et al.* (2015) demonstrates how to do multiple-species seed sourcing and collecting efficiently, and also highlights the common challenges of doing so. In a restoration project of the Araucaria Forest of southern Brazil, after compiling a list of target species for collecting, which included all the threatened, near threatened, and rare tree species ever recorded for this ecosystem, and mapping all the known occurrence records, the authors produced a list of potential

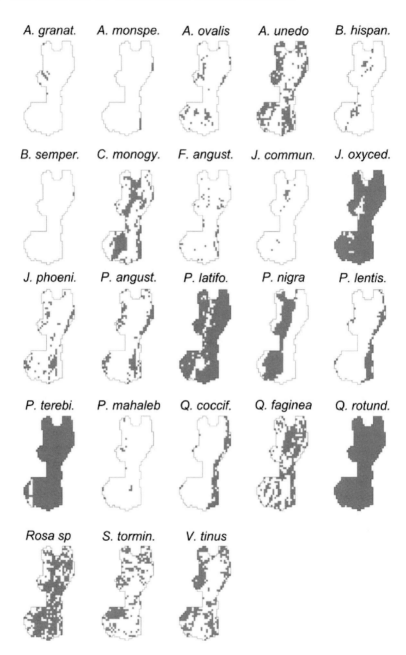

FIGURE 4.7 Maps of the predicted distribution for 23 species comprising a restoration species list, within a targeted forest restoration area (from Siles *et al.* 2010a). Cell size of the maps is 100 × 100 m.

forest remnants containing at least one target species. Then, because the chosen mother trees were supposed to be available for long-term seed collecting and monitoring, and to represent properly the species genetic diversity, the initial list of forest fragments was narrowed down to include only protected locations in different floristic associations and at different stages of forest succession. Each selected forest remnant was surveyed intensively to identify healthy (free of damage, infestation, or disease) and reproducing (showing presence of old fruits, seeds, or seedlings beneath the canopy) mother trees when possible. All attempts were made to have mother trees from at least three populations of each species and from at least 12 trees per population, or, when this was not possible, at least 20 trees per species in the whole study area. The selected trees of the same species were at least 50 m apart, and whenever possible >100 m apart. Exceptions were made only for the rare species found in a single cluster. Preliminary information about the target species reproductive phenology obtained from the literature was verified during 52 field trips throughout the study area, covering all seasons, each year during March 2011–March 2014. The data on onset and duration of flowering and fruiting were used to develop a seed collection calendar.

Despite the large search intensity (68.7 km of trails were surveyed), the authors failed to find any individual for 38 out of 71 targeted species, and located a sufficiently large number of mother trees (>12) in at least three remnants for only five species. For 23 of the target species, <20 mother trees were selected. These results demonstrate the difficulty of obtaining sufficient numbers of mother plants in a fragmented and degraded habitat, and the necessity of creating *ex situ* genetic depositories in a form of *quasi in situ* collections or seed orchards.

## 4.6 PLANTING DESIGN

Considering the effects of density in plant introductions, we must take into account that density of conspecifics can have both negative and positive effects on plant performance (Rathcke 1983). A negative

impact of conspecific seedling and adult neighbors, common in tropical and temperate forests (Harms *et al.* 2000; Comita and Hubbell
2009; Johnson *et al.* 2012), can be due to density-dependent effects
of host-specific natural enemies. Host specialist seed predators can
strongly affect the seed-to-seedling transition (Harper 1977), while
herbivore specialists and soil pathogens can affect the early seedling stage and transition to saplings (Augspurger 1983; Mangan *et al.*
2010). At the later developmental stages, both insects and insect-
transmitted pathogens can cause density-dependent mortality in
established seedlings and saplings (Wong *et al.* 1990; Gilbert *et al.*
2001). For example, Wong *et al.* (1990) showed that juveniles distant
from adults may have higher survival rates than juveniles close to
adults due to higher susceptibility to insect outbreaks. A recent study
of distance-dependent mortality in three tropical tree species showed
that multiple enemies (fungal pathogens, mammals, and insects)
affect seed-to-seedling transition and that source and severity of mortality at different stages of the life cycle is highly species and location
specific (Fricke *et al.* 2014). Conspecific negative density dependence
appears to be stronger for species less common as adults in the forest
community, and in species-rich regions than species-poor regions
(Comita *et al.* 2010; Mangan *et al.* 2010; Johnson *et al.* 2012; Zhu
*et al.* 2015).

A negative density-dependent effect can also arise from
autotoxicity effects (Li and Romane 1997; Cavieres *et al.* 2007). For
example, experiments with endangered *Nyssa yunnanensis* revealed
the inhibitory effect of root extract on seed germination and seedling
growth (Zhang *et al.* 2015).

There is a decline in the strength of conspecific negative
density dependence across life stages: from strong negative effects
at early life stages to weak positive effects for adults (Zhu *et al.*
2015). However, because individuals can spend decades in the seedling bank, weak neighborhood effects may accumulate over time
(Comita and Hubbell 2009). This implies that seedlings of rare and
endangered species must be planted at a distance from each other and

from the adult trees to prevent negative density effects, but this does not negate various positive density effects which become evident at later stages of the life cycle.

A positive density dependence in population growth rate at low densities is known as an Allee effect (Allee 1931). This effect can be caused by various genetic, demographic, and ecological factors, e.g., by increased levels of inbreeding depression, skewed sex ratios, reduced availability of mates, or failure to satiate predators. In plants, the latter phenomenon is apparently widespread in small and fragmented populations of species that require animal pollination (Aizen and Feinsinger 1994a; Agren 1996; Ghazoul 2005; Xia *et al.* 2013). Pollinators usually exhibit frequency-dependent foraging behavior when they forage more in dense patches in order to reduce the inter-patch travel time. A reduction in the number of individuals in a population decreases the number of flowers per unit area. Thus, plants growing at low densities may experience reproductive decline or failure due to difficulties in attracting pollinators (Allison 1990; Feinsinger *et al.* 1991; House 1992; Kunin 1992, 1993; Lamont *et al.* 1993; Agren 1996; Roll *et al.* 1997; Ghazoul *et al.* 1998; Bosch and Waser 2001; Forsyth 2003; Le Cadre *et al.* 2008). Genetic effects can also be significant due to the relationship between the outcrossing rates and the density of reproducing adults.

Another process contributing to an Allee effect is the lack of improved habitat that otherwise would be induced by the presence of a sufficiently large number of conspecific individuals. Examples are the hemlock *Tsuga heterophylla*, which can acidify soil and sequester water in the upper soil only at sufficiently high density (Ferson and Burgman 1990), and a floating aquatic macrophyte *Stratiotes aloides*, which can buffer against ammonium toxicity only in high- but not low-density stands (Harpenslager *et al.* 2016).

The major consequence of an Allee effect, either genetic, pollinator mediated, or environment facilitated, is the existence of a density threshold below which the aggregation unit (population) is likely to go extinct. For example, small remnant populations of

*Banksia goodii* exhibited complete reproductive failure below a certain threshold patch size (Lamont *et al.* 1993). As introductions of rare and endangered species usually have small to intermediate propagule sizes, a possible Allee effect on introduction success must be considered (Forsyth 2003; Deredec and Courchamp 2007; York *et al.* 2013; Abeli *et al.* 2016). The mechanisms underlying Allee effects suggest that population persistence is dependent on population size and the spatial distribution of flowering plants. The probability of reintroduction success is positively related to the number of introduced individuals (Abeli *et al.* 2016). In flowering plants, sufficiently large numbers of individuals spaced not too far from each other are needed to ensure high pollinator visitation rates (Groom 1998; Colas *et al.* 2001; Hackney and McGraw 2001; Le Cadre *et al.* 2008; Dauber *et al.* 2010), and the density appears to be more important for attracting pollinators than the population size (Kunin 1997).

It is also important that the introduced plants have different ages to reduce the susceptibility of a new population to pests, diseases, and wind storms. As was shown by Dani Sanchez *et al.* (2018), two-phase planting in groups of 25 saplings 2 m apart was a preferred restoration strategy for the Caicos pine *Pinus caribaea* var. *bahamensis*, a threatened endemic keystone species in the highly endangered pine forest ecosystem of the Bahaman Archipelago. The scheme of planting in year 1 and then in year 6 in the same plot resulted in a mosaic of trees of different ages, with the created stands less susceptible to damage and uprooting during strong winds.

The above considerations, plus results of long-term planting density experiments (if they exist) (e.g., York *et al.* 2013), must serve as the basis for decisions about number, density, and age structure of created populations. Relatively small but aggregated or high-density populations comprised of differently aged individuals appear to be a better design than larger populations with sparsely distributed individuals representing a single cohort.

## 4.7 BIOTIC INTERACTIONS

### 4.7.1 *Reestablishment of the Integrity of Disrupted Interactions Crucial for Ecosystem Functioning*

Restoration focusing only on focal species introduction has a much higher chance of failure than one considering reestablishment of the integrity of disrupted interactions crucial for ecosystem functioning. To restore seed dispersal, pollination, nutrient cycling, and the food web, one may need to introduce or control a suite of interacting species such as soil biota, herbivores, seed predators, and frugivorous vertebrates (e.g., Bond and Slingsby 1984; Brown and Heske 1990; Vander Wall 1994; Chapman and Onderdonk 1998; Traveset and Riera 2005; Beyer *et al.* 2007; Kaiser-Bunbury *et al.* 2010; Traveset *et al.* 2012; Grégoire Taillefer and Wheeler 2013; Anderson *et al.* 2014) and environmental engineers (McColley *et al.* 2012; Zysk-Gorczynska *et al.* 2015).

Many authors have demonstrated that the establishment success of introduced plants may require inoculation of seedlings with mycorrhizal fungi and soil bacteria (Barroetavena *et al.* 1998; Fisher and Jayachandran 2002; Gemma *et al.* 2002; Zandavalli *et al.* 2004; Zubek *et al.* 2009; Shen and Wang 2011; Ferrazzano and Williamson 2013; Fajardo *et al.* 2014). For example, inoculation with arbuscular mycorrhizal fungi improved growth and development of the translocated plants of endangered *Abronia macrocarpa* (Ferrazzano and Williamson 2013).

The most important mutualistic systems, as discussed in detail below, are fleshy-fruited plants and frugivores (Brodie and Aslan 2012) and coevolved insect pollinators and flowering plants (Aizen *et al.* 2012). Support for colonization or combined translocation of these interacting species is most likely to be necessary to restore lost ecosystem function. Depending on each other, coevolved taxa have rarely been considered in species restoration plans but must be incorporated into restoration planning and implementation to prevent coextinction.

Reestablishing a viable ecosystem may require environmental modifications needed for vertebrates, such as availability of perches or structural complexity of the vegetation (Beyer *et al.* 2007; Seddon *et al.* 2014; Fraser *et al.* 2015). Similarly, to be functional, an ecosystem may require the reintroduction of top predators. Examples of such reintroductions are wolves in Yellowstone National Park, which have improved the regeneration of arboreal vegetation and increased avian diversity by limiting deer population growth (Ripple and Beschta 2004), and beavers in France, which have helped to stabilize riverbanks by rejuvenating riparian forests (Fustec *et al.* 2001). Thus, reintroduction, or translocation of threatened animals, which can perform a functionally important role in a restored ecosystem, can simultaneously serve several important goals: to improve the conservation status of the introduced animal species, restore the degraded environment, and enhance populations of the threatened local plant species. For example, introduction of endangered giant tortoises (*Geochelone nigra hoodensis*) to one of the Galápagos Islands had a positive impact on an arboreal cactus *Opuntia megasperma*, which is itself endangered and a keystone resource for many animals on the island (Gibbs *et al.* 2008).

The maintenance of the network of plant–animal interactions must be recognized as a cornerstone of conservation policy (Montoya *et al.* 2008). Practical recommendations, for the areas where functionally important animals (e.g., seed dispersers or grazers) have become extinct, is to reintroduce these species from other regions, or, when this is not feasible because the species are globally extinct or have critically declined in number, to use functionally equivalent, at the community level, substitutes from the local or regional fauna (McConkey *et al.* 2012). For example, introduced tortoises have proved to be efficient extant substitutes for extinct beaked grazers/browsers in many island contexts (Gibbs *et al.* 2008; Griffiths *et al.* 2011; Pedrono *et al.* 2013; Burney and Burney 2016).

### 4.7.2    *Reestablishment of Pollination Services*

Disruption of pollination services is a common problem of degraded and altered ecosystems, and restoration plans must consider plant–pollinator interactions and networks that support those interactions to reestablish pollination processes and vectors that underpin the plant reproductive continuity of a restored ecosystem (Dixon 2009; Menz *et al.* 2011). Loss of pollinators and changes in pollinator assemblages servicing a species may have such negative consequences as low or zero seed set, population declines in associated species, and even coextinction (Bond 1994; Biesmeijer *et al.* 2006; Dixon 2009; Moir *et al.* 2010). Given the complexity of plant–pollinator networks and ubiquitous importance of pollination in ecosystem functioning, restoration planning cannot assume that plant–pollinator interactions will reestablish themselves (Handel 1997; Forup *et al.* 2008; Williams 2011). After disappearance from the Rondevlei Nature Reserve, South Africa, due to fragmentation, the obligate pollinator of the orchid *Disa draconis*, the long-tongued horsefly, pollination of this species does not occur naturally (Milton *et al.* 1999).

Some pollination systems are more vulnerable than others. Island pollination systems with a high degree of endemism and low taxonomic diversity are especially fragile because they evolved from rare colonization events and subsequent evolution of limited number of taxa. Because of the lack of coevolution with continental predators and competitors, they are highly vulnerable to invasive species (Traveset and Richardson 2006; Kaiser-Bunbury *et al.* 2010). For example, in Hawaii, the coevolution of honeycreepers and bees with local flora has occurred in the absence of social insects, and the endemic pollinators could not develop the appropriate competitive and defensive mechanisms. The historical absence of social insects in Hawaii explains the extent of their impact on this island. On the other hand, good examples of how eradication of exotic species can help restoration of plant–pollinator mutualisms also come from the islands. Ecosystem restoration of dwarf forest plant communities of isolated, rocky mountaintops in the Seychelles via removal of all exotic woody plants resulted in a marked increase in pollinator species, visits to flowers, and network interaction diversity. This led to distinctly improved reproductive performance of the 10 most abundant native plant species (Kaiser-Bunbury *et al.* 2017). The large-scale removal of an invasive nectar thief and arthropod predator, the social wasp *Vespula pensylvanica*, increased fruit production of a functionally important endemic Hawaiian tree species, *Metrosideros polymorpha* which provides a critical nectar resource and habitat for a large number of endemic biota. This happened due to a significant increase in the visitation rates to *M. polymorpha* by a substitute pollinator for the extinct honeycreepers and threatened bees *Hylaeus* spp., that is, the introduced *Apis mellifera* (Hanna *et al.* 2013).

Pollinator colonization and persistence in restored sites can be achieved only by satisfying the pollinators' needs which are essential for the completion of their life cycle either within the restoration site or within the pollinators' foraging distance. For example, solitary bees strongly depend on the availability of nest sites (Gathmann and Tscharntke 2002; Steffan-Dewenter and Schiele 2008) and butterflies on the availability of both larval host plants and nectar resources as adults (Dennis *et al.* 2003). Thus, the presence of habitat patches with pollinator-rich communities in proximity to restoration sites will facilitate the reestablishment of pollinator activities there. Establishing restoration sites in a highly fragmented landscape as stepping stones should enhance the connectivity among the habitat patches and facilitate dispersal of pollinators. In restoration sites embedded within an ecologically hostile matrix wide-ranging generalists such as bumblebees and the honeybee may succeed as colonizers, while nonflying or restricted-range pollinators, such as cursorial mammals, lizards, and manyinvertebrates, will stand little chance of colonizing these sites by themselves. Knowledge of the degree of specialization, colonization capability, and minimum habitat area requirements of the crucial pollinator groups is essential. In general, in species-rich plant communities, plants are more likely to exhibit higher levels of pollinator specialization. For example, many tropical orchid species use sexual deception to attract a single pollinator species with little sharing of pollinators among the species (Hoffman and Brown 1992; Phillips *et al.* 2009; Gaskett 2011). Species-rich ecosystems with highly specialized pollinator associations present the greatest challenges for restoration and will require detailed knowledge of the ecological requirements for both the plants and their pollinators.

A major recommendation from the studies that compared plant–pollinator networks of restored and unrestored communities (Forup and Memmott 2005; Forup *et al.* 2008; Williams 2011) is to prioritize restoration of generalist and strongly interacting species because the latter promote redundancy of pollination among plants and thus the creation of species-rich and structurally complex communities. Because pollinators contribute ecological functions that are critical to the ecosystems, the exact identities of these species are often are less important than their functional role. Thus, if crucial native pollinators are extinct, substitution by extant species can be a viable option. For example, in Hawaii the extinct bird pollinators of *Freycinetia aborea* and *Metrosideros polymorpha* have been replaced by an introduced bird *Zosterops japonica* (Cox 1983) and a honeybee *Apis mellifera* (Hanna *et al.* 2013), respectively.

In addition to addressing direct plant–pollinator mutualism, reestablishing pollination service in the restored site may require the introduction of facilitating plant species serving one of the following roles: framework species (species that provide a major nectar or pollen source); bridging species (plants that provide resources over resource-limited times); and magnet species (plants with attractive flowers associated with species with unattractive or small flowers) (Dixon 2009). Once a candidate framework, bridging, and magnet plants are identified, the best choice will be those species supporting the greatest local abundance and diversity of pollinators.

### 4.7.3 *Reestablishment of Frugivory*

The functioning of many ecosystems is impossible without another plant–animal mutualism, such as frugivory. Many plants have evolved edible seed coverings or appendages to attract frugivores, which will ingest the seeds together with the flesh, and then regurgitate, or defecate the viable seeds. Seed consumption followed by their release after passage through the gut can have two advantages for the plant: dissemination of the seeds away from the conspecifics and enhancement of seed germination.

Animal seed dispersal is a predominant form of dispersal in the tropics, with over 70% of tree, shrub, and vine species in tropical and subtropical rainforests having seeds enclosed in fleshy fruit (Howe and Swallowed 1982). Seed dispersal may be necessary for successful recruitment at different spatial scales by enabling seeds and seedlings to escape the high mortality that occurs in proximity to parent plants (i.e., Janzen–Connell effect) (Gilbert *et al.* 1994; Packer and Clay 2000; Mangan *et al.* 2010; Neghme *et al.* 2017), to colonize new sites, and maintain gene flow between populations. In the absence of dispersers, the seeds and seedlings may remain near parents and die due to competition with the adults and higher susceptibility to enemies when at high density.

The second positive effect of seed consumption by vertebrates can occur through an increase in germination percentage or rate as a result of gut treatment. This can be due to mechanical and/or chemical scarification of the seed coat, separation of seeds from the pulp, and the effect on germination and/or future seedling growth that results from fecal material surrounding the seeds (Traveset 1998). Two reviews of seed passage through the digestive tract of vertebrate frugivores in more than 200 plant species revealed a predominant enhancement effect, with ingested seeds germinating in greater numbers and more rapidly than uningested seeds (Traveset 1998; Traveset and Verdu 2002). Although no experimental evidence of this exists, germination failure in some species can be a result of local or global extinction of the associated frugivores.

Some frugivores, such as many small-sized birds and mammals called pulp feeders, consume the fruit pulp but not the seeds, thus leaving them in the vicinity of the parents. Nonetheless, by consuming part or most of the fruit pulp, such pulp thieves can provide an essential service to the plants by releasing seeds from the pericarp's inhibitory effect (Robertson *et al.* 2006). Fedriani *et al.*

(2012), studying the interaction between *Pyrus bourgaeana* and its pulp feeders, showed experimentally that pericarp removal had a consistently strong and long-lasting positive effect on seed performance (e.g., lower rotting and higher germination percentages) and seedling fate (greater emergence, growth, and survival to 2 years old). In this study, the cumulative probability of establishment for depulped seeds was 4–25 times higher than for seeds in intact ripe fruits.

Thus, frugivores can have a variety of positive effects on the plants they feed on. A spectrum of the fruit consumers can be roughly predicted from the fruit type: the species bearing fruits with one or several large seeds are visited by a few large frugivores, while those producing small or large soft fruits with many small seeds are visited by a wide spectrum of frugivores (Corlett 1998; Kitamura *et al.* 2002). The overlap between diets of different groups of large frugivores is usually small, especially between bird and mammal groups (Kitamura *et al.* 2002), and even frugivores with similar diets differ in dispersal methods, distances they travel, and microhabitats into which they disperse seeds. For example, gibbons disperse seeds via defecation while macaques disperse seeds via their cheek pouches (Kitamura *et al.* 2002), and small passerine birds disperse most seeds over short distances and into covered microhabitats, while mammals and medium-sized birds disperse seeds over long distances and mostly into open microhabitats (Jordano *et al.* 2007). The latter differences increase the chance that a seed will be disseminated into a favorable habitat with locations distributed randomly over space. Thus, if one frugivore group disappears its loss cannot be compensated by another group.

Rainforest clearing and fragmentation, on the one hand, and poaching and bushmeat harvest, on the other, have resulted in a dramatic decrease in the abundance or extirpation of many frugivorous vertebrates around the world. The disappearance of large fruit-eating birds and animals altered seedling banks by favoring seeds dispersed by bats, small birds, and wind (Wright *et al.* 2007; Terborgh *et al.*

2008; Brodie *et al.* 2009), and reduced or stopped recruitment in species dependent on large frugivores (e.g., Balcomb and Chapman 2003; Cordeiro and Howe 2003; Nunez-Iturri *et al.* 2008; Terborgh *et al.* 2008; Sethi and Howe 2009; Vanthomme *et al.* 2010; Wotton and Kelly 2011; Effiom *et al.* 2013; Hawes and Peres 2014; Beaune 2015; Pérez-Méndez *et al.* 2015). Defaunation has been recognized as a very significant conservation problem for both animals and plants (Redford 1992; Peres and Palacios 2007; Terborgh *et al.* 2008; Beaune *et al.* 2013; Harrison *et al.* 2013; Ripple *et al.* 2015; Culot *et al.* 2017), and must be addressed in restoration actions directly or indirectly. Without a strategic goal of restoring once existing but disrupted plant–animal interactions by protecting, attracting, or reintroducing frugivore populations, the habitat will never become the one it used to be. When information on fruit–frugivore interactions for a given community does not exist, a priority for conservation, reintroduction in the majority of cases should be the large-bodied, large-gaped, and wide-ranging frugivorous taxa (McConkey *et al.* 2012), because these animals and birds usually have the largest impact on ecosystem functioning (Donatti *et al.* 2011), but are first to disappear due to the highest hunting pressure. Before it is too late, it is time for tropical countries to consider complete hunting bans on particularly important seed-dispersing frugivores such as primates, hornbills, large pigeons, and fruit bats (Brodie and Aslan 2012).

Because of the important role that frugivorous animals play in maintaining ecosystem biodiversity, the plant species chosen for restoration plantings must include those producing broadly palatable fruits with long or multiple annual fruiting seasons (Howe 2016).

## 4.8    ADDRESSING RECRUITMENT PROBLEMS

### 4.8.1    *Causes of Recruitment Failure*

The natural regeneration cycle includes a sequence of stages, starting with the flowering of a mature adult and ending with the establishment of a new reproducing individual. Among the events associated

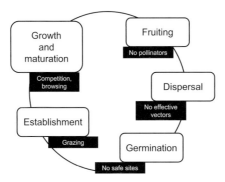

FIGURE 4.8 Ecological processes crucial for recruitment and possible causes of their failure in interaction with the biotic and abiotic environment.

with these stages, several are of decisive importance for regeneration success, as shown in Figure 4.8. They include the processes of pollen production, dispersal, and oviposition on stigmas, development of viable seeds, and their arrival at safe sites, seed germination, seedling survival, and establishment. Collectively, species requirements for successful completion of a regeneration cycle are known as the species ontogenetic niche. Decoupling between the environmental conditions and the species ontogenetic niche can lead to recruitment failure. Restoring recruitment requires removing the obstacles to seed production and their successful germination and growth (e.g., lack of pollinators, pre- or postdispersal seed predation, lack of germination cues, altered/unfavorable canopy or substrate conditions).

Spatial heterogeneity in environmental conditions can act as a filter, causing nonrandom germination and the survival and growth of individuals that differentially affect plant species (Harper 1977). Niche-based models recognize abiotic (e.g., topography and soil) and biotic (e.g., competition and facilitation) filters as important determinants of community assembly (Keddy 1992; Weiher and Keddy 1999; Chase and Leibold 2003). The habitat specialization of coexisting species serves as a framework for explaining the spatial distribution of plant species (reviewed in Rees *et al.* 2001; Wright

2002), although neutral processes based on dispersal limitation leading to local species sorting can also be important (Bell 2001; Hubbell 2001). Environmental filtering is the predominant driver of species assembling in temperate forests, whereas in the tropics dispersal limitation becomes equally or even more important (Myers *et al.* 2013 but see Lebrija-Trejos *et al.* 2010). However, although the contribution of neutral processes to shaping biodiversity patterns can be significant at intermediate and small geographic scales (Webb and Peart 2000; Harms *et al.* 2001; Condit *et al.* 2002; Baraloto and Goldberg 2004; Laliberté *et al.* 2009), from an ecosystem restoration perspective, environmental filtering is more important (regardless of its proportional contribution) because it is more constraining. When ecologically similar species are likely to coexist, this is because their similarities are primarily related to ecological traits that make them fit this environment (Leibold and McPeek 2006).

Although habitat specialization is predicted to operate throughout plant ontogeny, it became recognized, after a seminal paper by Grubb (1977), that the crucial life stage that defines a species niche, and to large extent creates niche separation among species, is the period from germination to establishment. For many plant species the niche at this stage of the life cycle is much narrower than at the adult stage and this explains why adults often thrive in a variety of habitats but without recruitment (see Figure 2.4). Particular sets of conditions favoring seedlings of different species thus function as regeneration niches that promote species coexistence (Grubb 1977; Huston 1994).

Soil conditions (moisture, pH, litter thickness), light, temperature, and biotic interactions limiting seed germination and seedling survival are among the strongest filters on recruitment for many taxa (Clark *et al.* 1998; Kobe 1999; Fine *et al.* 2004; Engelbrecht *et al.* 2007; Baldeck *et al.* 2013). The limiting role of these factors varies among species and can have either a synergetic or interacting effect (Denslow *et al.* 1998; Beckage *et al.* 2000; Coomes and Grubb 2000; Beckage and Clark 2003). Suitable soil conditions may require the presence of specific biota (microbes, fungi, and arthropods) (Haselwandter 1997; Turnau and Haselwandter 2002; Fahselt 2007). Leaf litter can be an important recruitment filter, inhibiting or enforcing seed germination through its effect on local humidity and infestation/predation rate (Schupp 1988; Molofsky and Augspurger 1992; Facelli 1994; Santos and Válio 2002), and enhancing or retarding early seedling survival (Ibanez and Schupp 2002; Wendelberger and Maschinski 2009).

Existing vegetation, soil texture, and relief microenvironments can be important for trapping seeds and enhancing their germination (e.g., Eldridge *et al.* 1991; Tsuyuzaki *et al.* 1997; Garcia *et al.* 2000; Rey and Alcantara 2000; Castro *et al.* 2004; Jones and del Moral 2005). Topographic microenvironments created by treefall (pits and elevated microsites such as logs, stumps, or root mounds) can be necessary for seedling establishment (Putz 1983; Harmon *et al.* 1986; Nakashizuka 1989; Peterson *et al.* 1990; Kuuluvainen and Juntunen 1998). In many coniferous and some broad-leaved forests, rotting logs act as safe sites, in which seeds are preferentially intercepted and retained, and seedlings have favorable conditions for their development (Harmon and Franklin 1989; Lack 1991; Narukawa and Yamamoto 2003; Zielonka 2006; Svoboda *et al.* 2010).

Beside topography and soil, local heterogeneity generated by canopy openings creating much higher light intensities than in the understory can also be related to the regeneration niche. Seedlings of many species, including nonpioneers, are heavily dependent on gaps and are termed "gap specialists" (Denslow 1987). However, in many forest types, the spatial patterns of tree diversity are better explained by seed limitation than by canopy gap availability (Hubbell *et al.* 1999; Brokaw and Busing 2000). Although gaps promote regeneration by creating opportunities for seedling recruitment, local seed pools determine which species will establish in a given gap (Hubbell *et al.* 1999; Curran and Webb 2000). Seed limitation applies first and foremost to species with large gravity- or animal-dispersed seeds and to sites where species occur at low densities (McEuen and Curran 2004).

Reliable determination of the species regeneration niche is not easy. When the factors determining species distribution are inferred from seedling, sapling, or adult spatial distributions, the observed patterns in many cases can be equally well explained by topographic–edaphic variables or space-limited seed dispersal. Observational and experimental evidence suggests that in both tropical and temperate forests there are often more safe sites than the seeds reach (e.g., Clark *et al.* 1998; Hubbell *et al.* 1999; McEuen and Curran 2004). Thus, observations on species reproductive biology, analysis of population demography, and tests of seed germination and seedling performance across microsites may be necessary to identify the regeneration limiting factor(s). Identification of the factors acting as "barriers" as well as those that can be "facilitators" to regeneration is a major step in any species recovery program. Once these factors, either physical or biological, are known, they can be manipulated to restore/improve regeneration. Thus, appropriate interventions will be those eliminating barriers, enforcing facilitating factors, or both. For threatened species, infertility, limited seed dispersal, interspecific competition, and poor soil conditions are the most important barriers for regeneration.

### 4.8.2   Seed Limitation

Recruitment limitation includes two components (Eriksson and Ehrlen 1992; Clark *et al.* 1998, 1999): (1) seed limitation, the failure of seeds to arrive at available sites, which can be due to a low density of adults, reduced adult fertility, or limited seed dispersal, and (2) establishment limitation, the failure of seeds to germinate or to develop into a reproducing individual, which can result from inadequate environmental conditions.

The importance of seed limitation is obvious – natural regeneration can occur only if the seeds are produced, dispersed, and able to germinate. If no seedlings are observed in a population, before testing whether the germination conditions are met, we must be sure that the seeds are produced and not consumed by predators before they can germinate, and remain viable at least some time after dispersal.

Anthropogenic alterations of habitats can be a cause of seed limitations. The transition of a continuously distributed population to spatially isolated small patches of plants may have numerous negative demographic, ecological, and genetic consequences. Degradation and fragmentation of habitats may result in reduced density and fecundity of parent trees (Ghazoul *et al.* 1998), decreased pollinator abundance, visitation rates, and pollen deposition (Jennersten 1988; Aizen and Feinsinger 1994b; Cunningham 2000; Quesada *et al.* 2003; Montero-Castano and Vila 2012), and impeded dispersal following frugivores' decimation (Stoner *et al.* 2007; Terborgh *et al.* 2008).

Outcrossing rates of wind-pollinated conifers (Perry and Knowles 1990; Boyle *et al.* 1991) and animal-pollinated trees (Murawski *et al.* 1990; Murawski and Hamrick 1991, 1992; Hall *et al.* 1994; Boshier *et al.* 1995; Fukue *et al.* 2007; Naito *et al.* 2008; Kamm *et al.* 2011) and shrubs (Franceschinelli and Bawa 2000; Gonzalez-Varo *et al.* 2009) positively correlate with the density of flowering individuals. Pollinators arrive less frequently in isolated populations, but visit most flowers in the patch when they do come. This implies that the majority of pollen transfer in isolated patches will be between members of the same patch. Therefore, in isolated patches, success of pollination is highly dependent on availability of mates, e.g., of genotypes representing different flower morphs or S-alleles, or a correct proportion of males and females in the patch. For these reasons, small patches face not only a higher extinction risk, but often suffer from reduced fruit/seed set or even lack of sexual reproduction (Percy and Cronk 1997; Groom 1998). Self-incompatible and predominantly outcrossing self-compatible species are more vulnerable to the negative genetic consequences of reducing plant density fragmentation than self-compatible and selfing species. For self-incompatible species in particular, beside inbreeding and genetic erosion, this may result in the loss of low-frequency self-incompatibility alleles (S-alleles) making mating between individuals impossible.

Besides pollen limitation, isolated patches with low within-patch density are also likely to suffer from reduced seed set due to the transfer of the wrong species pollen (e.g., Silander and Primack 1978), particularly if the plants are surrounded by other flowering species (Feinsinger *et al.* 1991; Aizen and Feinsinger 1994b). In some, apparently rare cases, if the density of conspecifics continues to increase after reaching an optimum, it may have the opposite effect on seed set. For example, in *Nierembergia linariifolia*, a shrub pollinated by solitary bees, seed set increased with increasing plant density up to a maximum value near 12 plants/m$^2$, after which it decreased with increasing plant density (Nattero *et al.* 2011).

There are species-specific thresholds in plant density below which negative ecological and genetic processes start to operate. For example, the reproductive output of a dipterocarp tree *Shorea siamensis* was not affected when population density was reduced from 85 to 51 flowering trees/ha but declined sharply when density was reduced further to 9 trees/ha (Ghazoul *et al.* 1998). These thresholds will be particularly important for species that are most sensitive to pollinator-mediated density-dependent effects, self-incompatible, and pollinated by small insect species. Examples are the thrip-pollinated species of *Shorea* (Ghazoul *et al.* 1998; Lee *et al.* 2006; Tani *et al.* 2009) and beetle-pollinated species of *Magnolia sensu* lato (Hirayama *et al.* 2005; Isagi *et al.* 2007; Lopez-A *et al.* 2008; Setsuko *et al.* 2013; Chen *et al.* 2016; Wang *et al.* 2017). *Magnolia* is a striking example of a taxon exhibiting depressed or completely lacking recruitment in natural populations that resulted from fragmentation-related disruption of the species reproductive ecology. Although the trees produce protogynous hermaphroditic flowers and occasionally set fruit, spatial isolation of the small populations and low density of trees lead to lack of recruitment in many *Magnolia* species. Another example is *Nesohedyotis arborea*, which is a tree endemic to St. Helena island. In this species, visitation of trees by syrphid flies, its major pollinators, was found to decline sharply at isolation distances exceeding 50 m, with a reduction of fruit set by more than half at trees isolated by more than 100 m (Percy and Cronk 1997). In addition, this species is dioe-cious, with a sex ratio 3:2 in favor of males, which also contributes to reduced fruit set. Therefore, the recommended actions include supplementation planting of opposite-sex individuals into single-sex subpopulations, and additional planting in equal sex proportions to link subpopulations by gene flow (Percy and Cronk 1997).

Density thresholds for successful recruitment exist also for short-lived plants. A study of an annual herb, *Clarkia concinna*, revealed that seed set dropped to zero when isolation distance among small patches exceeded 26–104 m depending on the patch size (Groom 1998). In contrast, sufficiently large patches attracted pollinators regardless of their degree of isolation. Thus, in this species, for patches smaller than 10 individuals, successful pollination requires the inter-patch distances to be less than 25 m, whereas for larger populations, the safe inter-patch distance is 100 m (Groom 1998). For small groups of plants that are dependent on insect pollinators, like *C. concinna*, clustering of subpopulations at the proper distance from each other is critical to long-term success.

All the above suggests that restoration of recruitment in insect-pollinated species exhibiting limited seed set and spatial segregation of reproducing plants requires reinforcement of the remaining wild populations with genetically variable seedlings or saplings, planted at a distance from each other within migrating distance of effective pollinators. The introduced groups of plants must be sufficiently large and not too far from each other to enable inter-group gene flow. This consideration must be explicit in projects augmenting existing, and especially creating, new populations, because in the latter the introduced population sizes are usually small and attrition during the establishment phase is inevitable.

Seed limitation can be caused by a combination of factors, which must be addressed simultaneously. For example, southern bog forests of Northern Patagonia have suffered broad-scale anthropogenic burning in the last 200 years. As a result of these fires, the forest composition has been altered and formerly dominant conifer *Pilgerodendron uviferum* has become a threatened species. Although burning stopped more than 70 years ago, recovery of this ecosystem has been very slow due to the low number of reproductive female *P. uviferum* trees (0.3 trees/ha) often being too far from the male trees, and limited seed dispersal (<20 m). To accelerate natural regeneration, combining passive and active restoration in *P. uviferum* forests is recommended. The recommended actions are planting male *P. uviferum* individuals near to female trees without regeneration, planting dispersed small groups comprising male and female trees, or sowing of pretreated seeds in disturbed forests. At the same time, it is recommended that the area inside a radius of 20–30 m from seed trees with regeneration is left to recover naturally (Bannister *et al.* 2014).

### 4.8.3   Establishment Limitation

Anthropogenic alterations of a habitat's physico-chemical parameters, such as soil moisture, nutrients, or light availability, can cause negative changes in biotic interactions with predators, pathogens, or competitors, and a reduction in suitability for seed germination and seedling establishment of microsites.

Besides seed limitation caused by low fecundity or poor seed dispersal as described above, seed predation can be another factor preventing seedling emergence, and increased levels of seed predation is often a consequence of habitat degradation (Silander 1978; Asquith *et al.* 1997; Tallmon *et al.* 2003; Donoso *et al.* 2004; Stephens *et al.* 2012). Recruitment in species suffering from intensive postdispersal predation depends upon the release of seeds into locations where the probability of predation is reduced (e.g., microsites with dense litter and vegetation cover) (Crawley 2000; Reed *et al.* 2006; Denham 2008). Recruitment of such species can be improved by increasing topographic heterogeneity, with, for example, artificially created pits acting as catchments for runoff, litter, and seeds.

In general, small-scale spatial heterogeneity, not only topographic but also in soil fertility, light availability, etc., is important because many species have very narrow conditions for seed germination and establishment. Within a habitat, the degree of habitat suitability at a small scale can vary greatly. Only those microsites that can support germination and plant establishment are called "safe sites" (*sensu* Harper *et al.* 1961; Fowler 1988). When parameters of safe sites are reliably determined, locations suitable for species introduction will be those with a sufficiently large number of microsites that can support natural species regeneration. Many empirical plant microhabitat studies have focused on the relationship between microsite variation and seed germination/seedling establishment. Identification of environmental factors (e.g., soil moisture, presence of mycorrhizae, available light) associated with seedling emergence and survival can be used to determine suitable and unsuitable sites for seeding/planting and then either (1) limit seeding/planting to the suitable microsites only, greatly reducing loss of the valuable plant material and restoration cost, or (2) choose the most appropriate treatments to attenuate these limitations. Vital knowledge can be obtained either through analysis of the pattern of seedlings' distribution or via experimental introductions. Let's have a closer look at these approaches.

Determination of favorable microhabitats can be based on comparison of occupied versus unoccupied locations. For example, analysis of microsites in Florida occupied by threatened *Lilium catesbaei* versus randomly selected unoccupied microsites revealed that this species prefers open, high light locations with minimal litter. Based on these results the recommended management techniques to provide more habitat for *L. catesbaei* were those eliminating litter and increasing light availability (prescribed fire and roller-chopping) (Sommers *et al.* 2011). Similarly, determination of key microhabitat predictors of persistence in a fire-mediated savanna for *Aloe peglerae*, an endangered South African plant, involved the analysis of aloe-occupied and aloe-vacant microhabitats. The key findings were that this succulent primarily survives in very rocky areas of low surrounding vegetation providing increased fire-protection (Arena *et al.* 2015).

Besides comparison of locations currently occupied and never occupied by a species, those locations from which it was extirpated can also be used to determine habitat characteristics essential for establishment. This has been done to determine the favored microhabitat for an endangered Florida perennial *Dicerandra immaculata* var. *immaculata*. Access to sunlight provided by gaps in the canopy and low competition from shrubs at ground level were found to be important for the species survival and recruitment. The management recommendations for gap creation were prescribed burning and/or mechanical removal of competing plants (Richardson *et al.* 2013).

It is also possible to compare the species population demography (total size, number of juveniles and reproducing adults) across occupied microsites to determine the safe sites. To identify microhabitat preferences of *Silene douglasii* var. *oraria*, a threatened perennial herb, Kephart and Paladino (1997) measured and mapped individuals in randomly selected permanent quadrats within rocky and grassy habitats, and related long-term demographic variability over 10 years to spatiotemporal environmental factors. Higher population densities, lower adult mortality, and more juvenile recruitment were observed in rocky areas, and there was a positive association between S. *douglasii* abundance and both poor soil development and high light availability. For *Pulsatilla patens*, a hemicryptophyte plant threatened in northeastern Europe, the favorable conditions for regeneration were determined via testing relationships between such habitat factors as cover of ground layer and amount of litter and the population structure of P. *patens* in 48 microsites. The total population size and number of juveniles were highest in sites with intermediate values of ground layer, and the flowering was negatively affected by dense moss layer and abundant litter. The recommended management involved opening the closed vegetation by either manually removing trees and breaking the ground layer or by burning (Kalliovirta *et al.* 2006). Several studies using this approach were performed by Menges and colleagues on endemic

Florida scrub herbs. The occupied microsites were used to study density and reproductive output of the perennial herb *Polygonella basiramia* in response to the amount of open space, presence of litter, and the associated species composition. Openness, which integrates shrub cover, herb cover, and ground cover (leaf litter), was identified as a necessary condition for the establishment and growth of *P. basiramia* individuals (Hawkes and Menges 1995). Also in *Eryngium cuneifolium*, openness was positively related to survival, growth, and fecundity of plants over a 4-year period (Menges and Kimmich 1996).

Principally different from the above approaches is the determination of favorable microhabitats via experimental introduction by comparing the establishment of introduced plants across microsites. For example, more than 1000 plants of *Dicerandra immaculata* var. *savannarum*, another endangered Florida perennial, were introduced into four types of microhabitats that differed in sunlight and presence of leaf litter. Survival and reproduction of introduced plants, and recruitment of new plants, was higher in microhabitats in full sun and no leaf litter and lower in partially shaded habitats. The recommended management was the same as for *D. immaculata* var. *immaculata* (see above): gap creation through prescribed burning and/or mechanical removal of competing plants (Peterson *et al.* 2013). Atondo-Bueno *et al.* (2016) assessed the microenvironmental conditions for *Oreomunnea mexicana*, a highly endangered relict tree species of Mexico and Central America, through seeding in 60 microsites. From the observed positive correlation of seedling emergence with soil moisture and negative correlation with vegetation cover, it was recommended that seeds are sown in sites with soil moisture above 50% and an overhead vegetation cover of between 60 and 70%.

Both seeds and seedlings were experimentally introduced into 40 plots to study natural regeneration in *Abies guatemalensis*, an endangered Central American conifer. Based on analysis of germination and transplant survival, the authors recommended limited selective thinning of the canopy and small-scale litter removal to promote this species regeneration (Kollmann *et al.* 2008).

Experimental introductions were conducted to determine the favorable microsite conditions for many commercially important conifer species, e.g., *Larix occidentalis* (Oswald and Neuenschwander 1993), *Picea abies* (Hornberg *et al.* 1997), *Pinus ponderosa* (Coop and Givnish 2008), *Pinus jeffreyi* (Alpert and Loik 2013), *Abies lasiocarpa*, *Picea engelmannii*, and *Pinus contorta* (Newsome *et al.* 2016). Although the motivation of these studies was to increase timber extraction rather than restoration of the natural populations, this information can be very useful for conservation-oriented restoration.

Analysis of a regeneration niche can utilize a combination of approaches, as for example in a study of threatened Florida shrub *Amorpha herbacea* var. *crenulata* (Wendelberger and Maschinski 2009). To identify the microsites required for successful seedling establishment, the authors used geographical information systems, observations on the establishment of wild seedlings, and testing germination of experimentally introduced seeds. Only microsites with litter depths less than 3 cm were found to be suitable for this species, and such microsites constituted 9% of habitat occupied by the largest population. To dispose of the litter accumulated due to fire suppression, and to improve the probability of seedling establishment, removing the litter and debris manually was recommended.

As some of the above examples demonstrate, beside topography and soil, spatial heterogeneity generated by canopy openings is a very important component of a species regeneration niche. Seedlings of many species, including nonpioneers, benefit from the formation of gaps in the forest canopy creating much higher light intensities than in the understory, either through rapid germination or increased survival and growth rates (Denslow 1987; Gray and Spies 1996; Rueger *et al.* 2009). Large openings created by disturbances such as fire and windstorms or small openings created by a single tree or branch fall are necessary for the regeneration of slow-growing shade-tolerant tree species (e.g., Fowells 1965; Whitmore 1975).

Formation of a sufficiently large canopy gap also enhances opportunities for recruitment from a juvenile to an adult. There is a group of trees, including many threatened species, with relatively high survival but very slow growth in the understory. These species maintain a juvenile bank of shade-tolerant and very slow-growing plants (Whitmore 1989). *Abies amabilis* and *Tsuga heterophylla* in the US Pacific Northwest can survive with little growth for many decades under closed canopy and resume growth after disturbance (Gray and Spies 1996). Other examples are several highly threatened old-lived species of Araucariaceae. New Caledonian *Araucaria laubenfelsii* is very slowly growing, able to tolerate low light levels (i.e., 8% of full sunlight), and has residency times of up to 90 years in the seedling stage (Rigg *et al.* 2010). Similarly, the shaded seedlings of New Zealand's *Agathis australis* have been reported to have a very slow or extended establishment phase for seedlings and saplings of more than 50 years (Steward and Beveridge 2010).

But juvenile persistence in the understory has a tradeoff. Juveniles are much more vulnerable than adults to mortality through abiotic (drought, fire, frost, or herbivory) and biotic agents (herbivores, pathogens, and litterfall) (Bond 1989); and prolonged persistence in the juvenile stage results in longer exposure to these risks. High juvenile mortality may have contributed to the long-term decline of currently critically endangered species such as *Wollemia nobilis*, in which successful recruitment from seedling to juvenile is infrequent (Zimmer *et al.* 2016), apparently due to juvenile susceptibility to pathogens.

Thus, threatened (and inferior as competitors) species regenerating in gaps can suffer from competition in two ways. First, their seedling growth can be suppressed by herbs, shrubs, or non-threatened tree species, thereby prolonging the juvenile phase. In turn, a prolonged stay in the juvenile stage can result in that slowly maturing species being eliminated before it reaches adulthood when it is immune to disturbance, grazing, or pathogens. Adults of many conifers can survive a fire but seedlings and saplings cannot (e.g., *Sequoiadendron giganteum*; York *et al.* 2010).

Limited or no species regeneration can be due to multiple constraints involving both seed and establishment limitation. For example, forests composed of cork oak (*Quercus suber*) were once widespread in the Iberian Peninsula, but became transformed by human management into a mosaic landscape of remaining small forest patches, oak savannas, shrublands, and grasslands. Poor or no tree regeneration across the Iberian Peninsula has been reported not only in the latter three ecosystems, but also in oak-dominated patches. To restore oak forests management actions need to simultaneously relieve low seed availability, high rates of seed predation, low rates of seed germination, and low rates of seedling survival due to thermal and water stress (Pulido and Diaz 2005; Acacio *et al.* 2007). This can be done through a combination of such actions as removing competing vegetation, adding acorns, planting nurse shrubs to facilitate oak seedling germination and survival, and introducing rodent predators.

## 4.9 MONITORING AND ASSESSMENT OF SUCCESS

Evaluating success of restoration efforts in terms of number of seedlings planted, their survival, or growth, as well as by short-term changes in community structure is inappropriate. Clearly, a criterion for evaluating restoration success is how closely the ecosystem has shifted toward a reference. However, successful reestablishment or adjustment toward the desired reference ecosystem can only be truly evaluated in the long term, and is evident in a number of parameters, both demographic and community. Although conservation-oriented restoration differs from traditional restoration in many aspects, all nine key attributes of successful restoration identified by SER (2004) (Table 4.3) apply to it. These attributes cover three general ecological outcomes indicative of the ecosystem health: vegetation structure, species composition and abundance, as well as ecological processes (e.g., nutrient cycling or certain level of disturbance) (Noss 1990; Ruiz-Jaen and Aide 2005a, b). These nine attributes were recently

Table 4.3 *Nine attributes of restored ecosystem for determining when restoration can be considered accomplished (SER 2004)*

| No. | Attribute |
| --- | --- |
| 1 | The restored ecosystem contains a characteristic assemblage of the species that occur in the reference ecosystem and that provide appropriate community structure |
| 2 | The restored ecosystem consists of indigenous species to the greatest practicable extent |
| 3 | All functional groups necessary for the continued development and/or stability of the restored ecosystem are represented or, if they are not, the missing groups have the potential to colonize by natural means |
| 4 | The physical environment of the restored ecosystem is capable of sustaining reproducing populations of the species necessary for its continued stability or development along the desired trajectory |
| 5 | The restored ecosystem apparently functions normally for its ecological stage of development, and signs of dysfunction are absent |
| 6 | The restored ecosystem is suitably integrated into a larger ecological matrix or landscape, with which it interacts through abiotic and biotic flows and exchanges |
| 7 | Potential threats to the health and integrity of the restored ecosystem from the surrounding landscape have been eliminated or reduced as much as possible |
| 8 | The restored ecosystem is sufficiently resilient to endure the normal periodic stress events in the local environment that serve to maintain the integrity of the ecosystem |
| 9 | The restored ecosystem is self-sustaining to the same degree as its reference ecosystem, and has the potential to persist indefinitely under existing environmental conditions |

reorganized (McDonald *et al.* 2016) into six: (1) absence of threat (previous 7–9); (2) reinstated physical conditions (previous 4); (3) restored species composition (previous 1–3); (4) restored ecosystem functionality (previous 5); (5) restored external exchanges (previous

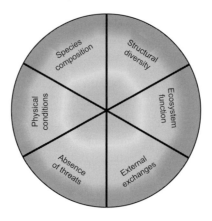

FIGURE 4.9 Six ecosystem attributes important for assessing restoration
success (McDonald *et al.* 2016) and the three possible ecosystem
states denoted by different colors (historical, hybrid, and novel) (see
Figure 4.2 for explanation). The success of a restoration project will
be a function of how close the restored ecosystem attributes approach
the historical (reference) ecosystem. *A black and white version of this
figure will appear in some formats. For the color version, please refer
to the plate section.*

6); and (6) restored structural diversity (Figure 4.9). The last attribute
is an important new addition to those nine listed in SER (2004) and
emphasizes as important tasks in restoration, reinstatement of faunal
food webs, the preexisting strata in forest structure, and habitat spatial heterogeneity. In a properly restored habitat, the structural and
trophic complexity and spatial patterning should closely resemble
the reference ecosystem.

Gaston *et al.* (2008) recognized three possible ways to assess
PA efficiency: how features have changed within PA through time (a
HereThen-HereNow analysis), how the state of features within a PA
compares with that outside it (a HereNow-ThereNow analysis), or
how the state of features has changed within a PA compared with how
it has changed outside it (a HereThenNow-ThereThenNow analysis).
These approaches can be useful for assessing restoration success,
with reference sites being the outside areas used for comparison. To
evaluate changes in vegetation structure, species abundances, and

ecosystem functioning through time and in comparison with the reference, a restoration project must necessarily have an effective monitoring and evaluation system. This system should (1) establish a baseline and a set of indicators that relate to the specific object- ives of restoration; (2) measure the progress toward a set object- ives throughout the project; and (3) cover all its main stages (e.g., from establishment of outplants to their regeneration). Only prop- erly organized continuous monitoring and assessment, especially if designed to test specific hypotheses that are relevant to policy and management questions, will allow for refined management practices where necessary (Noss 1990; Walters and Holling 1990). An example of such detailed monitoring protocol is one that has been developed and recently released for Brazilian Atlantic Forest restoration (Viani *et al.* 2017). As the authors note, it can serve as a model to inform the development of similar protocols in other tropical regions.

An iterative process of project monitoring – assessment – corrective management should include the following major steps (Vallauri *et al.* 2005):

1. Confirmation of the hypotheses used to develop the restoration plan (e.g., about the major threats) and that defined goals are achieved within the specified time frame (e.g., that specific damaged functional components of the ecosystem are restored and have started to function).
2. Fine-tuned management actions that correct problems encountered during restoration (e.g., lower survival of seedlings than expected or poor reproduction) or incorrect treatment choices (e.g., too low thinning intensity).
3. Adjusting management interventions to changes along a restoration trajectory that can take decades, and especially those that could not be forecasted when the project was initiated (e.g., increased human population pressure or climate change).

Because the iterative process of project assessment – corrective management uses continually updated monitoring data it can provide greater predictive power in decision-making (Murray and Marmorek 2003; Moore *et al.* 2011; Larson *et al.* 2013; Brudvig *et al.* 2017).

This method is well established in ecosystem monitoring and management in the United States and several other countries. Adaptive management has been used in land and wildlife management in forests of the US Pacific Northwest (Gray 2000) and British Columbia (Nyberg 1999), restoration of Wisconsin pine and oak barrens (Power and Haney 1998), and reintroduction of native grassland vegetation at Organ Pipes National Park in Australia (McDougall and Morgan 2005). This approach is also used by the US Fish and Wildlife Service's Native Prairie Adaptive Management Initiative to assess the ecosystem effects of different management actions, such as prescribed burning and grazing (Hunt *et al.* 2015).

When the number of introduced or naturally regenerating species in a restoration project is large, evaluation of project success may require the use of techniques that allow objective comparisons of the diversity and structure of the restored communities subject to different restoration actions with the reference(s). A technique of choice for controlling the effects of abundance and sampling effort in restoration assessments is rarefaction, standardizing species richness in compared communities to a common sampling effort (Rey *et al.* 2009). Another promising approach in restoration assessments is incorporation of metrics of network structure (Tylianakis *et al.* 2010; Kaiser-Bunbury and Bluethgen 2015).

# 5    Conservation-Oriented Restoration Silvicultural Toolkit

## 5.1    ACTIVE INTERVENTIONS IN RESTORATION

Natural succession processes can be facilitated with minimal intervention (e.g., land protection) or with intensive practices involving changes to the soil and/or remaining vegetation and planting. A range of silvicultural techniques has been developed in forestry for managing secondary forests that can be used in conservation-oriented restoration. They are, in increasing order of intensity and associated cost: (1) passive restoration via protection; (2) additional protection from fire during the dry season or, in contrast, periodic burning; (3) liberation cutting to reduce competition and promote growth of immature trees; (4) thinning the overstory to promote seed production and growth of desirable species; (5) enrichment planting without thinning; (6) enrichment planting after thinning; (7) replacement planting. Some more specific interventions may be needed to create the necessary conditions for early plant developmental stages such as small-scale topographic heterogeneity or presence of facilitating species. Restoration of ecosystems invaded by non-native invasive plants will require such interventions as removal and control of the invaders (Florens *et al.* 2010; Hudson *et al.* 2013; Bellingham *et al.* 2018), and many species, including threatened ones, can recover dramatically as a consequence of the sole removal of invasive alien plants (Baider and Florens 2011). How necessary particular methods are for the restoration of a target area will depend upon the history of past ecosystem management, the degradation level, and the desired outcome (i.e., reference conditions).

Specific features of the existing practices of active intervention, their limitations, and circumstances under which they can be expected to provide the greatest benefits are discussed below.

### 5.1.1 Facilitation by Benefactor Species

A favorable microhabitat providing necessary regeneration conditions can be created through facilitation when one plant species enhances the establishment of another species. This positive influence of the adult plants on young plants of another species is called "nurse plant syndrome" (Niering et al. 1963). A benefactor or nurse plant makes the physical environment under its canopy more suitable for the beneficiary, by modifying light, temperature, soil moisture, or nutrient regime, protecting it from herbivores, providing a space free from competitors, or accumulating its seeds (e.g., McAuliffe 1986; Franco and Nobel 1989; Valiente-Banuet and Ezcurra 1991; Callaway 1992, 1995; Guevara et al. 1992; Belsky 1994; Pugnaire et al. 1996; Garcia et al. 2000; Shumway 2000; Holl 2002; Bakker et al. 2004; Duarte et al. 2006; Filazzola and Lortie 2014). The nurse plant syndrome is more common in environments where the constraining role of abiotic factors or herbivory on plant performance is stronger, such as arid (Flores and Jurado 2003) or alpine habitats (Cavieres et al. 2006; Butterfield et al. 2013). In arid regions many studies have reported a facilitative effect of shrubs and other low-stature vegetation on seedlings of succulents through shading, thereby reducing their mortality due to temperature extremes (e.g., Jordan and Nobel 1979; Nobel 1980; Franco and Nobel 1989; Leirana-Alcocer and Parra-Tabla 1999; Drezner 2006; Landero and Valiente-Banuet 2010). Similarly, a number of studies reported the attenuating effect of low temperatures, strong winds, and excessive radiation on many species growing within cushion plants in alpine habitats (Alliende and Hoffmann 1985; Pyšek and Liska 1991; Nunez et al. 1999; Arroyo et al. 2003; Akhalkatsi et al. 2006; Antonsson et al. 2009; Catorci et al. 2011; Butterfield et al. 2013).

In Mediterranean ecosystems, the regeneration niche of the majority of woody species is positively associated with shrubs (Herrera *et al.* 1994; Rousset and Lepart 1999; Garcia *et al.* 2000; Rey and Alcantara 2000; Castro *et al.* 2004). Some species, such as the threatened *Taxus baccata* and *Acer opalus* subsp. *granatense*, critically depend on the presence of prickly shrubs that protect the recruits against ungulate herbivores (Garcia and Obeso 2003; Gomez-Aparicio *et al.* 2005b). In the presence of large herbivores, seedlings of *Quercus robur* can survive only under spiny *Prunus spinosa* shrubs (Bakker *et al.* 2004). Also the mortality of *Quercus douglasii* seedlings from grazing is much higher in open grassland than under a canopy of shrubs *Salvia leucophylla* and *Artemisia californi* (Callaway 1992). Likewise, *Quercus pubescens* and *Fagus sylvatica*, tree species that play critical roles in the mid- and late-succession stages of forest dynamics in Mediterranean Europe, usually regenerate under unpalatable shrubs *Buxus sempervirens* and *Juniperus communis* protecting them from herbivores (Rousset and Lepart 2000; Dolezal *et al.* 2004). However, *F. sylvatica* can survive in a narrower area under a shrub than *Q. pubescens* due to its higher susceptibility to competition with herbs (Kunstler *et al.* 2006). Thus, for *F. sylvatica* the nurse shrubs also provide a competition-free space. For other species nurse plants can provide more favorable soil moisture and protection from excessive radiation. For planted seedlings of mid-successional *Prunus serotina* and *Sassafras albidum* in the coastal environments of the North American Atlantic Coast, mortality was lower beneath *Juniperus virginiana* than in exposed sites (Joy and Young 2002).

Considering direct physical protection from herbivory provided by the nurse plant to the beneficiary, we should take into account that an external defense mechanism depends on foraging behavior decisions of the herbivore and palatability of a target plant individual relative to the surrounding vegetation (associational avoidance, Milchunas and Noy-Meir 2002). The degree of protection offered by the nurse plant increases as its palatability relative to the palatability of a beneficiary decreases. While an unpalatable shrub will reduce the probability of browsing a seedling/sapling growing beneath it, a highly palatable shrub will only promote it (Rousset and Lepart 2003; Baraza *et al.* 2006). Growing in an open area will be advantageous for the young plants whose palatability is lower than most of the shrubs in the area, and therefore these young plants will not attract herbivore attention when growing alone.

Besides palatability of the species in a benefactor–beneficiary pair, herbivory pressure also affects the degree of young plant protection by nurse plants. Baraza *et al.* (2006) proposed a conceptual scheme of interactions between large herbivores, nurse shrubs, and seedlings/saplings (Figure 5.1). Under high herbivore pressure, only unpalatable shrubs can protect palatable saplings, while for unpalatable saplings their growth in the proximity of shrubs is disadvantageous. The situation is quite different under moderate and low herbivore pressure. When herbivore pressure is low, shrubs of varying palatability can protect palatable saplings from herbivores, and unpalatable saplings are safe in any microhabitat. Under intermediate herbivore pressure, shrubs of intermediate or low palatability can protect palatable saplings from herbivores but palatable shrubs cannot, and unpalatable saplings will start to undergo damage when growing under palatable shrubs (Figure 5.1). Thus, a safe site for recruitment is community context dependent, and is determined by the characteristics of the neighbors and the intensity of herbivore pressure. Consequently, the landscape can change from being, at a low herbivore level, a high-quality environment with many suitable microhabitats for recruitment to becoming, at a high herbivore level, a low-quality environment with the only safe sites being the unpalatable shrubs.

Reforestation techniques ignoring the dependence of tree seedlings on shrubs as a microhabitat for successful development frequently show low establishment success because of the high mortality of planted seedlings through water stress or herbivory. The improved chances of seedling establishment provided by nurse plants to late-successional species accelerate succession

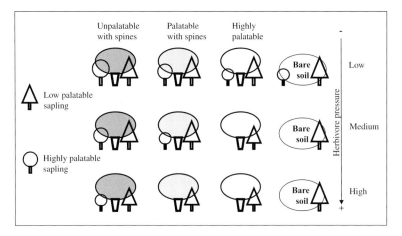

FIGURE 5.1 A conceptual scheme showing the relationship between sapling establishment (damage) probability, sapling palatability, palatability of nursing shrubs, and herbivore pressure (from Baraza *et al.* 2006). Probability of establishment is proportional to the size of a sapling symbol in the figure. Under high herbivore pressure (bottom row) palatable saplings escape browsing only when growing under unpalatable shrubs, while for unpalatable saplings their growth in the proximity of shrubs (including unpalatable ones) can increase the probability of attack. Under low herbivore pressure (upper row), palatable saplings can occupy all microhabitats, and damage probability decreases with a decrease in palatability of shrubs, while for unpalatable saplings the damage probability is close to zero for all microhabitats. Under intermediate herbivore pressure (middle row), palatable saplings can grow only under unpalatable or low palatability shrubs. In this case the proximity of palatable shrubs can increase damage to unpalatable saplings.

and therefore utilization of nurse plants has been recognized as an important silvicultural tool in restoration (Padilla and Pugnaire 2006; Gomez-Aparicio 2009 and references therein). New techniques that alleviate drought stress and/or protect seedlings from the large herbivores using nurse plants have shown promise for restoration in Mediterranean landscapes (Maestre *et al.* 2001; Castro *et al.* 2004, 2006; Gomez-Aparicio *et al.* 2004; Siles *et al.* 2008; Rey *et al.* 2009; Torroba-Balmori *et al.* 2015). For example, use of nurse shrubs in restoration of *Quercus pyrenaica* forests, once abundant in the western

Mediterranean, but nowadays highly degraded or cleared, was shown to greatly increase the establishment success of planted seedlings (Castro *et al.* 2006). Spiny tall shrubs abundant in many native plant communities of the Mediterranean were proposed as keystone species for restoration in this region due, in addition to their protective role for many late-successional trees, to the ease of establishment and ability to attract vertebrate seed dispersers and nucleate the arrival of other fleshy- fruited tall shrubs and trees (Verdu and Garcia-Fayos 1996; Siles *et al.* 2010b).

Use of shrubs or trees as nurse plants is a method of vegetation restoration gaining popularity in other semiarid regions besides the European Mediterranean (Maschinski *et al.* 2004; Blignaut and Milton 2005; van Zonneveld *et al.* 2012; Talamo *et al.* 2015), temperate regions (Smit *et al.* 2006; Tones and Renison 2016) and the tropics (Vieira *et al.* 1994; Hooper *et al.* 2002, 2005b; Rigg *et al.* 2002; Sanchez-Velasquez *et al.* 2004; Griscom and Ashton 2011; Montes-Hernandez and López-Barrera 2013). For example, *Quercus insignis*, an endangered species of Central American tropical montane forests, was experimentally introduced into abandoned grassland with small remnant forest patches. Analysis of 4-year-old seedling survival and growth revealed the protective role of shrubs to these seedlings from herbivores. Based on these results, it is recommended that *Q. insignis* seedlings be introduced under shrubs and isolated trees (Montes-Hernandez and López-Barrera 2013).

Because nurse plants act as habitat modifiers creating microhabitats necessary for the establishment of many species, increasing ecosystem structural complexity and species richness, they can be considered ecological and conservational keystone species in their environments (Suzan *et al.* 1996; Tewksbury and Lloyd 2001; van Zonneveld *et al.* 2012; Filazzola and Lortie 2014). As was shown by Rey *et al.* (2009), restoration utilizing existing vegetation as nurses for the planted seedlings significantly increases the diversity of planted trees, life-form diversity, late-successional species cover and resilience in the restored community, and accelerates the approach of the reference community in species composition.

## 5.1.2   Enrichment Planting

Passive recovery of severely degraded forests takes a very long time. Recovery of just the above-ground biomass regardless of the species composition takes close to 100 years (Martin *et al.* 2013; Poorter *et al.* 2016; Wheeler *et al.* 2016). In addition, the trajectory of passive forest recovery is highly stochastic and not always predictable (Norden *et al.* 2015). Secondary forests can attain the species richness of pre-logged forests relatively rapidly, but almost inevitably will differ substantially in forest structure and floristic composition, with a higher proportion of pioneers and undeveloped understory. Restoration of these habitats requires enrichment with late-successional species and absent functional groups, e.g., large-fruited or insect-/bird-pollinated trees (Martinez-Garza and Howe 2003; Garcia *et al.* 2015).

Enrichment planting (or underplanting) is the introduction of target species to the ecosystem without the elimination of the individuals already present (Figure 5.2). This is a method of choice in areas where natural regeneration is insufficient and for reintroducing species that have disappeared. Enrichment planting is popular in restoration of degraded and secondary forests, with a recognized utility in recovery of native and rare forest species (e.g., Sovu *et al.* 2010; Millet *et al.* 2013; Garcia *et al.* 2015). There is no alternative to this method for the establishment of species that, under natural regeneration, are unable to colonize the site or regenerate by themselves (Ashton *et al.* 2001; Martinez-Garza and Howe 2003; Lamb *et al.* 2005). In conservation-oriented restoration, enrichment can be done with threatened species that are either locally present but have regeneration problems, or locally absent but with good prospects of establishment upon introduction.

Underplanting has a long history of use in forestry for promoting regeneration of shade-tolerant timber species in different

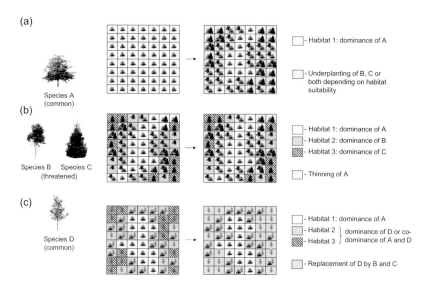

FIGURE 5.2 Restoration of the threatened species habitats. (a) A common species A dominates the landscape. The recommended restoration action is underplanting of the threatened species B and C in habitats suitable for them. (b) Two threatened species B and C dominate two of the three site habitats. The recommended restoration action is thinning of the nonthreatened species A in habitats where it co-occurs with species B and C. (c) The two threatened species B and C do not occur in the same habitats as above at another site because of dispersal barriers or former extirpation. The recommended restoration action is replacement of nonthreatened species D in habitats where it occurs alone or co-occurs with species A. *A black and white version of this figure will appear in some formats. For the color version, please refer to the plate section.*

climatic zones and forest types (e.g., Weaver 1987; Adjers *et al.* 1995; Montagnini *et al.* 1997; Maas-Hebner *et al.* 2005; Erefur *et al.* 2011; Navarro-Cerrillo *et al.* 2011; Dey *et al.* 2012). Some of the successfully underplanted tree species are those that became threatened as a result of overharvesting for timber (e.g., *Bertholletia excelsa* and *Swietenia macrophylla*; d'Oliveira 2000; Lopes *et al.* 2008). The vast experience gained by forestry can be adopted and applied in conservation-oriented restoration.

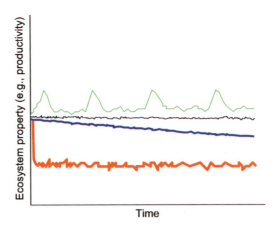

FIGURE 1.1 Four possible ecosystem trajectories under anthropogenic disturbance. Line width corresponds to the probability of occurrence: no change (black), periodic successional changes (green), gradual directional reversible change (blue), and sudden irreversible change (red). *A black and white version of this figure will appear in some formats.*

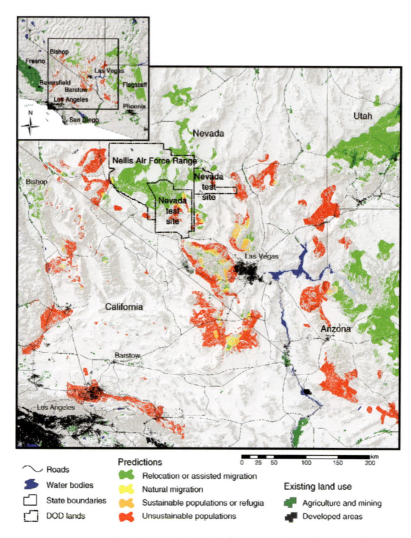

FIGURE 2.3 Assisted migration (= colonization) as a solution to the effect of climate change on Joshua tree (*Yucca brevifolia*) (from Cole *et al.* 2011). Four different colors designate areas where SDM predicts existing populations will become unsustainable (red), remain viable (orange), migrate naturally and persist (yellow), or conditions are suitable for assisted colonization in protected areas (green). DOD stands for the US Department of Defense. *A black and white version of this figure will appear in some formats.*

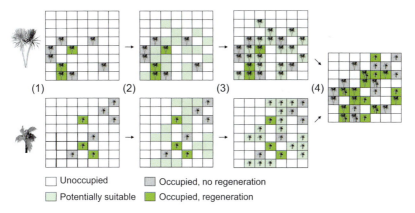

| | Unoccupied | | Occupied, no regeneration |
|---|---|---|---|
| | Potentially suitable | | Occupied, regeneration |

FIGURE 2.5 Necessary steps of conservation-oriented restoration using threatened plants: (1) mapping and population demographic survey; (2) species distribution modeling and analysis of regeneration niche; (3) multiple site introduction trials; (4) monitoring of establishment. The introduction trials may result in no establishment, establishment without regeneration, and establishment with regeneration. Only the introductions that resulted in establishment with regeneration can be considered successful. *A black and white version of this figure will appear in some formats.*

FIGURE 2.6 The Makauwahi Cave restoration project in the Hawaiian Archipelago. (a) and (b) show an aerial view of the Makauwahi Cave reserve with the restored area in the center, and a section of a nature trail providing information to visitors about the restoration techniques and the plants used, as it winds through the restorations. (c) The former sugar cane field in front of Mt. Haupu that has been restored using native Hawaiian threatened and rare plants. Despite use of heavy equipment for site preparation and planting in rows like crops, mixing many species and several growth forms in each row allows the restored site to become indistinguishable from a naturally regenerating forest in a decade or less. Photos by Ellen Coulombe, Lida Pigott Burney, and David Burney. *A black and white version of this figure will appear in some formats.*

FIGURE 2.7 Recreation of the tropical forest using native threatened and rare plant species on Rodrigues (a, b) and Mauritius (c) islands of the Mascarene Archipelago. The upper photos show the landscape view (a) before and (b) during restoration. Photos by Arnaud Meunier and Christine Griffiths. *A black and white version of this figure will appear in some formats.*

FIGURE 2.10 Six PMRs, each with the main plant species it protects (inset) and the country location. (a) Cap d'Or (*Silene hifacensis*), Spain; (b) Lavajo del Tio Bernardo (*Marsilea strigosa*), Spain; (c) Chryssoskalitissa (*Phoenix theophrasti*), Crete; (d) Lyubash Mounts (*Lathyrus pancici*), Bulgaria; (e) Metropolitan Geawargios Haddad (*Iris bismarckiana*), Lebanon; (f) Mystero (*Ophrys kotschyi*), Cyprus. Photo of the PMR Metropolitan Geawargios Haddad by Magda Bou Dagher Kharrat. Other photos by Emilio Laguna. *A black and white version of this figure will appear in some formats.*

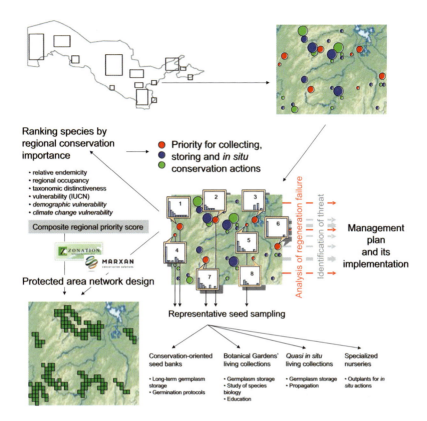

FIGURE 3.1 A scheme of regional conservation planning (from Volis 2018). Each colored circle denotes a population of one of three species with the circle size and color corresponding to a population size and species identity, respectively. All populations of one species (in red) are provided with size class distributions. In size class distribution histograms the *x*- and *y*-axes are size classes and plant density per unit area, respectively. *A black and white version of this figure will appear in some formats.*

| A. Population size reduction. Population reduction (measured over the longer of 10 years or 3 generations) based on any of A1 to A4 | | | |
|---|---|---|---|
| | **Critically Endangered** | **Endangered** | **Vulnerable** |
| A1 | ≥ 90% | ≥ 70% | ≥ 50% |
| A2, A3 & A4 | ≥ 80% | ≥ 50% | ≥ 30% |
| A1 Population reduction observed, estimated, inferred, or suspected in the past where the causes of the reduction are clearly reversible AND understood AND have ceased. A2 Population reduction observed, estimated, inferred, or suspected in the past where the causes of reduction may not have ceased OR may not be understood OR may not be reversible. A3 Population reduction projected, inferred or suspected to be met in the future (up to a maximum of 100 years) [(a) cannot be used for A3]. A4 An observed, estimated, inferred, projected or suspected population reduction where the time period must include both the past and the future (up to a max. of 100 years in the furure), and where the ceased causes of reduction may not have ceased OR may not be understood OR may not be reversible. | based on any of the following: | (a) direct observation [except A3] (b) an index of abundance appropriate to the taxon (c) a decline in area of occupancy (AOO), extent of occurrence (EOO) and/or habitat quality (d) actual or potential levels of exploitation (e) effects of introduced taxa, hybridization, pathogens, pollutants, competitors or parasites. | |

| B. Geographic range in the form of either B1 (extent of occurrence) AND/OR B2 (area of occupancy) | | | |
|---|---|---|---|
| | **Critically Endangered** | **Endangered** | **Vulnerable** |
| B1. Extent of occurrence (EOO) | < 100 km$^2$ | < 5 000 km$^2$ | < 2 000 km$^2$ |
| B2. Area of occupancy (AOO) | < 10 km$^2$ | < 500 km$^2$ | < 200 km$^2$ |
| AND at least 2 of the following 3 conditions: | | | |
| (a) Severely fragmented OR Number of locations | =1 | ≤ 5 | ≤ 10 |
| (b) Continuing decline observed, estimated, inferred or projected in any of: (i) extent of occurrence; (ii) area of occupancy; (iii) area, extent and/or quality of habitat; (iv) number of locations or subpopulations; (v) number of mature individuals | | | |
| (c) Extreme fluctuations in any of: (i) extent of occurrence; (ii) area of occupancy; (iii) number of locations or subpopulations; (iv) number of mature individuals | | | |

| C. Small population size and decline | | | |
|---|---|---|---|
| | **Critically Endangered** | **Endangered** | **Vulnerable** |
| Number of mature individuals | < 250 | < 2 500 | < 10 000 |
| AND at least one of C1 or C2 | | | |
| C1. An observed, estimated or projected continuing decline of at least (up to a max. of 100 years in future) | 25% in 3 years or 1 generation (whichever is longer) | 20% in 5 years or 2 generations (whichever is longer) | 10% in 10 years or 3 generations (whichever is longer) |
| C2. An observed, estimated. projected or inferred continuing decline AND at least 1 of the following 3 conditions | | | |
| (a)   (i) Number of mature individuals in each subpopulation | ≤ 50 | ≤ 250 | ≤ 1 000 |
| (ii) % of mature individuals in one subpopulation = | 90–100% | 95–100% | 100% |
| (b) Extreme fluctuations in the number of mature individuals | | | |

| D. Very small or restricted population | | | |
|---|---|---|---|
| | **Critically Endangered** | **Endangered** | **Vulnerable** |
| D. Number of mature individuals | < 50 | < 250 | D1. < 1 000 |
| D2. Only applies to the VU category. Restricted area of occupancy or number of locations with a plausible future threat that could drive the taxon to CR or EX in a very short time. | - | - | D2. typically: AOO < 20 km$^2$ or number of locations ≤ 5 |

| E. Quantitative Analysis | | | |
|---|---|---|---|
| | **Critically Endangered** | **Endangered** | **Vulnerable** |
| Indicating the probability of extinction in the wild to be: | ≥ 50%  in 10 years or 3 generations, whichever is longer (100 years max.) | ≥ 20% in 20 years or 5 generations, whichever is longer (100 years max.) | ≥ 10% in 100 years |

FIGURE 3.3  Five criteria (A–E) used to evaluate if a taxon belongs to a threatened category (Critically Endangered, Endangered, or Vulnerable) (Standards Petitions Working Group 2017). *A black and white version of this figure will appear in some formats.*

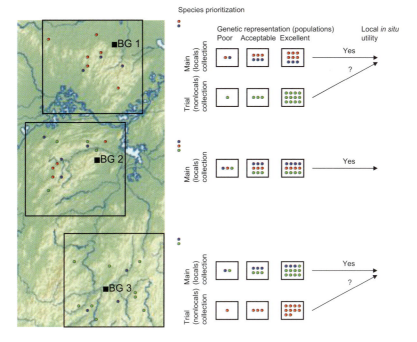

FIGURE 3.12 A proposed unified management strategy of botanical gardens' living collections exemplified by collections in three hypothetical botanic gardens (BG1-3) (from Volis 2017). Colored circles denote populations of three different species in three ecoregions denoted by rectangles. Species in each ecoregion are prioritized based on a set of criteria. Representative collections are created for species with the highest regional priority. Only collections representing all known populations in the region (excellent representation) or at least three populations (acceptable representation) can be used for *in situ* actions. In addition to the main collections of regionally local species, trial collections of nonlocal threatened species can be used for regional *in situ* actions based on plants' performance and SDM predictions. *A black and white version of this figure will appear in some formats.*

FIGURE 3.13 A *quasi in situ* living collection of *Iris atrofusca* created in Tel Beer Sheva National Park, Israel, 5 years after planting. Note the large number of fruits produced by the plants. Photo by the author. *A black and white version of this figure will appear in some formats.*

(a)
(b)
(c)

FIGURE 3.14 The forestry seed orchards preserving genetic diversity of native woody species of Flanders, Belgium: (a) *Crataegus manogyna*, (b) *Malus sylvestris*, (c) *Prunus padus*. Photos by Kristine Vander Mijnsbrugge. *A black and white version of this figure will appear in some formats.*

FIGURE 3.15 Managed breeding program for critically endangered Hawaiian endemic *Argyroxiphium kauense*. The photos show (a) the silversword plants cultivated at the Volcano Rare Plant Facility and (b) in the remnant population that acted as both mother plants and pollen donors; (c) hand pollination of a mother plant; (d) seedlings produced for outplanting at the Kahuku site in Hawaii Volcanoes National Park; and (e) the reintroduced plants. Photos by Robert Robichaux and David Boyle. *A black and white version of this figure will appear in some formats.*

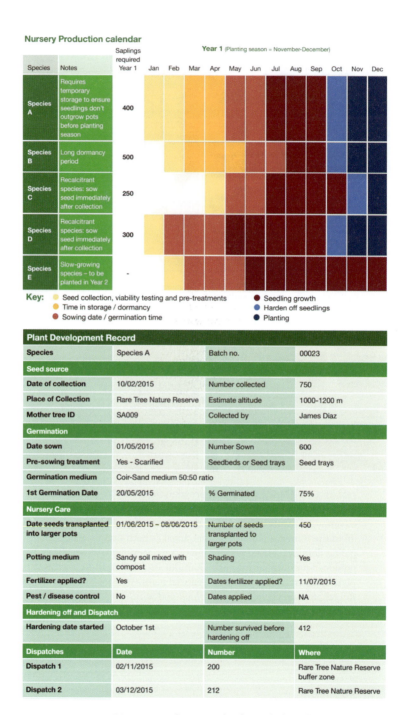

FIGURE 3.18 Nursery production calendar and plant development records (from Stott and Gill 2014). *A black and white version of this figure will appear in some formats.*

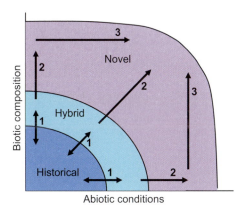

| Species-targeted conservation | Conservation-oriented restoration | Traditional restoration |

Strength of local adaptation
Breadth of potential niche
Vulnerability to climate change
Genetic diversity
Biotic interactions

Structural complexity
Species diversity
Resilience
Functional groups
Seed and pollen flow

Ecosystem services
Productivity

Historical      Hybrid      Novel

Ecosystems

FIGURE 4.1 Three approaches to conservation of species and ecosystems, attributes of their targets on which the approach actions are focused, and the range of ecosystem degradation they can be successfully applied to (terms are *sensu* Hobbs *et al.* 2013, for explanation, see Figure 4.2). *A black and white version of this figure will appear in some formats.*

FIGURE 4.2 Types of ecosystems under increasing levels of change in abiotic conditions (*x*-axis) and biotic composition (*y*-axis) (after Hobbs *et al.* 2009). A historical ecosystem remains within its historical range of variability; a hybrid ecosystem is biotically and/or abiotically dissimilar to its historical ecosystem but is capable of returning to the historical state; novel ecosystems are biotically and/or abiotically dissimilar to the historical state and have passed a threshold such that they cannot return to the historical state. Pathways represent possible directions of change: (1) shifts from historical to hybrid ecosystems that are reversible; (2) nonreversible shifts from historical or hybrid ecosystems to novel ecosystems; and (3) further possible biotic and abiotic shifts within novel ecosystems. *A black and white version of this figure will appear in some formats.*

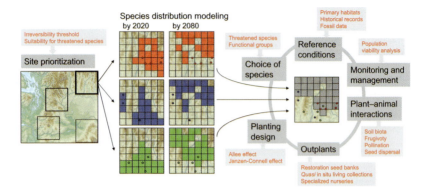

FIGURE 4.3 Stages of conservation-oriented restoration with major issues and appropriate methodology to solve them (from Volis 2018). At the site prioritization stage, four quadrats denote candidate sites with border line width corresponding to the priority rank. As a result of ecological niche modeling for the three species for which extant populations are denoted by colored circles, the small-scale planting sites are defined (grid cells in gray). *A black and white version of this figure will appear in some formats.*

FIGURE 4.4 (a) The nursery workers at Camara Nursery in São Paulo State, Brazil, are assembling the mixtures of native tree species from Atlantic Forest. (b) Boxes, each with a mixture of 50 native tree species, are ready to be transported to restoration sites. Photos by Robin Chazdon. *A black and white version of this figure will appear in some formats.*

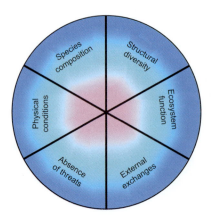

FIGURE 4.9 Six ecosystem attributes important for assessing restoration success (McDonald *et al.* 2016) and the three possible ecosystem states denoted by different colors (historical, hybrid, and novel) (see Figure 4.2 for explanation). The success of a restoration project will be a function of how close the restored ecosystem attributes approach the historical (reference) ecosystem. *A black and white version of this figure will appear in some formats.*

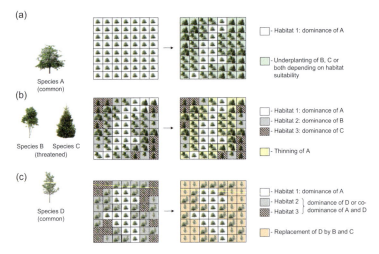

FIGURE 5.2 Restoration of the threatened species habitats.
(a) A common species A dominates the landscape. The recommended restoration action is underplanting of the threatened species B and C in habitats suitable for them. (b) Two threatened species B and C dominate two of the three site habitats. The recommended restoration action is thinning of the nonthreatened species A in habitats where it co-occurs with species B and C. (c) The two threatened species B and C do not occur in the same habitats as above at another site because of dispersal barriers or former extirpation. The recommended restoration action is replacement of nonthreatened species D in habitats where it occurs alone or co-occurs with species A. *A black and white version of this figure will appear in some formats.*

FIGURE 6.6 Thinned Norway spruce forest plantation in Sweden enriched with broad-leaved species. Photo by David Lamb. *A black and white version of this figure will appear in some formats.*

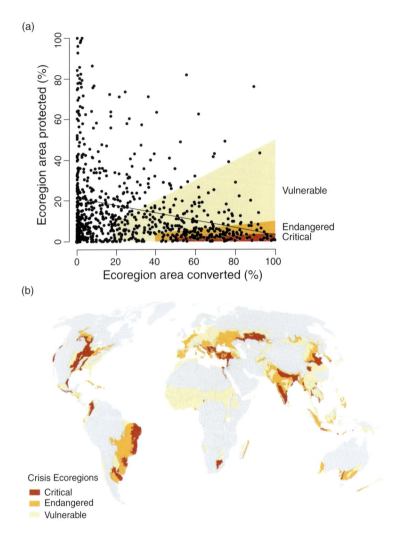

FIGURE 7.3 Classification of crisis ecoregions as Vulnerable, Endangered, and Critically endangered (a), and the world map of crisis ecoregions (b) (from Hoekstra *et al.* 2005). Scatterplot shows the relationship between percentage area converted and percentage area protected in each of 810 terrestrial ecoregions (slope = –0.18). Ecoregions with >50% habitat conversion and Conservation Risk Index (CRI) >25 are classified as Critically Endangered (red); ecoregions with >40% conversion and CRI >10 are classified as Endangered (orange); and those with >20% conversion and CRI >2 are classified as Vulnerable (yellow). CRI for each ecoregion was calculated as the ratio of percentage area converted to percentage area protected. *A black and white version of this figure will appear in some formats.*

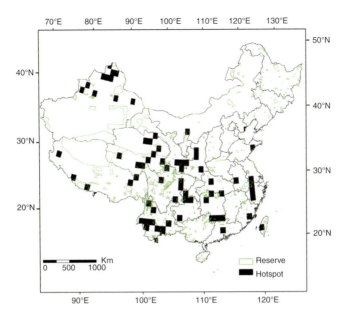

FIGURE 7.6 The identified putative refugia (black squares) and the network of Chinese protected areas (green contour lines). *A black and white version of this figure will appear in some formats.*

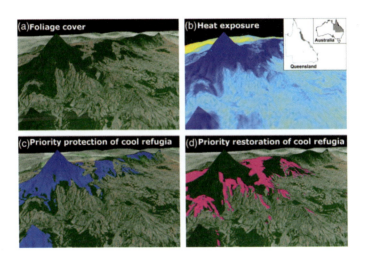

FIGURE 7.7 Cool rainforest refugia within a modified landscape (from Shoo *et al.* 2011). Patterns of foliage cover (green, a) have a strong influence on buffering heat exposure (i.e., maximum temperature of the warmest period, blue to red, b). To increase the local extent of cool rainforest refugia, it is proposed to supplement the extant protected areas (blue, c) with formerly supporting cool habitat but currently deforested areas after restoration of their foliage cover through reforestation (purple, d). *A black and white version of this figure will appear in some formats.*

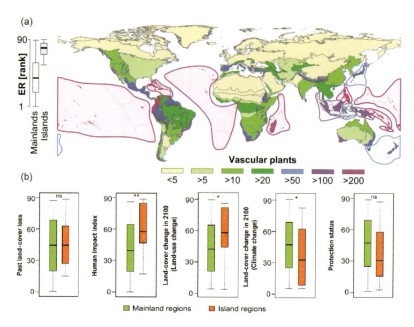

FIGURE 7.11 A global assessment of endemism and species richness across island and mainland regions (extracted from Kier *et al.* 2009). (a) Global patterns of endemism richness (ER; range equivalents per 10 000 km²) for vascular plants across 90 biogeographic regions. Map legends were classified using quantiles, i.e., each color class contains a comparable number of regions. Box-and-whisker plots illustrate rank-based differences in ER between mainland (*n* = 76; white boxes) and island regions (*n* = 14; gray boxes). Boxes mark second and third quartiles; whiskers mark the range of the data. (b) Comparison of key conservation features (expressed as ranks) for mainland (green) and island regions (orange), including past land-cover loss (year 2000), human impact (year 2000), future land-cover loss (projections for the year 2100) because of land-use change and climate change, and current protection status (proportion covered by protected areas). Significance codes indicate differences (Mann–Whitney U-test) between mainlands and islands: **, *P* < 0.01; *, *P* < 0.1; ns, not significant. *A black and white version of this figure will appear in some formats.*

FIGURE 7.12 Experimental translocation of critically endangered *Wollemia nobilis*. The photos show: (a) the outplants produced at the Australian Botanic Garden Mount Annan, (b) the translocation site, (c) a small introduced outplant, and (d) a larger outplant. Photos by Haidi Zimmer. *A black and white version of this figure will appear in some formats.*

FIGURE 7.13 Mount Jinfo Nature Reserve. (a, b) The reserve landscape, (c) depressed growth of two juvenile *Cathaya argyrophylla* trees (marked by arrows) under the canopy of a large evergreen tree, (d) *C. argyrophylla* sapling growing in the canopy gap, and (e) clear-cutting of understory during the bamboo shoot season. Photos by Yongchuan Yang. *A black and white version of this figure will appear in some formats.*

FIGURE 7.14 Restoration of a native Hawaiian forest at Limahuli Preserve. (a–c) A time series of photographs taken at 6-month intervals: (a) was taken immediately after initial tree planting, but a few days before outplanting of groundcover species. The next two photos show good establishment and vigorous growth of both common and endangered species 6 months (b) and 1 year after planting (c). (d–e) The reestablished native plant community that resulted from restoration, using two different methods, small (0.2 ha) and large (2.0 ha) sections of forest that had been completely taken over by invasive species. Thirteen native plant species including three endangered tree species (0.2 ha, d) and 14 native plant species including four endangered tree species (2.0 ha, e) are identifiable in these pictures. Photo credit: National Tropical Botanical Garden. *A black and white version of this figure will appear in some formats.*

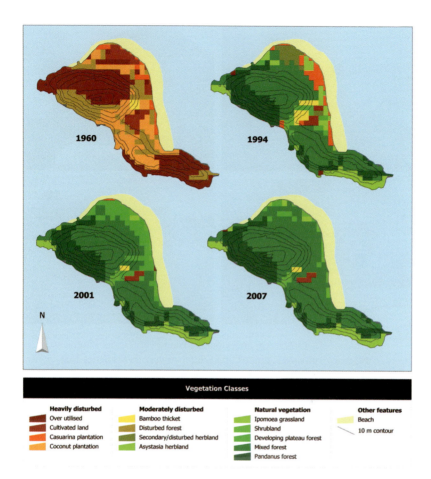

FIGURE 7.16 (a–d) Recovered Cousine Island vegetation. Photos by
Michael Samways. *A black and white version of this figure will appear
in some formats.*

Intensive experimentation with underplanted timber species performed in the last 50 years suggests that to optimize growth and survival of introduced seedlings, it is usually necessary to manipulate the uppermost and sometimes secondary forest canopies. Experimental planting of seedlings of several conifer species (*Abies grandis*, *Picea sitchensis*, and *Tsuga heterophylla*) underneath unthinned *Pseudotsuga menziesii* stands turned out to be lethal, with all seedlings dead after 2 years (Maas-Hebner *et al.* 2005). In contrast, planting in thinned stands revealed an increase in survival rates with higher light (Maas-Hebner *et al.* 2005). Similarly, experiments with enrichment planting in mixed dipterocarp tropical forests suggest that the optimal conditions for the establishment of the majority of species, including shade-tolerant ones, are within gaps (Adjers *et al.* 1995; Kuusipalo *et al.* 1997; d'Oliveira 2000; Ashton *et al.* 2001; Romell *et al.* 2008; Schulze 2008; Schwartz *et al.* 2013). On the other hand, the successful establishment of shade-tolerant species needs some protection from direct sunlight, as well as the milder temperature and moisture conditions provided by the understory (Ashton *et al.* 1997; Parrotta *et al.* 1997; Kobe 1999; Paquette *et al.* 2006). Besides, in canopy gaps the underplanted individuals will face competition with grasses and herbs that immediately colonize the gaps upon creation. For these reasons, the survival and growth of the outplants usually improve as stand density decreases to an intermediate level, below which they either drop or stabilize (Paquette *et al.* 2006). Improved establishment and better growth of underplanted seedlings can be achieved by leaving only the uppermost canopy layer, thus minimizing mortality from competition with weeds at early stages of development and, later, opening the canopy to expose the established plants to the full sun to maximize their growth.

The planted individuals may suffer from intense competition with established and naturally recruiting seedlings, and therefore will require postplanting management of competing vegetation for several years (Adjers *et al.* 1995; Montagnini *et al.* 1997;

Navarro-Cerrillo *et al.* 2011; Schwartz *et al.* 2013). Although not without exceptions (e.g., Navarro-Cerrillo *et al.* 2011), because the mortality of the outplants and maintenance cost can be substantial, it is better to use well-developed tall seedlings for planting that are stronger competitors. An example of successful underplanting of a threatened species represented by a single extant population with fewer than 100 adults is given in Zimmer *et al.* (2016) and will be discussed in Section 7.2.

### 5.1.3   Thinning the Understory: Liberation Cutting and Selective Thinning

Thinning is the silvicultural technique of removing competing vegetation to redistribute site resources (light, water, and nutrients) to the more valuable trees. In forestry, this intervention is applied at sites where adults and/or juveniles of desirable species are already present. One particular form of thinning is early-age liberation cutting that releases young growth from competition with less desirable species, mostly by enhancing the degree of crown illumination. Such removal of understory competitors is an important silvicultural treatment in commercial forestry to improve establishment of target timber species (Zeide 2001; Cameron 2002). However, in forestry this silvicultural treatment is usually implemented through the removal of large undesirable stems (e.g., >40 cm DBH) with no upper size limit for cutting (e.g., Wadsworth and Zweede 2006), which is clearly inapplicable in conservation-oriented restoration. If the goal is restoration of degraded or logged forest, large trees that would be undesirable for timber production will, in many cases, be very desirable. Therefore, liberation applied within a restoration framework must have a primary focus on low-stature vegetation such as smothering seedlings and saplings vegetation: lianas, bamboos, and dense shrubs. For example, lianas are well known for their strong effect on the development of old-growth tree seedlings and saplings (Schnitzer *et al.* 2000; Perez-Salicrup 2001; Schnitzer and Bongers 2002; Schnitzer and Carson 2010; Wright *et al.* 2015; Lussetti *et al.* 2016; Martinez-Izquierdo *et al.* 2016). Bamboos have similar effects impeding tree regeneration through their ability to physically suppress the growth and establishment of juvenile trees in both temperate and tropical forests (Taylor and Qin 1988; Oliveira-Filho *et al.* 1994; Tabarelli and Mantovani 2000; Griscom and Ashton 2003; Taylor *et al.* 2004; Guilherme *et al.* 2004; Wang *et al.* 2006; Lima *et al.* 2012; Rother *et al.* 2016).

Besides smothering nonarboreal vegetation such as climbers, bamboos, and shrubs, small stems of some tree species in the understory or subcanopy can also be cut (selective thinning; Swinfield *et al.* 2016), if they are target species competitors. This low-intensity liberation will only subtly elevate light conditions and therefore will not stimulate recruitment of fast-growing colonizers. This recommendation resonates with a suggestion from a group of researchers who studied the effect of liberation cutting on the development of seedling stock in mixed dipterocarp forest of Indonesia – to create artificial gaps by removing shading trees over patches of abundant stock of dipterocarp seedlings, while leaving other parts of the forest untouched (Kuusipalo *et al.* 1996, 1997; Tuomela *et al.* 1996).

## 5.1.4 *Thinning the Overstory: Individual and Group Selection*

Thinning to stimulate regeneration is applied in forestry not only to the understory vegetation, as described in the previous section, but also to the overstory vegetation. Many commercial timber species, after germinating in the understory, require some opening in the canopy for growth and transition from a juvenile to adult (Denslow 1987). Thinning of the uppermost canopy is more efficient than density reduction of low-value, light-demanding species in the understory or subcanopy in terms of increase in the amount of light available to high-value species. Another positive effect of canopy thinning, besides enhanced growth and accession to the canopy of the target trees, is increased stand structural complexity that increases the number of available niches for plants and animals.

Gaps in the overstory are important for fruit production in understory plants and therefore for associated nectariferous and frugivorous birds (Levey 1988). A meta-analysis by Verschuyl *et al.* (2011) revealed generally positive or neutral effects of canopy thinning in North American forests on the diversity and abundance across all studied animal taxa including birds, mammals, reptiles, amphibians, and invertebrates. In Europe, thinning of dense Mediterranean woodland had a similar effect on the bird community through increased woodland structural diversity (De la Montana *et al.* 2006).

Overstory thinning is known to be effective in promoting the growth of timber species. Increased vigor in response to thinning has been observed in old trees of, among others, *Sequoiadendron giganteum* (York *et al.* 2010), *Pseudotsuga menziesii* (Newton and Cole 1987; Latham and Tappeiner 2002), *Pinus strobus* (Bebber *et al.* 2004), *Pinus contorta* (Waring and Pitman 1985), *Pinus ponderosa* (Latham and Tappeiner 2002; McDowell *et al.* 2003), *Picea glauca* (Youngblood 1991), and *Pinus lambertiana* (Latham and Tappeiner 2002).

When applied to young trees, growth induced by the thinning growth response, as a rule, is observed immediately after thinning (e.g., Guariguata 1999), but rarely lasts for more than a few years (Oren *et al.* 1987). In contrast, in old trees, the response to decreases in stand density is usually delayed for several years (Youngblood 1991; Latham and Tappeiner 2002; Bebber *et al.* 2004), but improved tree growth can be seen over decades following thinning. Analysis of the response of old-growth stands of boreal *Picea mariana* to partial harvest revealed a 2-year phase of no response, increase in years 3–9, and then a stage of decline in growth increment in years 10–12 after harvest (Thorpe *et al.* 2007).

In boreal conifers, an increase in basal area growth of the individual trees was evident 14 years after thinning a 174-year-old *Picea glauca* stand in Alaska (Youngblood 1991), and at least 10 years after thinning old-growth (older than 120 years) stands of *Pinus strobus* in Ontario (Bebber *et al.* 2004). In temperate coniferous species, the effect of thinning appears to last even longer. In the US Pacific Northwest, a strong positive effect of thinning on the growth in diameter of *Pseudotsuga menziesii, Pinus ponderosa,* and *Pinus lambertiana* trees ranging in age from 158 to 650 years lasted for at least 20 years (Latham and Tappeiner 2002), and in 15- to 20-year-old stands of *Thuja plicata* for more than 25 years (Devine and Harrington 2009). For angiosperm tree species of the US Midwest, a period of increased growth following thinning lasted 10–12 years in *Quercus velutina* and *Quercus coccinea* in 32-year-old mixed stands of these two species (Cutter *et al.* 1991), and at least 10 years in *Acer rubrum* in its 50-year-old stand (Strong and Erdmann 2000). Similarly, increase in stem diameter of thinned old-growth stands of Mediterranean *Quercus ilex* persisted for at least 12 years (Mayor and Roda 1993). And in *Liriodendron tulipifera* from the Appalachian Mountains, improved basal growth was evident 40 years after thinning (Keyser and Brown 2014).

Thinning creates openings in the canopy with higher irradiance as compared with the surrounding stand. The light environment within a gap can be measured directly with sensors or estimated by determining the proportion of unobscured sky over a point. However, the easiest way to compare openings of different sizes is by using the ratio of the gap diameter to the surrounding stand height. Both light intensity and soil moisture in the center of the gap increase as this ratio increases, leveling off when this ratio approaches 2 (Runkle 1982).

The two kinds of thinning silvicultural treatments that either reduce densities or create distinct canopy gaps are called single-tree selection and group selection, respectively (Smith *et al.* 1997; Nyland 2002) (Figure 5.3). In single-tree selection, trees are removed either singly or in small groups at more or less regular intervals, creating canopy gaps that cover an area equivalent to the crown spread of one to several large trees. Group selection is done through the creation of openings of up to 0.5 ha spaced by at least 50 m.

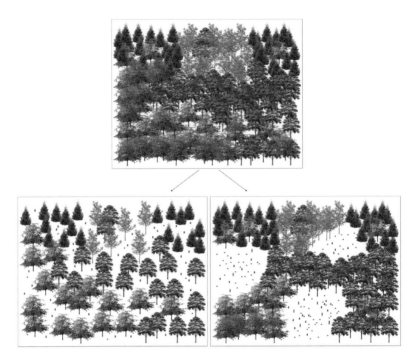

FIGURE 5.3 A forest stand managed through single-tree selection (left) and group selection (right).

For conservation-oriented restoration, both existing thinning treatments can be used, but in different situations. The single-tree selection better suits relatively intact secondary forests with well-developed understory that need either compositional adjustments in favor of threatened or functionally important species, or their growth enhancement. This intervention can be vital for many critically endangered species that are restricted by superior competitors to the most unfavorable or inaccessible microhabitats (e.g., Tang and Ohsawa 2002; He *et al.* 2015; Qian *et al.* 2016, 2018). By reducing the density of the dominant species, thinning will allow population growth of the imperiled species, thus improving the prospects of their survival (Figure 5.2). In addition, because dominance is related inversely to species diversity (Huston 1979), thinning of the dominants will not only preserve the threatened species but also maintain high species diversity within a community.

In contrast to single-tree selection, distinct canopy gaps of a large size that characterize group selection can be effective in the restoration of structural and/or biological complexity of naturally colonized abandoned pastures, severely degraded forests, and even-aged forest plantations. Group selection can be especially effective if followed by planting those species that have high conservation value in the gaps.

It is known that under natural processes, the development of a late-successional structure with several canopy layers, a wide range of tree sizes, standing dead trees, and abundant coarse woody debris can take 100–200 years or longer (Franklin *et al.* 2002). Thinning can accelerate the development of the late-successional forest structure in young or recovering stands by enhancing the growth of large trees, and creating vertical (multilayered canopies) and horizontal heterogeneity (a mosaic of canopy gaps and areas of high tree density) (Zenner 2004). Therefore, the goal of thinning either single or small groups of trees applied within a restoration framework should be the creation of uneven-aged stands with complex spatial canopy gap distributions, necessary for maintaining both plant and animal biodiversity (Thysell and Carey 2001; Carey 2003; O'Hara *et al.* 2010; Dodson *et al.* 2012; Blakey *et al.* 2016) (Figure 5.4).

## 5.1.5  *Replacement Planting*

Enrichment planting, even if supplemented by canopy thinning, will not always be the optimal strategy for conservation-oriented restoration. The establishment of slow-growing threatened species, which are inferior as competitors, may require more radical interventions. Some unique or rare habitats partly degraded as a result of human activity and subsequently colonized naturally or afforested by aggressive and fast-growing tree species, or simply invaded by exotics, may need complete replacement of their dominant and/ or subdominant vegetation, especially if the latter has a nonlocal

FIGURE 5.4 Mechanically thinned (right) and unthinned (left) dry mixed conifer forest, Lassen Volcanic National Park, California (from Falk 2017). The photo is courtesy of Calvin Farris, NPS, and is used by permission of Missouri Botanical Garden Press, St. Louis.

origin. For example, in the Azores Archipelago, introduced woody *Pittosporum undulatum* colonized a wide range of habitats including nature reserves and other protected areas, and negatively impacted native flora and fauna. Regional conservation planning is considering replacing *P. undulatum* with Macaronesian species, such as *Morella faya*, *Picconia azorica*, *Prunus azorica*, and others, in those locations where they are most likely to establish and persist (Costa *et al.* 2012). For such a process, I propose to use the existing term "replacement planting." The current usage of this term is limited to cases when protected trees have been removed for some reason and must be replaced by new trees. I think that for such cases the term "compensatory planting" is better suited.

Replacement planting, when undesirable or low conservation value vegetation is removed and threatened or functionally important nonthreatened species are planted instead, can be a solution for critically endangered species that have only a few extant populations and especially those that are poor competitors. For such species, if unoccupied suitable sites exist, an option is replacement planting in favorable habitats (Figure 5.2).

Endangered tree species of China (of which many are Tertiary relics; Thorne 1999; Manchester *et al.* 2009) are examples of species with extremely high conservation value, very narrow distribution, and small population sizes (López-Pujol *et al.* 2006). Although many of the extant populations of these species are protected, these populations are too small and often have very poor regeneration. In addition, protection often does not prevent legal and illegal detrimental activities of the nature reserves' local residents in the "protected" environment (Liu *et al.* 2001; Qian *et al.* 2018). As a result, for many endangered species, undisturbed reference habitats do not currently exist. For example, cultivation of profitable plants in the understory, selective cutting, and harvesting of firewood resulted in disruption of natural regeneration in the only natural population of *Metasequoia glyptostroboides* (Tang *et al.* 2011b). Similarly, a few remaining populations of *Michelia coriacea* all suffer from very poor regeneration due to past logging, invasion of exotic weeds, and cutting for firewood (Tang *et al.* 2011a). Thus, ensuring a future for these species is possible only through their introduction into multiple suitable habitats both inside and outside their known range accompanied by strong restoration interventions applied to the habitats. In many cases, control of competing species by the creation of canopy gaps and liberation cutting will not be enough. Management of populations of many relict tree species, e.g., *Cathaya argyrophylla* (Qian *et al.* 2016), *Thuja sutchuenensis* (Tang *et al.* 2015), and *Taiwania cryptomerioides* (He *et al.* 2015), may require removal of currently dominant tree species and replacement with the target threatened species.

## 5.1.6 *Utilizing Deadwood to Improve Regeneration*

Processes naturally occurring in unmanaged forests, such as continuous small-scale disturbances and occasional large-scale disturbances, lead to a spatial mosaic representing a full variety of

FIGURE 5.5 Typical examples of natural deadwood habitats in (a) boreal, (b) temperate, (c) subtropical, and (d) tropical biomes. From Seibold *et al.* (2015).

tree developmental phases. For this reason, the characteristic elements of a mature natural forest are not only old and very large trees, but also standing dead trees and their decaying partial remains (snags, stumps, and logs) (Harmon *et al.* 1986; Peterken 1996; Hunter 1999; Seibold *et al.* 2015). The abundance of this structural component progressively decreases in logged forests with a number of rotations. In unmanaged forests in Fennoscandia, the proportion of deadwood in the total wood volume is 30–40%, but this declines to about 20% after one rotation, and to about 1% after several rotations of intensive wood extraction (Angelstam 1997). The abundance of deadwood is often low even within reserves because of the previously practiced removal of dead trees in these areas. This is unfortunate, because deadwood plays an important role in all forest biomes (Figure 5.5) modifying the microclimate and microtopography and supporting diverse fauna either directly or indirectly. For amphibians, reptiles, and mammals, it provides shelter; for birds, it provides foraging and nesting sites; and for many fungi and invertebrates, it serves as a substrate and food (Rose *et al.* 2001; Bunnell and Houde 2010). Not surprisingly, a radical decline in the amount of deadwood may lead to the disappearance of associated saproxylic species (Siitonen 2001).

Besides this general role, large woody debris lying on the forest floor (e.g., logs) in various forest ecosystems creates favorable conditions for seed germination, and the survival and growth of tree seedlings of many species, mainly conifers but also broad-leaved trees (Harmon *et al.* 1986; Christie and Armesto 2003; Bellingham and Richardson 2006; Lonsdale *et al.* 2008). Representing only a small proportion of the available microsites, fallen wood accounts for a disproportionate percentage of established seedlings. For example, 98% of *Tsuga heterophylla* seedling recruitment in the western Cascades (United States) was observed on decayed logs although the latter usually cover only 4–11% of the forest floor (Christy and Mack 1984; Spies *et al.* 1988). Similarly, 43% of the saplings of *Picea abies* in the western Carpathians were found growing on decaying logs and stumps which covered only 4% of the forest floor (Zielonka 2006). This phenomenon, i.e., differences in species composition and densities of tree seedlings growing on logs and the forest floor, is well documented for a variety of forest types, but especially cool climate boreal and subalpine forests (e.g., Harmon and Franklin 1989; Hofgaard 1993; Szewczyk and Szwagrzyk 1996; Siitonen *et al.* 2000; Takahashi *et al.* 2000; Christie and Armesto 2003; Narukawa *et al.* 2003; Svoboda *et al.* 2010).

The presence of deadwood of necessary attributes (size and species identity) in the target restoration areas can be achieved through either passive or active management. If the forest has some standing old dead trees and/or logs, the only management required is just to leave them to decay. If the forest has some overmature live trees but no dead trees, the necessary management actions can be the creation of snags and high stumps by girdling or topping, and the creation of lying deadwood by felling, pulling down, or uprooting trees. If, however, the stand is too young or has been too intensively logged in the past, a solution can be the introduction of logs. In any case, damaging and felling selected individual trees will accelerate the formation of dead wood that would naturally require long forest successional development. Accumulation of deadwood following abandonment of forest activities is a very slow process; it takes decades for a fallen tree to become a suitable substrate for seedlings. For spruce, for example, seedling establishment starts approximately 10 years after tree death and colonization by seedlings reaches a maximum when the logs are 30–60 years old (Zielonka 2006).

### 5.1.7   Use of Ecotechnology-Based Tools

The importance of small-scale heterogeneity in surface soil properties has so far received insufficient attention in restoration programs. Only in arid and semiarid environments has the creation of spatial heterogeneity been recognized as important for improving land productivity (e.g., Shachak et al. 1998). In these environments, runoff generated by sloped surfaces covered with water-impervious soil crust is spatially redistributed toward adjacent vegetated patches collecting water, sediments, and nutrients. As a result, the vegetation is markedly patchy, leading to the development of "resource islands" or "islands of fertility" with vegetative patches surrounded by relatively infertile soil (Garcia-Moya and McKell 1970; Garner and Steinberger 1989; Ludwig and Tongway 1995; Pugnaire et al. 1996; Moro et al. 1997). These fertile patches are usually occupied by shrubs with a layer of annual and perennial herbs underneath their canopies and are points of high biological activity where facilitation is an important interaction between plant species. Herbs growing under the canopy usually benefit from higher soil moisture and nutrient availability.

The restoration of degraded drylands throughout the world is achieved by planting tree/shrub seedlings to initiate or accelerate natural recovery of the vegetation. The fundamental problems of traditional reforestation in Mediterranean and other dryland areas are early plant mortalities due to excessive radiation, high temperature, and low rainfall causing severe summer drought, small-scale heterogeneity in the distribution of soil resources, and high rates of grazing.

Summer drought occurring during the first years after planting is the most limiting factor for the establishment and growth of woody seedlings in dryland environments. Ecotechnological tools can be used to alleviate the problem of low water availability by mimicking the natural processes of the formation of "resource islands" with improved microclimatic conditions.

In arid lands, the success of restoration can be increased by mechanical preparation of the area prior to planting to increase the volume of available soil and improve the capture and retention of water (Eldridge *et al.* 2002; Ferrandiz *et al.* 2006; Palacios *et al.* 2009; Villar-Salvador *et al.* 2009; Pineiro *et al.* 2013). This involves the creation of contour furrows and water-catchment basins around each plant (micro-catchment) to direct the runoff toward the planted seedlings. The topographic modifications are often followed by the application of complementary ecotechnological techniques mitigating stressful environmental conditions and increasing the availability of resources for planted individuals (e.g., artificial shading, stone, woody, or plastic mulches, small waterproof sheets, dry wells, buried clay pots, and addition of fertilizers and gels that absorb and very slowly release water), and reducing seedling predation (screen cages or tree shelters against grazing) (Bellot *et al.* 2002; Leroy and Caraglio 2003; Rey Benayas and Camacho-Cruz 2004; Pausas *et al.* 2004; Bainbridge 2007; Devine *et al.* 2007; Li *et al.* 2008; Chirino *et al.* 2011; Padilla *et al.* 2011; Bakker *et al.* 2012; Valdecantos *et al.* 2014). The other emerging ecotechnological tools are the preconditioning of seedlings in nurseries (including mycorrhizal inoculation and nutrient and drought hardening; Requena *et al.* 2001; Vilagrosa *et al.* 2003; Barea *et al.* 2011; Trubat *et al.* 2011).

In a degraded soil, the mycorrhizal component may disappear or, at least, be severely depleted, and restoration may require an appropriate inoculation. Similarly, degraded soils can be at least partly restored by applying methods which improve soil quality, e.g., the addition of organic amendments to soil. Both addition of composted organic residue and mycorrhizal inoculation improved the establishment of *Olea europaea* (Caravaca *et al.* 2002), and mycorrhizal inoculation alone greatly increased seedling growth of three other native shrubs, *Pistacia lentiscus*, *Retama sphaerocarpa*, and *Rhamnus lycioides* (Caravaca *et al.* 2003) in semiarid southern Spain.

Analyzing the available literature on the application of ecotechnological tools in the restoration of degraded drylands, Pineiro *et al.* (2013) came to the conclusion that the inoculation with mycorrhizal fungi (mycorrhization) in the nursery (alone, or in combination with organic amendments) and the use of tree shelters were the most effective treatments for enhancing both the survival and growth of planted seedlings.

Usage of ecotechnological tools in restoration is not limited to drylands. In less water-limited environments, plastic mulching can both increase soil moisture and also reduce competition of planted trees with grasses (Barajas-Guzmán *et al.* 2006; González-Tokman *et al.* 2018).

## 5.1.8 Use of Non-Native Species in Planting

Fast-growing exotic tree species (e.g., *Pinus* and *Eucalyptus*) are often used in large-scale restoration projects either alone or interplanted with natives, due to their ability to facilitate the establishment of native species and to help offset restoration costs (Figure 5.6). Their seedlings are cheap, require lower maintenance interventions than native pioneer species, and can provide income from timber harvesting after just 5–7 years. However, care must be taken in the selection of non-native plants for use in restoration in general, and especially in conservation-oriented restoration. Once established, invasive species can prevent recolonization by native species for centuries, if not forever (Mascaro *et al.* 2008; Ostertag *et al.* 2008), and encroach on adjacent intact ecosystems (Richardson and Brown 1986; Ledgard 2001; Abella and MacDonald 2002; Williams and Wardle 2005; Catling and King 2007; Florens *et al.* 2016). Less dramatic examples show that plantation monocultures composed

FIGURE 5.6 Plantation of exotic *Eucalyptus* with mid- and late-successional native trees in the Atlantic Forest of Brazil (from Brancalion and van Melis 2017). Photo by Pedro Brancalion, and is used by permission from Missouri Botanical Garden Press, St. Louis.

of exotic species are more difficult for native species to colonize (Harrington and Ewel 1997; Keenan *et al.* 1997; Mascaro 2011).

There are classifications of alien species by their environmental impact (e.g., Blackburn *et al.* 2011, 2014) that can be used to rank them by undesirability. But the candidate alien species must also be evaluated according to two other properties: invasiveness and ease of complete eradication from the restoration site, if necessary. The first attribute determines the probability of the species spreading into adjacent areas and establishing self-sustaining populations there, but what is also very important from conservationists' point of view is the probability and cost of eradicating the exotic species from the site where it was planted but is no longer wanted. Global lists of invasive alien trees are available (Richardson and Rejmanek 2011; Rejmanek and Richardson 2013), but species differ substantially in the degree of invasiveness, and also in the probability of persistence. Prolific seed production, adaptations for long-distance seed dispersal, and early maturation are traits indicating potential invasiveness. However, vegetative reproduction by means of suckering or resprouting is a trait that does not contribute to long-distance spread and the colonization of new habitats, but greatly increases persistence within the already colonized habitat (Lloret *et al.* 2005). Species capable of vegetative propagation and producing numerous seeds are very difficult to eradicate even locally.

Thus, in decisions of whether or not to use exotic species, it is critical to predict with certainty: (1) whether their populations will persist once established or be replaced by native species over time; (2) if persistent, the expected negative impact on native biota; and (3) if such impact is probable, the expected cost of their local eradication.

Based on the above considerations, when the long-term impact is unknown, the non-native tree species can be ranked in their suitability for conservation-oriented restoration in the following way: introduced trees that do not reproduce; they do reproduce but the populations are not self-sustaining; the populations are self-sustaining but reproduction is only through seeds and the seed bank is short-lived. Species that can readily establish self-sustaining populations and are capable of resprouting or producing a persistent seed bank must be excluded from consideration.

Several studies have reported the creation of plantations of exotic trees in a restoration project that has conservation goals. In Vietnam, conversion of degraded areas dominated by aggressive grasses such as *Imperata cylindrica* into old-growth forest turned out to be an easier task if the exotic nitrogen-fixing and fast-growing *Acacia auriculiformis* was used to improve the soil and shade the grasses. Later, after closure of the canopy and disappearance of grasses, *A. auriculiformis* stands were thinned and underplanted with a variety of native late-successional shade-tolerant species, including threatened ones. Periodic thinning of *A. auriculiformis* in this system ensures development of the native species (McNamara *et al.* 2006). Similarly, four threatened dipterocarp species were underplanted in a 3-year-old plantation of *Leucaena leucocephala* in Thailand. After thinning the *L. leucocephala* stand after 4 years, the dipterocarp trees started to overtake *L. leucocephala* in height over the next 6 years (Sakai *et al.* 2014). Unfortunately, the authors of these two papers did not report the prospects of complete eradication of the exotic nurse species after reestablishment of the pre-disturbance forest, which is absolutely necessary in a conservation-oriented restoration project. In fact, such prospects seem to be doubtful given that there are "no clearly documented cases of the eradication of an alien tree" (van Wilgen and Richardson 2014).

Summarizing the above, when choosing material for plantings, a safe approach would be to give preference to indigenous species whenever possible (see also Hartley 2002; Rodrigues *et al.* 2009; Bremer and Farley 2010). Slower growth and the higher cost of their seedlings in the majority of cases will be greatly compensated by there being no risk of the appearance of a persistent novel ecosystem.

## 5.2   MAJOR PRINCIPLES OF FOREST RESTORATION

### 5.2.1   *Secondary Forest Succession*

Secondary forest succession is the woody vegetation that re-establishes after complete forest clearance for timber or pulp, or conversion to agricultural field or pasture. Discontinuity of forest cover is the defining feature of secondary forest. Vegetation clearance is followed by a spontaneous gradual process of the recovery of the structure, function, and composition of the pre-disturbance ecosystem, in which changes in vegetation are accompanied by changes in fauna and soil biota. A gradual change of the regrowth structure and composition during succession, however, occurs at very different rates, being much faster for the former. Structural changes in the recovering ecosystem can be summarized as four successional stages each having distinct features (Table 5.1) common to the majority of forest ecosystems, embracing tropical, subtropical, temperate, and boreal forests (Peet 1992; Finegan 1996; Guariguata and Ostertag 2001). These stages are initial vegetation colonization, dominance of short-lived, and then longer-lived woody pioneer species, and finally replacement by mature forest.

Table 5.1 *Hypothesized sequence for processes occurring during secondary forest succession, in reference to old-growth conditions (Guariguata and Ostertag 2001)*

| Process | Canopy vegetation | Time scale (years) | Notes |
|---|---|---|---|
| Initial colonization | Grass, herbs, ferns | 1–5 | Factors that affect initial colonization include:<br>1. Landscape features (distance to forest, topography)<br>2. Climate and microclimate<br>3. Presence/absence of past vegetation (seed bank, resprouts, remnant trees, and shrubs)<br>4. Site characteristics (nutrient availability, soil compaction, mycorrhizae, type, and intensity of past land use)<br>5. Multi-species interactions (seed predation, herbivory, perch availability, pathogens, competition, phenological stage) |
| Early forest development | Short-lived pioneers | 5–20 | Early stages of forest development include:<br>1. Canopy closure<br>2. Fine root biomass levels become similar to mature forest<br>3. Stemflow and throughfall levels become similar to mature forest<br>4. High rates of litterfall, net primary productivity (NPP), and turnover of nutrients<br>5. Rapid accumulation of biomass, along with self-thinning, and the appearance of standing dead<br>6. Accumulation of species, and the possibility that understory species richness is similar to mature forest |

| Late forest development | Long-lived pioneers | 20–100 | Later stages of forest development include: |
|---|---|---|---|
| | | | 1. More frequent small-scale disturbances, particularly small canopy gaps |
| | | | 2. Prevalence of advanced regeneration |
| | | | 3. Greater storage of nutrients in biomass, perhaps lower rates of NPP and litterfall |
| | | | 4. Low spatial heterogeneity in understory light levels |
| Old-growth forest | Shade-tolerant trees | 100–400 | Old-growth forest characterized by: |
| | | | 1. Very diverse overstory tree species composition |
| | | | 2. Prevalence of large canopy gaps, other chronic disturbances |
| | | | 3. Very large trees |
| | | | 4. High spatial heterogeneity in understory light levels |

The initial colonization stage (Table 5.1) starts from germination of the seeds buried in the soil and dispersed from the surrounding stands; the former usually being the more important (Young *et al.* 1987; Garwood 1989). Therefore, depletion of the soil seed bank during prolonged land use after vegetation clearance (especially if the seed rain is limited due to dispersal limitations) can have detrimental consequences for the rate and trajectory of forest recovery. Seed availability and seed longevity in the soil determine the species composition of the regrowth, and for this reason stands of similar age and land-use history can have very dissimilar canopy composition (Guariguata and Ostertag 2001). In addition to lack of seeds, competition/interference exerted by exotic or weed plants is another major barrier to forest regeneration. Once the tree seedlings have emerged they have to compete with herbaceous vegetation (grass, forbs, and ferns) that rapidly responds to canopy opening. Remnant vegetation (including resprouting stumps and rootstocks), if present, can have a strong facilitating role in forest recovery through its effect on seed dispersal and tree seedling germination/establishment. The former is due to food supply of isolated fruit-bearing remnant trees and shrubs to frugivorous birds and bats, and the latter because conditions are more favorable for tree seedlings (higher soil moisture, less herbaceous vegetation) beneath these trees and shrubs than in the open vegetation.

In the tropics, the first 5 (or more) years of forest recovery after the secondary succession starts are characterized by vegetation dominated by grasses, forbs, ferns, and shrubs, which are eventually shaded out by short-lived and fast-growing light-demanding pioneer tree species.

Early- and late-successional tree species have different growth patterns. Early successional species exhibit a "height-growth type" that is characterized by a rapid increase in plant height, while late-successional species are termed "crown-growth type" because they allocate resources preferentially to the construction of lateral branches and foliage rather than to plant height growth (Boojh and

Ramakrishnan 1982a, b). This difference is paralleled by differences in shade-tolerance and gap-size preferences reflected in terms "large-gap specialists" and "small-gap specialists." The former species "require the high light intensities and temperatures of large gaps for germination and seedling establishment. Early growth is rapid, and saplings are able to reach the upper forest strata during the lifetime of a single gap" (Denslow 1987). And the latter species "germinate in the understory or in small clearings. Saplings are able to survive understory light conditions owing to low respiration rates and low light requirements at saturation, but they are dependent on some canopy opening for substantive growth and reproduction" (Denslow 1987).

The fast-growing but short-lived early-successional tree species usually produce numerous, widely dispersed seeds (generally also stored in the dormant soil seed bank) (e.g., *Macaranga* in Asia and *Cecropia* in tropical America). These pioneers have a much smaller species pool than shade-tolerant late-successional species. The creation of a closed canopy by these species marks the start of the second stage of succession. The canopy emergence then causes the disappearance of aggressive herbaceous vegetation. Early-successional trees tend to include species that are able to acquire nutrients from sources not accessible to most other species (e.g., through nitrogen fixation or deep root systems). These species often improve the surface soils and prevent erosion by dense surface rooting and abundant litter production, creating conditions for the establishment of other more site-demanding species.

At this stage, although the closed canopy allows the arrival and initial growth of other more slowly establishing species, light becomes a limiting factor for further stand development because the late-successional species cannot advance into the canopy. However, a few decades after the initiation of succession, rates of treefall gap formation greatly increase due to canopy senescence of early colonizing short-lived tree species. With the creation of canopy gaps, long-lived pioneers start to replace the early colonizers in the canopy and the latter become dominated by long-lived, taller-statured, but nevertheless shade-intolerant tree species. Because most of these species are unable to reproduce under their own shade, eventually they are replaced by the shade-tolerant species characteristic of old-growth forest. The duration of the latter stage is related to the life span of old-growth canopy species and, therefore, to their turnover rate (Finegan 1996; Guariguata and Ostertag 2001) (Table 5.1). Figure 5.7 shows the consecutive stages of forest development following disturbance.

What is especially important from the biodiversity conservation viewpoint, of course, is not the basal area, standing biomass, tree density, or even species richness, but how close the species composition is to the primeval forest. While species richness can approach the old growth within two decades, returning to a species composition resembling that of the old-growth forest is a much longer process that can take centuries (Matlack 1994; Dent *et al.* 2013; Goosem *et al.* 2016). Therefore, the conservation value of a secondary forest increases over time, as more late-successional species characteristic of old growth accumulate and grow in abundance. For example, in secondary forests of the Venezuelan Amazon, saplings of tree species characteristic of old-growth forest started to dominate in abundance no earlier than 60 years after the start of succession (Saldarriaga *et al.* 1988). In the Atlantic Forest of Brazil, the percentage of old-growth tree species in secondary forests increased from 12% in 10-year-old stands to only 42% in 40-year-old stands (Piotto *et al.* 2009), and, after 50 years, secondary forests in this area shared only 19% of the same species with nearby primary forest (Liebsch *et al.* 2008). Liebsch *et al.* (2008) concluded that it would take between 1000 and 4000 years to reach the endemism levels that exist in mature forest. In temperate forests, the recovery of understory was estimated to take at least 50 years (Metzger and Schultz 1984) or even considerably longer than 80 years (Duffy and Meier 1992). Slow recovery of species composition in secondary forests is mostly due to poor dispersal of late-successional species (Whitmore 1991; Ingle 2003), evident in the fact that many large-seeded vertebrate-dispersed species

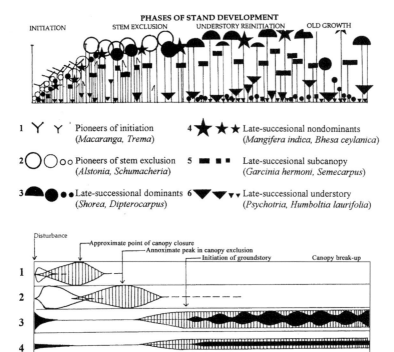

**PHASES OF STAND DEVELOPMENT**

INITIATION          STEM EXCLUSION          UNDERSTORY REINITIATION          OLD GROWTH

1  Y  Y  Pioneers of initiation
         (*Macaranga, Trema*)

4  ★ ★ ★  Late-succesional nondominants
          (*Mangifera indica, Bhesa ceylanica*)

2  ○ ○ ○ ○  Pioneers of stem exclusion
           (*Alstonia, Schumacheria*)

5  ■ ■ ■  Late-succesional subcanopy
          (*Garcinia hermoni, Semecarpus*)

3  ◖ ● ● ●  Late-successional dominants
           (*Shorea, Dipterocarpus*)

6  ▼ ▼ ▼ ▼  Late-successional understory
           (*Psychotria, Humboltia laurifolia*)

T I M E    L I N E

■ Juveniles recruited under canopy light conditions and considered as advanced regeneration (seedlings, seedling sprouts, root and stem suckers). The breadth of the bar represents amount of regeneration relative to other ecological guilds.

☐ Juveniles recruited under open conditions of full sun (buried seed, seed dispersed by wind or small animals into opening after disturbance).

▥ Stand canopy dominance. The breadth of the bar represents the degree of crown dominance in relation to other ecological guilds.

FIGURE 5.7  Six regeneration guilds, their recruitment frequency, and canopy dominance at different stages of forest development following disturbance (from Ashton *et al.* 2001). Each of the numbered horizontal graphs depicts the time of appearance and the changing abundance at the ground understory of each regeneration guild listed, with their respective numbers and crown symbols illustrated below the stand diagram. Examples of species for each guild are from Sri Lankan mixed dipterocarp forest.

persist in surrounding remnant forests but are rare or absent in secondary forests.

To understand succession, it is useful to group species into
guilds based upon ecological similarity of regeneration origin (initial or advance regeneration), stage of stand development (stand initiation, stem exclusion, understory reinitiation), and growth habit
(understory, subcanopy, canopy). These guilds, rather than stand
age, characterize successional stages. The six guilds recognized by
Ashton *et al.* (2001) are: (1) pioneers that dominate stand initiation;
(2) pioneers that dominate stand stem exclusion; (3) late-successional
canopy dominants; (4) late-successional canopy nondominants;
(5) late-successional subcanopy; and (6) late-successional understory
(Figure 5.7 and Table 5.2). Except for the first two guilds, the other
guilds depend on advance regeneration.

In a temperate climate, the process of woody colonization after
disturbance takes the same successional route of the initial dominance of herbaceous vegetation followed by the colonization of early-
and late-successional forest tree species (Smith and Olff 1998; Rebele
2013). Pioneer species of open habitats (e.g., *Betula*, *Alnus*, *Acer*,
*Pinus*) arrive early and are normally replaced by late-successional
tree species of such genera as *Fagus* and *Quercus* when a more or less
closed canopy has developed. However, lack of sources for bird- and
mammal-dispersed seed may result in early arriving pioneers (e.g.,
*Acer*, *Pinus*) also dominating late in the succession.

### 5.2.2   Land Use and Forest Regeneration Potential

Anthropogenic disturbances that lead to secondary succession
(forest clear-cut, agricultural use) fundamentally differ from natural disturbances such as treefall gaps in their much higher severity,
duration, and extent of impacted area (Janzen 1990; Myster 2004),
to which ecosystem keystone species are poorly adapted (Martinez-
Ramos *et al.* 2016). These disturbances include the elimination of the
original vegetation and replacement with exotic species grown for an
extended period of time, often with burning and recurrent weeding.

Table 5.2 *Six regeneration guilds recognized by Ashton* et al. *(2001) with characteristic life-history traits*

| Guild | Mode of dispersal | Other life-history traits |
|---|---|---|
| Pioneers of stand initiation | Wind, small birds, bats | Seed that is abundant and small; shade intolerant; short-lived; fast growing; often umbrella-crowned |
| Pioneers of stem exclusion | As above | Seed that is abundant and small; shade intolerant; medium-lived; fast growing; often columnar-crowned |
| Late-successional dominants | Gravity, hoarding mammals | Seeds of medium size; germination sporadic as a cohort; seedling bank in the groundstory; intermediate to shade tolerant |
| Late-successional nondominants | Large birds, primates | Seeds of large size; germination regular; shade tolerant |
| Late-successional subcanopy | As above; often capable of resprouting | As above |
| Late-successional understory | Small birds and rodents, vegetative propagation | Shade tolerant |

In contrast to the colonization of naturally disturbed areas, colonizers of degraded lands may inhibit or even divert succession to a different type of plant community (Uhl *et al.* 1988; Nepstad *et al.* 1990; Aide and Cavelier 1994; Cohen *et al.* 1995; Chapman and Chapman 1997; Endress and Chinea 2001; Brown and Gurevitch 2004; Bonnell *et al.* 2011; Piiroinen *et al.* 2017). Janzen (1990) listed a number of ways in which the recovery of human-affected habitats differs from the recovery of natural disturbances, such as treefall gaps. The local fauna, in general, and of functionally important groups (pollinators, seed dispersers), in particular, is likely to be substantially altered. Frugivores become less common and seed predators become more common. Wind-dispersed species start to dominate over animal-dispersed species and access to light will favor species that did not originally occur in the altered habitat. As a result, the pathway of succession becomes much less predictable.

A site's regeneration ability will depend on past management and disturbance history, and therefore the classification of the latter aligned with restoration methods is highly needed. Ashton *et al.* (2001) proposed distinguishing two types of forest degradation processes that result from human activity: (1) chronic (having continuous degradation impacts, e.g., recurrent fires, cattle grazing, selective logging), and (2) acute (one-time detrimental impact, e.g., land clearance). These processes cause various structural and functional changes in the ecosystem. Structural changes can be due to lack of seed source, or inability to establish (e.g., open conditions, susceptibility to browse, competition with grasses) but not because of fundamental changes to the system integrity. In contrast, functional changes are more severe, altering the soil fertility and texture, hydrology, and hence site productivity and nutrient cycling (Griscom and Ashton 2011).

Figure 5.8 reproduced from Ashton *et al.* (2001) shows four types of land uses representing either chronic or acute degradation. One type of chronic degradation is a bottom-up disturbance effect on the ecosystem through changes in the groundstory or understory (e.g., cultivation of crops such as cardamom, coffee, bamboo, etc., beneath the forest canopy). Weeding and grubbing associated with crop cultivation persistently eradicate emerging vegetation thus inhibiting regeneration of both understory and canopy, simplifying the forest structure, and preventing the replacement of the forest strata (Figure 5.8a). In contrast, the top-down disturbance effect in the ecosystem is caused by changes in the canopy (e.g., selective logging), but with similar consequences for the forest structure.

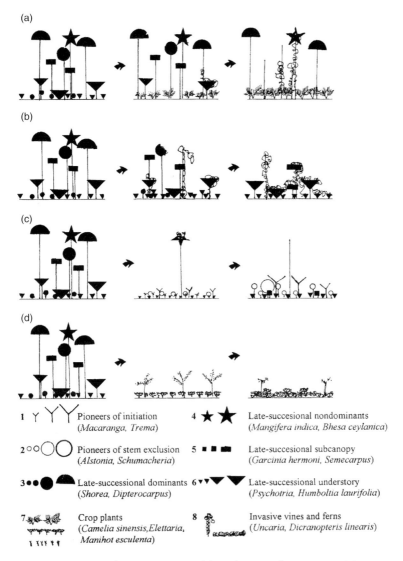

1 ⋎ Ⲩ Ⲩ  Pioneers of initiation
(*Macaranga, Trema*)

2 ∘∘ ◯◯  Pioneers of stem exclusion
(*Alstonia, Schumacheria*)

3 • • ● ◢  Late-successional dominants
(*Shorea, Dipterocarpus*)

7  Crop plants
(*Camelia sinensis,Elettaria,
Manihot esculenta*)

4 ★ ★  Late-succesional nondominants
(*Mangifera indica, Bhesa ceylanica*)

5 ▪ ▪ ▬  Late-succesional subcanopy
(*Garcinia hermoni, Semecarpus*)

6 ▾▾▼ ▼  Late-successional understory
(*Psychotria, Humboltia laurifolia*)

8  Invasive vines and ferns
(*Uncaria, Dicranopteris linearis*)

FIGURE 5.8 Chronosequences of degradation in forest composition
and structure following anthropogenic disturbances of a mature
rainforest (using Sri Lankan mixed dipterocarp forest as an example)
(from Ashton *et al.* 2001). The four types of degradation are: (a)
chronic degradation from continuous removal of the groundstory or
understory for nontimber forest product cultivation (bottom-up effect);
(b) chronic degradation from frequent removal of the largest timbers
in selective logging (top-down effect); (c) acute degradation from one-
time incomplete clearance for temporary crop cultivation; (d) acute
degradation from one-time complete clearance for permanent crop
cultivation with subsequent abandonment.

Periodic cutting of canopy trees, either targeting particular timber species or through diameter-limit cutting, will lead to gradual degradation of the forest structure, proliferation of vines, reduction of strata, and replacement of canopy species by those from the subcanopy (Figure 5.8b). Acute degradation is caused by one-time disturbances of different intensity. Temporary incomplete clearance occurs under agricultural practices with intensive site preparation and cultivation for less than 5 years such as swidden (or slash-and-burn) systems. These practices have little impact on soil and often leave stumps and rootstocks that are capable of resprouting after forest clearance. In such lands, regeneration after abandonment can be rapid, albeit that the pioneer species will predominate (Figure 5.8c). Permanent land clearance for cultivation and exploitation for a long time will result in reduced soil fertility and infiltration ability, as well as the complete eradication of all or almost all ecological legacies, making natural forest regeneration impossible. After abandonment, the land can be invaded by grasses, ferns, and vines, and become grassland (Figure 5.8d). Grassland that has resulted from forest clearance is the most extreme degradation type.

A process of succession following anthropogenic disturbance can be better understood by looking at a trajectory of secondary succession in the dry tropics as schematically presented in Figure 5.9 and described by Alvarez-Yepiz et al. (2008). A dry Pacific forest in Mexico, once it is cleared by slash-and-burn, is either used for a few years for cultivation of crops and, when soil fertility decreases, is transformed into exotic grassland, or is turned into pasture right after forest clearance. After the abandonment of management practices (i.e., fire, intensive grazing, and agriculture) in the deforested land, resprouting from roots or stumps, and seedling recruitment from the soil seed bank and seed rain, lead to the establishment of fast-growing pioneer species with the latter dominating for a period of about 5–50 years. Recruitment of primary forest species starts early during the dominant pioneer trees stage, and leads to the creation of mixed forest where primary tree species subdominate.

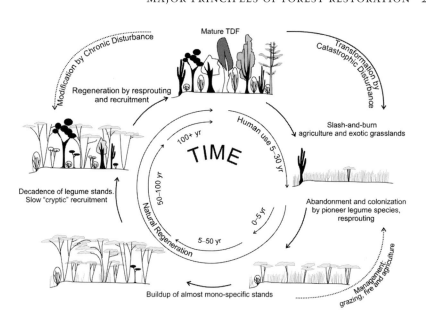

FIGURE 5.9 Secondary succession in tropical dry forests following anthropogenic disturbance through slash-and-burn (from Alvarez-Yepiz *et al.* 2008). Counterclockwise dotted arrows represent the responses from recurrent human impacts.

This stage lasts between 50 and 100 years after abandonment, and, if not interrupted, will proceed to the dominance of the late-successional primary species. However, chronic disturbance (i.e., grazing, selective logging, and fuelwood collection) will retard or prevent the achievement of the latter stage. Similarly, resumed disturbance (grazing, fire, and agricultural use) will return the pioneer forest to a stage of grassland (Figure 5.9).

The capacity to recover from human disturbances, i.e., ecological resilience, is higher for more recently and less modified ecosystems, with the consequences of chronic degradation being easier to remedy. In general, greater agricultural impact makes regeneration longer, more problematic, and a less likely process (Figure 5.10). However, the resilience of plant communities to disturbances depends not only on internal (within local site) factors, such as biological legacies and environmental conditions

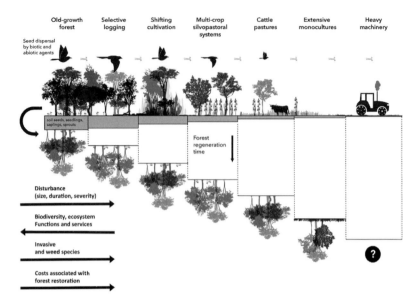

FIGURE 5.10 The relationship between agricultural land use and forest regeneration potential (from Martinez-Ramos *et al.* 2016). As size, duration, and severity of agricultural disturbance increases the forest regeneration potential decreases and forest recovery takes longer (as indicated by the length of the discontinuous lines). Under extreme disturbance conditions, natural forest regeneration is highly unlikely (uncertain time of recovery). With increasing agricultural disturbance seed dispersal limitation intensifies (indicated by the decrease in the size of the bird image; symbol in front of birds represents wind), establishment limitation increases, and biodiversity, ecosystem functions, and services of regenerating forests decline, while the costs required for restoration increase.

after disturbance, but also on external (within surrounding landscape) processes, such as the sources for colonization in the remaining landscape matrix (Chazdon 2003; Nascimento *et al.* 2006; Jakovac *et al.* 2015; Crouzeilles *et al.* 2016; Arroyo-Rodriguez *et al.* 2017). As the intensity of disturbance inflicted by agricultural land use increases, the potential for forest regeneration in the abandoned fields decreases through both reduced suitability of the biophysical conditions and reduced availability of propagules (seed rain from proximate forest patches, seed, seedling, sapling, and resprouting banks) (Janzen 1990; Ashton *et al.* 2001; Holl and Aide 2011). Among regenerating forests, those closer to the remaining untouched forest have higher levels of diversity, and within those adjacent to the intact forests, species diversity usually progressively declines with distance from the boundary separating them (Matlack 1994; Brunet and von Oheimb 1998; Bossuyt *et al.* 1999; Singleton *et al.* 2001; Barnes and Chapman 2014). With increases in the size of the agricultural area and distance to the nearest forest patch, the representation of animal-dispersed species decreases in the seed rain (da Silva and Tabarelli 2000), while the increasing duration and severity of agricultural uses (duration of crop cultivation or cattle grazing, frequency of burning) more strongly depletes the propagule pools (Figure 5.10). At the same time, post-agricultural biophysical environmental conditions can support an increasingly smaller number of original species (with an increasingly higher proportion of pioneers) as agricultural land use intensifies, with fewer species dominating the altered community. All this will lead to reduced biodiversity and structural complexity of regenerating forest and increased costs of land restoration. Under extreme disturbance, ecosystem degradation can cross a threshold above which restoration of pre-disturbance conditions is impossible or, if possible, time to regeneration is uncertain with most expensive treatments (Figure 5.10).

Thus, sites with the highest regeneration potential and hence conservation value are those that have the least modification by agriculture, where remnant trees and seed banks composed of native species persist, and which are surrounded by well-preserved biodiversity-rich native forests covering a significant part of the landscape. Such areas converge relatively quickly, within a few decades, to the community attributes of nearby reference forests. Nevertheless, such convergence is usually evident for stand structural attributes, although not for species composition, as even under mild land uses, such as selective logging or shifting cultivation, and high forest cover in the landscape, significant differences in species composition between secondary forests and reference areas are commonly observed (Guariguata and Ostertag 2001; Chazdon *et al.* 2007; Arroyo-Rodriguez *et al.* 2017). So, for the question "How predictable is a forest succession pathway?" we should distinguish predictability of recovery in structure and recovery in composition. The recovery of forest structure is mostly determined by agricultural-use history, while the recovery of species composition is determined by landscape patterns, with the extent of an area covered by old-growth forest up to 10 km around a disturbed site being the most important (Crouzeilles and Curran 2016).

With increasing landscape alterations, the heterogeneity of landscapes also increases due to higher fragmentation of the forest cover. This heterogeneity makes pollination and frugivory, and thus seed production, predation, and dispersal among forest patches, more variable, increasing differentiation in species composition (Arroyo-Rodriguez *et al.* 2013; Puettker *et al.* 2015) and reducing the predictability of the successional pathways in particular locations. An increase in landscape disturbance up to a certain point is also associated with an increase in variation in biotic and abiotic conditions among regenerating stands, and more compositionally diverse successional pathways (Figure 5.11). Intermediate forest cover (from 20% to 50% of remaining forest cover) shows the greatest variability in the degree of fragmentation (e.g., number of forest patches) and in the total forest edge (Villard and Metzger (2014), and therefore at this level of landscape alteration successional pathways should be the most variable, and hence be less predictable. Crouzeilles and Curran (2016), in their meta-analysis of the influence of forest cover on restoration success, revealed that when the percentage of the forest cover surrounding a disturbed site falls below 50%, restoration success becomes increasingly uncertain. With further decrease in forest cover, however, the landscape becomes more homogeneous (i.e., dominated by nonforested areas), biologically more impoverished and uniform, and with much more predictable successional pathways (e.g., hyper-dominance of

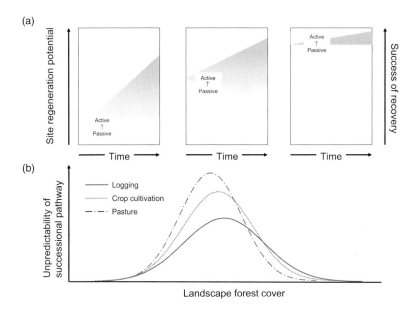

FIGURE 5.11 (a) Variability in successional pathways and recovery success as a result of a site's regeneration potential determined by the level of landscape antropogenic disturbance and the restoration method (based on Bechara *et al.* 2016). Recovery success is measured by functional and structural resemblance to the old-growth forests (reflected in intensity of the grey color). Importance of active restoration decreases with the site regeneration potential. Under intermediate regeneration potential, it is difficult to predict the recovery trajectory under both passive and active restoration.

(b) The hypothesized relationship between the landscape anthropogenic disturbance level and the predictability of successional pathways (based on Arroyo-Rodriguez *et al.* 2017). Succession is expected to be more predictable (i.e., convergent with the vegetation of nearby old-growth forests) for a more recently modified land, and the higher remaining forest cover in the landscape, which is illustrated with three predominant in the landscape land-use types (see text for further explanations).

disturbance-adapted species). Eventually, landscape disturbance above a certain threshold will limit, and even interrupt, the eco-logical resilience of the system (Folke *et al.* 2004; Ortega-Pieck *et al.* 2011; Banks-Leite *et al.* 2014; Jakovac *et al.* 2015), greatly reducing the number of potential pathways and making restoration to the pre-disturbance state very problematic or impossible (Figure 5.11).

Fragmentation of the natural habitat in the landscape should always be seen as extremely undesirable, because, among other negative effects, the high extent of forest edges will cause a hyper-proliferation of short-lived pioneer trees, particularly along the edges (Laurance *et al.* 2006), while large-seeded, long-lived, late-successional tree species will be discouraged at the landscape scale. Some authors have hypothesized that, as a result of landscape-scale diffusion of pioneers and elimination of many large-seeded, old-growth tree species, such pioneer-dominated assemblages may approach a near-equilibrium, quasi-final stage rather than a transient successional stage (Santos *et al.* 2008; Tabarelli *et al.* 2008). The probability of res-toration success measured in terms of the basal area, standing biomass, and canopy structure, in contrast to the predictability of a particular succession pathway (i.e., species composition), linearly increases along a gradient of contiguous forest cover due to an increase in the immigration rates (Crouzeilles *et al.* 2014) (Figure 5.11).

### 5.2.3   Forest Restoration Pathways

There are different approaches to the restoration of forest ecosystems, and their utility for conservation-oriented restoration depends on several factors with major ones being the initial state of the forest or land degradation they can be applied to and the outcomes they can achieve (Chazdon 2008; Holl and Aide 2011).

As was shown in the preceding section, a site's potential for recovery is determined by the three main characteristics of the agri-cultural land use before abandonment: field size, duration, and land-use severity. Thus, some quantitative approach incorporating the measures of these characteristics in an overall estimate of ecological disturbance and its effect on forest regeneration can be a useful tool in restoration. This tool already exists (Holl 2007; Zermeno-Hernandez *et al.* 2015) and can be used for choosing the most appropriate restor-ation method for the target area.

As for the methods of forest restoration, there has been great progress in the development of technical approaches for restoration

Table 5.3 *Suitable conservation-oriented forest restoration approaches and their merits (based on Shono* et al. *2007a)*

| Restoration approach | Costs | Biodiversity | Time for forest development | Research input required |
|---|---|---|---|---|
| Passive natural regeneration | Low | Low to medium | Slow | Low |
| ANR[a] without enrichment planting | Low | Low to medium | Slow to medium | Low |
| ANR with enrichment planting | Low to medium | Medium | Medium | Low to medium |
| Framework species method | Medium to high | Medium | Medium | High |
| Dense planting of many species | High | High | Fast | High |
| Mixed-plantation method | High | Low to medium | Fast | Low to medium |
| Nucleation and cluster planting methods | Low to medium | Low to medium | Medium | Low to medium |

[a] Assisted Natural Regeneration

(Ashton *et al.* 2001; Mansourian *et al.* 2005), and these approaches (Table 5.3) will be presented and discussed in the next sections. The existing methods of forest restoration can be grouped into three broadly defined types of restoration pathways. All three assume protection from further anthropogenic impact, be it logging, nontimber forest product cultivation, collection of firewood, etc. They are:

1. no or minor intervention into naturally occurring succession;
2. enrichment planting of the late-successional, functionally important, or threatened species;
3. use of plantation analogs to facilitate secondary succession.

Application of these methods in a conservation project should be based on three main principles (Rodrigues and Gandolfi 2007; Rodrigues *et al.* 2009): (1) reconstruct species-rich functional communities; (2) stimulate whenever possible the potential for self-recovery still existing in the area (resilience); and (3) plan restoration actions from a landscape perspective. Within these principles, a project generally should have the following site-level goals: increase species richness (first of all rare and threatened species); support local fauna; and eliminate or control invasive exotic species.

## 5.3   EXISTING METHODS OF FOREST RESTORATION

### 5.3.1   *Passive Natural Regeneration*

Passive natural regeneration is considerably less costly than restoration based on planting trees, and can potentially be applied over much larger areas. Therefore, it is important to have a set of criteria to identify areas for which natural regeneration is the most appropriate and cost-effective restoration option (Rodrigues *et al.* 2011). Site regeneration potential is inversely related to the intensity of land clearance and pre-abandonment use. The latter is controlled by two factors: (1) the fertility of the soil – the greater the fertility, the greater the intensity of land clearance to maximize productivity; and (2) the site topography – the greater the degree of rough terrain (steeper slopes, more pronounced depressions), the lower is the intensity of land clearance. Thus, sites with rough terrain, and lower soil fertility, because of their low agricultural value, have a higher probability of abandonment due to marginal profits and migration of local residents to urban areas (Chazdon and Guariguata 2016), and have more remnant forest fragments (Griscom and Ashton 2011).

The speed and trajectory of natural forest regrowth is determined by processes at both local and landscape scales as described above, and the most important factors for successful natural regeneration are close proximity to mature forest patches, low levels of soil disturbance, and abundant seed-dispersing fauna (Aide *et al.* 2000; Griscom and Ashton 2011; de Rezende *et al.* 2015; Sloan *et al.* 2016; Arroyo-Rodriguez *et al.* 2017). Traits that facilitate regeneration are resprouting ability (Guimaraes and Proctor 2007), root suckering (Scowcroft and Yeh 2013), and a long-lived seed and seedling bank (De Steven *et al.* 2006; Fernández-Lugo *et al.* 2015). Rapid regeneration requires the presence of legacy trees, seeds, rootstocks, stolons, and/ or resprouts below the soil surface and/or seeds dispersed from local or surrounding plants (Chazdon and Guariguata 2016). These and other features presented in Table 5.4 are either internal (within the local site) or external (within the surrounding landscape) ecological legacies left by the preceding land use, and they are good indicators of the site's capacity for natural regeneration. External legacies determine the potential for seed dispersal, whereas internal legacies determine the potential for local regeneration within the site. The greater the extent of old-growth forest remaining in the landscape, the greater the seed flow into the regenerating forest. Because 50–90% of tree species in the tropics are dispersed by birds and mammals (Howe and Swallowed 1982), the landscape context is important for both faunal conservation and forest regeneration (Reid *et al.* 2015).

It must be noted, however, that even sites with high regeneration potential (i.e., those with intact soil, old-growth stands in close proximity, etc.) are unlikely to recover the full complement of species present in the original ecosystem, especially the endemic and rare species (e.g., Matlack 1994; Wunderle 1997; Van Gemerden *et al.* 2003; Lugo and Helmer 2004). Using data for 410 of Brazil's Atlantic Forest tree species, Liebsch *et al.* (2008) estimated that it should take 100–300 years after a disturbance to reach the proportion of animal-dispersed species (80%

Table 5.4 *Internal (within the local site) and external (within the surrounding landscape) ecological legacies that can be used to predict the site capacity for natural regeneration (Chazdon and Guariguata 2016)*

| Indicator | Internal | External |
|---|---|---|
| Presence of topsoil and soil organic matter | X | |
| Soil seed bank | X | |
| Presence of rootstocks | X | |
| Abundance and cover of shrubs | X | X |
| Abundance of remnant trees | X | X |
| Abundance of animal-dispersed trees | X | X |
| Living fences/hedgerows | X | X |
| Local avian abundance and diversity | X | X |
| Local mammal frugivore abundance and diversity | X | X |
| Remnant forest patches within 100 m | | X |
| Riparian vegetation within 100 m | | X |
| Large forest remnants or reserves within 200 m | | X |
| Regional avian abundance and diversity | | X |
| Regional mammal abundance and diversity | | X |

of the species), of nonpioneer species (90%), and of understory species (50%) found in mature tropical forests, while it will take between 1000 and 4000 years to reach the endemism levels (40% of the species) that exist in the mature forests of the region.

Thus, if the goal is to restore the original biodiversity, greater levels of management beyond initiating natural succession (enrichment planting, liberation cutting, creation of natural or artificial perches, etc.) are necessary (Zimmerman *et al.* 2000; Bertacchi *et al.* 2016).

### 5.3.2   Assisted or Accelerated Natural Regeneration

Assisted natural regeneration, originally proposed by Dalmacio (1987), is a method to accelerate, with relatively little human

intervention, the reestablishment of native tree species that naturally colonize sites with low to intermediate levels of degradation
but generally intact soil. This is done by protecting and nurturing
the mother trees and their offspring already present in the area and
removing or reducing barriers to regeneration. The latter is achieved
by mulching or applying fertilizer to improve soil quality, and liberating naturally regenerating seedlings or saplings from competition
with undergrowth (Hardwick *et al.* 1997, 2004; Dugan 2000; Shono
*et al.* 2007a). Besides suppressing or eliminating the weeds around
naturally established seedlings and saplings, fuel breaks and fences
can be established to protect them from fire and grazing. Where
two or more seedlings or saplings grow close to each other, the less
successful, or less desirable one is removed and, where appropriate,
transplanted to another location in the restoration site. Usually this
method implies no or minimal tree planting. Instead, to overcome
the problem of poor seed availability in degraded fields, artificial
perches can be used to attract birds and thus increase seed arrival
from the surrounding natural vegetation (Guidetti *et al.* 2016).

This strategy is applicable in areas with sufficiently dense stands
of remaining trees or patches of natural forest (500–800 wildings/ha),
even within a wider degraded landscape, as these trees provide propagation material or attract seed-dispersing animals (Shono *et al.* 2007a).
However, most species able to colonize these areas will be pioneer
fast-growing species. In areas with fewer than 500 wildings/ha and
reduced species richness, stands must be enriched with a variety of
species, first fast-growing and light-demanding species that create
shade in the understory and a habitat for late-successional species.
Once the planted and existing trees start to create the canopy, the
stand can be enriched with shade-tolerant species, including species
that provide fruit for birds, bats, and other animals that disperse
seed. Actually, planting late-successional species is necessary even
in areas with dense stands of remaining vegetation because otherwise
mostly fast-growing pioneer species will colonize the site. Although
this method can be used for a variety of goals, many of which have

nothing to do with nature conservation (Shono *et al.* 2007a), I present it here because it can be useful for the restoration of the threatened species habitat. However, for the latter it should be supplemented by the planting of functionally important and threatened species (Zimmerman *et al.* 2000).

### 5.3.3   *Restoration Plantings*

Despite widespread public perception, natural regeneration is not innately superior to artificial management of the ecosystem rehabilitation process. Lands that been repeatedly burned, overgrazed, planted with exotic grass species, and have lost topsoil and parent tree seed sources usually lack core functional attributes of site productivity (soil carbon, nitrogen, infiltration capacity) (Uhl *et al.* 1988; Lamb *et al.* 2005). If left alone, such lands may never return to the pre-disturbance state. For such degraded areas, restoration planting using several existing methods is a useful management tool to ameliorate soil structure and fertility, shade out grasses and weeds, diminish the fire hazard, and facilitate colonization of the site by a wider range of species (Parrotta *et al.* 1997; Lamb *et al.* 2005; Shono *et al.* 2006; de la Pena-Domene *et al.* 2013). Planting certain kinds of trees with or without postplanting management can facilitate native secondary growth and support diverse biota within a few decades of establishment (Reay and Norton 1999; Catterall *et al.* 2004; Kanowski *et al.* 2005; Shono *et al.* 2006). Restoration planting methods are similar to plantations in that they usually utilize a single cohort of seedlings in initial planting, but use natives only, and at least 20 or more species (Figure 4.4). For example, in Eastern Australia restoration plantings utilize up to 100 locally occurring species of rainforest plants (Goosem and Tucker 1995; Kooyman 1996).

A forest cannot be considered successfully restored without establishment of the canopy-dominating late-successional tree species because only the latter species create the environmental conditions characteristic of a mature forest, suitable for fauna and for many subcanopy plants. Most pioneer species are small seeded and wind dispersed, but in primary tropical forests large-seeded, animal-dispersed species predominate. Naturally regenerating secondary forests, even as old as 60 or more years, often lack these large-seeded, old-growth canopy tree species due to dispersal limitations (Martinez-Garza and Howe 2003). Ensuring that shade-tolerant, late-successional, animal-dispersed species reestablish in a community is therefore a priority for restoration. This can be done through restoration planting, but when planted on lands cleared of forest, these species usually demonstrate poor seedling establishment. Early growth and establishment of many late-successional tree species require protection from the desiccating effects of full sun, which can be implemented by planting them either in the open environment together with shade-intolerant pioneer species or in the canopy of the latter.

Thus, species used in restoration plantings can be mature forest species unable to recolonize the site by themselves and/or pioneer species facilitating their establishment. Restoration planting should always focus on adding functionally important species of trees that are unlikely to establish naturally and those species that are threatened (Shono et al. 2006).

Among other active interventions, planting tree seedlings is by far more effective for restoring species richness than adding litter, seeding, plowing, or grass removal (Sampaio et al. 2007). Several restoration methods involving planting tree seedlings have been developed to reforest degraded lands and they are presented below. They differ in density of planting, spatial design, and type of species used. Although they have higher cost in comparison with passive or assisted regeneration methods, which currently limits their application to relatively small-scale projects, these methods are the future of conservation-oriented restoration of severely degraded habitats.

Nevertheless, in large-scale restoration projects, balancing passive and active restoration is essential because of funding constraints. The resilience of lands targeted for restoration, on the one hand, and available amount of financial resources, on the other,

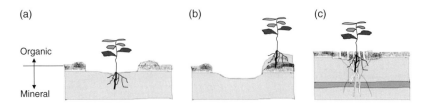

FIGURE 5.12 Three types of mechanical site preparation and their main effect on soil structure: (a) scarification, (b) mounding, and (c) subsoiling (from Lof *et al.* 2012). Dark gray area below mineral soil in (c) represents a hardened layer with impaired drainage. Planting spots are indicated by the seedling.

will dictate how much of the designated land will be subjected to active and how much to passive restoration.

To improve survival of planted seedlings, soil preparation may be necessary. Scarification of the soil (Figure 5.12a) removes vegetation and the upper organic layers, and uncovers bare soil. Scarification does not significantly influence soil structure but increases soil temperature and moisture, which may be necessary for improving seedling growth conditions. In contrast to scarification, mounding (Figure 5.12b) does change soil structure, creating elevated planting spots of inverted or mixed soil (Sutton 1993). This technique is effective in providing the seedling with conditions free from water logging and vegetative competition (e.g., Knapp *et al.* 2006; Zamith and Scarano 2010). Subsoiling or ripping (Figure 5.12c) fractures soil structure without mixing soil horizons. This technique is used for dry and compacted soils to improve root growth and plant development conditions by increasing soil aeration, soil water infiltration, and drainage, and reducing soil bulk density (e.g., Barbera *et al.* 2005). Compacted layers often develop after long-term grazing or agricultural land use. Therefore, when restoration of these systems is done without subsoiling, often the initially good growth of planted trees becomes stagnated.

## 5.3.4   *Miyawaki Method (Dense Planting of Many Species)*

This method is based on an idea that it is not necessary to closely simulate natural succession by initially planting only

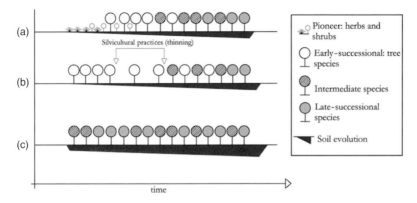

FIGURE 5.13 Successional stages under (a) natural regeneration, (b) traditional reforestation, and (c) the Miyawaki method (from Schirone *et al.* 2011).

early-successional species. In many, though not all cases, late-successional species can also grow during the early stages of community development, and the absence of such species in many natural successions is as much a result of seed limitation as physiological tolerances (Lamb 2000). This reasoning motivated a proposal to plant late-successional species together with pioneers (Miyawaki 1993).

In this approach the planting stock comprises seedlings of the maximum possible number of tree species that characterize the potential natural vegetation, from pioneer species to late-successional ones (Figure 5.13). Before planting, the soil is improved where necessary and feasible by adding topsoil, straw, and debris from neighboring woods. It is advisable to acclimatize the planting stock before planting for several weeks in the proximity to the restoration site under artificial or natural sheltering. Then seedlings are planted randomly at high density (2–5 individuals/m², i.e., more than 10 000 seedlings/ha) and mulched with straw or other organic materials. Weeding and watering are applied once or twice, if necessary. After the first 2 years, the planted forest is left to grow naturally. Development into late-successional forest takes, according to Miyawaki (2014), 40–50 years.

This method has proved to be successful not only in subtropical and tropical regions (Miyawaki 2014) but also in Mediterranean Europe (Schirone and Vessella 2014). The Mediterranean study deserves a closer look.

Since 1905, all attempts to reforest degraded natural forests in Sardinia, Italy, were done using low-density planting (300–2200 plants/ha) along contour lines after forming terraces by subsoiling, or across the slope in pits. In 1997, a modified Miyawaki method was applied at two sites on which several reforestation projects had failed. Planting was done with 22 and 23 tree and shrub species, respectively, of which 65% and 69% were late-successional ones. Seedlings were grown in plastic bags for 1 year before planting. Several modifications were made to the Miyawaki method. Instead of adding new topsoil to the restoration sites, the soil was tilled, and no weeding was done after planting. Twelve years after planting, early-successional tree species were well established, with stable populations, and the plots had a high level of plant biodiversity.

## 5.3.5 Framework Species Method (Close-Spaced Planting of Limited Numbers of Species)

This method is similar to the Miyawaki method in that both pioneer and climax tree species are planted together, and that it is suitable for seriously degraded environments with little or no remaining vegetation (Goosem and Tucker 1995; Elliott *et al.* 1998; Shono *et al.* 2007). The difference is that a selected group of species native to the restored site is planted rather than the maximum number.

In this method, the new forest is created with several dozen fast-growing trees that form a "framework" within which successional processes can operate. A mixture of both pioneer and climax tree species is planted in a single step at close spacing (>1000 seedlings/ha). Planted framework trees rapidly reestablish a multilayered closed forest canopy, eradicate weeds, and restore forest productivity and nutrient cycles. The plots of planted framework trees can act as a nucleus for forest restoration in a larger degraded landscape by attracting frugivorous birds and animals. The wildlife attracted disperses the seeds produced by the planted framework species into surrounding degraded areas, and also brings seeds of other (nonplanted) trees into planted areas. The forest created has improved soil and weed-free conditions, which favors germination and establishment of seedlings.

Essential characteristics of framework species are: (1) high survival and rapid growth when planted at degraded sites; (2) wide dense crowns that shade out competing weeds; (3) provision of resources that attract seed-dispersing wildlife (e.g., fruits, nectar, nesting sites, etc.) at a young age; and (4) reliable seed availability, and rapid and synchronous germination and growth of seedlings in containers to a plantable size in less than 1 year (Goosem and Tucker 1995; Elliott *et al.* 1998). For regions with frequently occurring fires, an ability to resist or recover after fire is also an important functional trait.

This method appears to work well when fragments of intact forest remain within about 10 km of restoration sites (Jalonen and Elliott 2014). Selecting candidate framework species requires extensive background studies including germination trials, monitoring of early seedling growth in a nursery, and observations of flowering and fruit production. Choosing 57 framework species for a trial plot in Doi Suthep-Pui National Park, Thailand, required testing 400 indigenous tree species for germination and growth, and more than 100 for flowering and fruit production (Elliott *et al.* 2003). After propagation in a nursery, seedlings are planted at an average spacing of 1.8 × 1.8 m (approximately 3100 trees/ha) and tended for at least 2 years after planting by regular weeding and other management practices if necessary (e.g., fertilizer application, protection from wildlife) (Jalonen and Elliott 2014). If biodiversity recovery is not evident 4–5 years after tree planting subsequent enrichment planting is recommended (Forest Restoration Research Unit 2006).

An advantage of this method is that in a degraded environment it allows the restoration, within a short time period, of the initial conditions needed for succession to start. However, successful recovery of the formerly degraded area into a forest with conservation value will depend to a large extent on the availability of particular species that can be dispersed into the site from the nearby forest. Tucker and Murphy (1997) reported that framework species plots established to restore moist tropical forest in Queensland, Australia, became colonized by 15–49 tree species within 5–7 years of planting. The majority of species recorded in the plots were fleshy-fruited early- and intermediate-successional taxa dispersed by birds.

Recently, this method has been tested as a tool to accelerate the forest recovery process within protected areas in the seasonally dry evergreen mountain tropical forest of northern Thailand. Because these protected areas represented sites formerly cleared by logging, they were subjected to intensive intervention via the planting of indigenous forest tree species to complement natural forest regeneration with the goal to restore the original forest ecosystems. According to Jalonen and Elliott (2014), in a trial plot in Doi Suthep-Pui National Park, with 57 planted framework tree species, the total tree species on the site 9 years after planting amounted to 130 species, equivalent to 85% of the total tree flora expected in an intact forest in a similar area under the same conditions. Most of the nonplanted tree species originated from seeds dispersed from nearby forest by birds, fruit bats, and civets. The bird species richness also increased from about 30 species before planting to 88 after 6 years, which is about 54% of bird species in nearby intact forest.

### 5.3.6   *Mixed-Plantation Method (Spatially Organized Planting)*

The publication of Ashton *et al.* (2001) was apparently the first to propose a formal scheme of spatially organized planting of pioneer and late-successional species. A spatial configuration was designed to conform to species differences in shade intolerance, self-thinning, and crown architecture. In the spatial arrangement of species, more shade-intolerant species surround less intolerant species, pioneers surround late-successional species, and the predetermined intra- and inter-specific spacing among trees is compatible with their known crown morphologies (Figure 5.14). Such plantation goes through three phases of stand development termed by Ashton *et al.* (2001) nurse, training, and mature phases (Figure 5.14). The nurse or stand initiation phase is the period during which the site conditions are improved by planted nurse trees to accommodate the survival and growth of the other planted species. Nurse trees are fast growing with umbrella-shaped crowns and are pioneer stem initiation species. They are widely planted to provide shade to the other species. Subsequent planting is done with late-successional, slow-growing, canopy-dominant species within the matrix of faster-growing pioneers. When the stem exclusion pioneers overtop the nurse trees, the plantation can be considered to be in the training or stem exclusion phase. Pioneers with small compact crowns dominate the canopy during this phase, but, once overtopped, readily self-thin. Thus, pioneers with expansive but shallow-layered crowns provide stem training for the shade-tolerant late-successional species. The moment when the late-successional canopy trees begin to overtop the pioneers of stem exclusion marks the start of the mature or under-story initiation phase (Figure 5.14).

The schemes proposed by Nave and Rodrigues (2007) can be considered as further development of this general approach. In these schemes species that differ in growth, shade tolerance, and other attributes, are planted in differently arranged "filling" and "diversity" planting lines (Figure 5.15). In the "filling" line, 15–30 fast-growing pioneer species are planted to ameliorate soil and/or suppress weeds by shading. The individuals planted in "filling" lines create an environment favorable for the development of individuals in the "diversity" lines and unfavorable for the colonization of the area by competing grasses, lianas, ferns, etc. The "diversity" line is composed of a larger number of tree species in comparison with the "filling" line, which comprises functionally important groups without fast-growing or good-cover characteristics. The maximum species diversity is used in each line. The method of filling diversity lines seems to have very good prospects for conservation-oriented restoration because it (1) allows rapid ground coverage; (2) uses a diversity of ecological

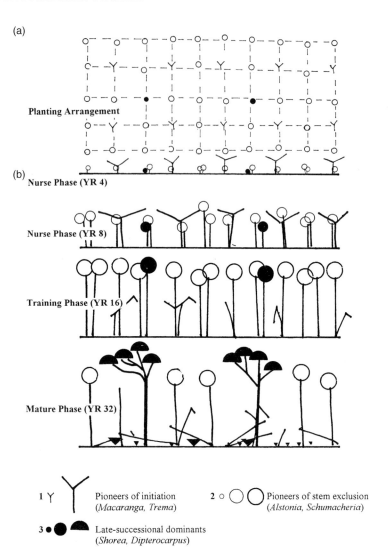

**1 Y** Pioneers of initiation
(*Macaranga, Trema*)

**2 ◯ ◯** Pioneers of stem exclusion
(*Alstonia, Schumacheria*)

**3 ● ● ▲** Late-successional dominants
(*Shorea, Dipterocarpus*)

FIGURE 5.14 Mixed-plantation forest restoration method applied to Sri Lankan abandoned agricultural land (from Ashton *et al.* 2001). (a) Initial planting arrangements between initiation pioneers and stem exclusion pioneers within which canopy tree dominants of late succession are planted. (b) The profiles of subsequent stages of forest development over time.

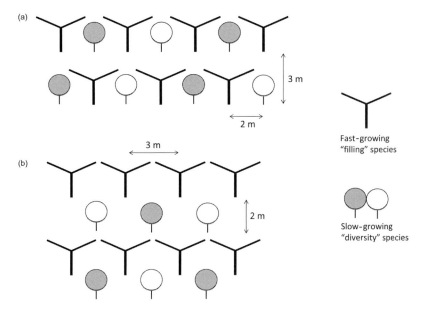

FIGURE 5.15 Planting design using (a) alternating planting groups in the same line and (b) alternating lines of the planting groups.

groups thus creating and maintaining structural and compositional complexity in the restored forest; and (3) uses as many species as possible in planting but with only a few individuals from each species, which is an essential feature of the mature forest.

## 5.3.7   Nucleation Method (Scattered Planting)

The idea of planting tree species in patches or islands for forest restoration is rapidly gaining popularity (Corbin and Holl 2012; Peterson *et al.* 2014; Piiroinen *et al.* 2015; Bechara *et al.* 2016). This idea stems from the nucleation theory of succession (Yarranton and Morrison 1974), a natural recovery process where patchily establishing woody pioneers facilitates the recruitment of nuclei for other species, expanding, and coalescing over time. The "woodland islets" (Rey Benayas *et al.* 2008) and "nucleation" (Cole *et al.* 2010) strategies, based on this idea, were proposed as low-cost

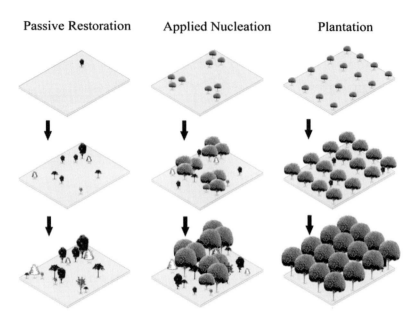

FIGURE 5.16 Dynamics of forest regeneration under three common strategies to restore forest cover: passive restoration, applied nucleation, and plantation (from Corbin and Holl 2012). Through time (from top to bottom), tree cover expands via tree growth and colonization. The resulting forest cover is lowest under passive restoration and highest under plantation, while the species diversity exhibits the opposite trend. Applied nucleation results in greater forest cover compared to passive restoration, and lower cover but a more diverse community compared to the plantation.

alternatives to continuous plantation-style restorations (reviewed in Corbin and Holl 2012). Accumulating evidence suggests that such scattered planting enhances seed rain and accelerates natural regeneration. It does so to a similar degree as plantation-style restoration (Cole *et al.* 2010; Zahawi *et al.* 2013; Holl *et al.* 2017), but produces less homogeneous habitat conditions (Holl *et al.* 2013) (Figure 5.16). Establishment of planted material is often followed by the arrival of birds and other animals that may introduce and disperse seeds of species for which they are the dispersal vectors (Cole *et al.* 2010; Corbin and Holl 2012; de la Pena-Domene *et al.* 2014), and dispersal and establishment of new recruits within and outside the nuclei (Albornoz *et al.* 2013; Zahawi *et al.* 2013; Peterson *et al.* 2014; Piiroinen *et al.* 2015). Thus, nucleation planting can improve the recruitment of resident populations by attracting the pollinators and seed dispersers needed to restore regeneration and facilitate the recruitment of other species through enhanced seed dispersal and improved establishment conditions. Gradual spatial expansion of planted nuclei should lead to rich and highly heterogeneous species composition across the contiguous forest landscape (Corbin and Holl 2012). The minimum critical island size below which nucleation planting does not enhance seed rain and tree recruitment appears to be about 100 m$^2$ (Zahawi and Augspurger 2006; Zahawi *et al.* 2013). The appropriate distance separating seedlings in planted islands, from the available literature, appears to be more than 1 m but less than 3 m (Rey Benayas *et al.* 2008; Holl *et al.* 2011; Zahawi *et al.* 2013).

It is not only planted tree islands or remnant trees embedded in a deforested or degraded landscape that can serve as nuclei (Guevara *et al.* 1986; Toh *et al.* 1999; Slocum 2001; Carriere *et al.* 2002a, b; Schlawin and Zahawi 2008; Holl *et al.* 2011), the small remaining populations of endangered species in less degraded habitats can also act as nuclei. Supplementation of the latter nuclei with planted conspecifics can boost regeneration and lead to population growth and expansion. Alternatively, the planted clusters of endangered woody species can serve as nuclei in sites where the species does not have extant populations.

### 5.3.8   *Group or Cluster Planting Method (Spatially Organized Scattered Planting)*

Somewhat similar to the nucleation silvicultural method is group or cluster planting, developed in commercial forestry for the artificial regeneration of oaks (Saha *et al.* 2017 and references therein). In this method, the target tree species (oaks) are surrounded by trainer trees (e.g., *Carpinus betulus*, *Tilia cordata*, and *Fagus sylvatica*; Saha *et al.* 2014) that are likely to facilitate their growth and protect them from

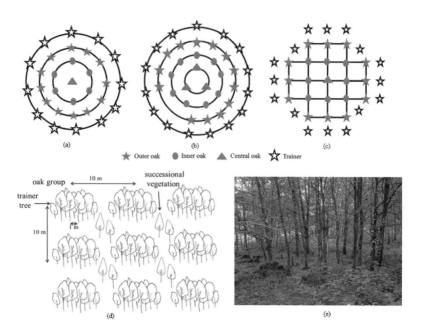

★ Outer oak   ● Inner oak   ▲ Central oak   ☆ Trainer

FIGURE 5.17 Group (or cluster) planting design by Gockel (1995) with three variants (a, b, c); spatial configuration of planted groups (d); and 20-year-old group planting in Lerchenfeld-Schwarzenborn, Hesse, Germany (e) (from Saha *et al.* 2017).

browsing at the early stage of development. The clusters are spaced at a distance apart that represents the optimal density of mature target trees (Figure 5.17). The current practice in southwestern Germany, for example, is to plant 40 to 70 clusters/ha (Saha *et al.* 2017). The spaces left unplanted between clusters are subsequently occupied by naturally regenerating early-, mid-, and late-successional woody plants, and the diversity, density, and basal area of naturally regenerated trees all increase with the size of unplanted area between the clusters (Saha *et al.* 2013). Cluster planting especially suits areas embedded in a matrix of existing forests due to availability of propagules for colonization, but, like the nucleation method, can also be used for reforestation of degraded lands with little remaining vegetation nearby and therefore fewer tree propagules. The "woodland islets" design resembling the low-density cluster planting design was successfully used for afforestation of degraded land in Spain with *Quercus ilex* (Rey Benayas *et al.* 2008).

Originally developed for oaks, cluster planting can be equally suitable for regeneration of other broad-leaved and conifer species, and this method has been successfully trialed in several forest restoration projects, for example, the restoration of threatened *Pilgerodendron uviferum* in Chile (Bannister 2015) and the afforestation of Swiss high-elevation mountains, using various coniferous species (Schönenberger 2001).

## 5.3.9   Level of Habitat Degradation and the Appropriate Restoration Method

Elliott *et al.* (2013) categorized restoration sites by their degree of degradation and described five levels of degradation, each of which requires a different restoration strategy. The categories can be distinguished by several critical "thresholds" of degradation, both internal (within the local site) (1–3) and external (within the surrounding landscape) (4–6):

1. The density of trees is so low that herbaceous weeds suppress tree seedling establishment.
2. Sources of forest regeneration (the remnant trees, seed, or seedling bank, live snags and stumps) are too insignificant for reestablishment of climax forest tree species.
3. Soil is so degraded that poor soil conditions limit the establishment of tree seedlings.
4. The remaining patches of climax forest in the landscape are too small and too far from the restoration site to provide the seed flow of the climax forest species.
5. Populations of seed-dispersing animals in the landscape are so diminished that seeds are no longer transported in sufficiently high densities to the restoration site to reestablish the climax forest species.
6. The cover of combustible herbaceous weeds in the landscape has achieved a level at which survival of established trees is unlikely due to fire risk.

Recognizing which of the five stages of degradation a site has reached will determine which restoration method is most appropriate

(Table 5.3). A rapid site assessment with clear quantitative criteria to determine the site degradation level was proposed by Elliott *et al.* (2013). The assessment can be done using circular sample plots of 5 m radius, and recording the number of (1) trees >10 cm DBH and >1.3 m height, (2) saplings taller than 50 cm but having <10 cm DBH, and (3) live tree stumps/snags within the circle. The number of "regenerants" per circle is the total number of trees in all three of these categories (Elliott *et al.* 2013). Information about the number of regenerants and the number of species they represent, together with a rough estimation of soil condition, the percentage cover of exposed soil, grasses, and herbaceous weeds, and the tree seedling density across the plot can be used for the rapid determination of the site's degradation stage (Table 5.5 and Figure 5.18).

For all site degradation levels, the goals of the initial restoration activities are (1) to achieve a sufficiently high density of regenerants (i.e., ≥3100/ha) and tree species richness (i.e., ≥10% of the tree species characteristic of the target climax forest or ≥30 tree species if the species richness of the climax forest is unknown); and (2) to remove factors preventing forest regeneration (e.g., fire, grazing, hunting of seed dispersers). The density of 3100 regenerants per hectare (which corresponds to an average tree spacing of 1.8 m apart) is necessary to shade out the weeds. Canopy closure with this planting density can be achieved in the majority of tropical forests within 2–3 years of planting.

Stage 2 of degradation differs from stage 1 mainly in grass and herb cover, and therefore restoration of stage 2 degradation sites requires canopy closure acceleration measures. If ANR in a form of weeding, fertilizer application, and/or mulching is not efficient because of a poor seed/seedling bank or limited influx of seeds from the nearby forest remnants, then enrichment planting is necessary. Under stage 3 degradation conditions, prevention of fire is important due to the abundance of highly flammable herbaceous vegetation. Because of the low density of natural regenerants at this degradation stage, protection and ANR must be complemented with the planting of framework tree species. The framework tree species should

Table 5.5 *Simplified guide to choosing a restoration strategy (from Elliott et al. 2013)*

| Landscape critical thresholds | | | Suggested restoration | Site-critical thresholds | | |
|---|---|---|---|---|---|---|
| Forest in landscape | Seed dispersal mechanisms | Fire risk | strategy | Vegetation cover | Natural regenerants | Soil |
| Remnant forest remains within a few km of the restoration site | Mostly intact, limiting the recovery of tree species richness | Low to medium | Protection | Tree canopy cover exceeds herbaceous weed cover | Natural regenerants exceed 3100/ha with more than 30* | Soil does not limit tree seedling establishment |
| | | Medium to high | Protection + assisted natural regeneration (ANR) | Tree crown cover insufficient to shade out herbaceous weeds | common tree species represented | |

(continued)

Table 5.5 (cont.)

| Landscape critical thresholds | | | Suggested restoration | Site-critical thresholds | | |
|---|---|---|---|---|---|---|
| Forest in landscape | Seed dispersal mechanisms | Fire risk | strategy | Vegetation cover | Natural regenerants | Soil |
| Remnant forest patches very sparse or absent from the surrounding landscape | Seed-dispersing animals rare or absent such that the recruitment of tree species to the restoration site will be limited | High | Protection + ANR + planting framework tree species | Herbaceous weed cover greatly exceeds tree crown cover | Natural regenerants sparser than 3100/ha with fewer than 30* common tree species represented | |
| | | | Protection + ANR + maximum diversity tree planting | | | |
| | | Initially low (soil conditions limit plant growth); higher as the vegetation recovers | Soil amelioration + nurse tree plantation, followed by thinning and gradual replacement of maximum diversity tree planting | Herbaceous weed cover limited by poor soil conditions | | Soil degradation limits tree seedling establishment |

* Or roughly 10% of the estimated number of tree species in the target forest, if known.

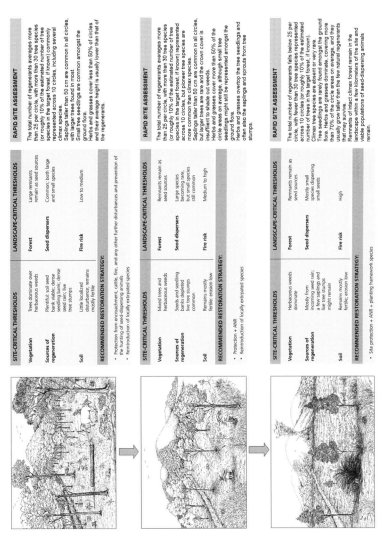

**SITE-CRITICAL THRESHOLDS**

| | |
|---|---|
| Vegetation | Trees dominate over herbaceous weeds |
| Sources of regeneration | Plentiful soil seed bank viable; dense seedling bank; dense seed rain; live tree stumps |
| Soil | Little localized disturbance; remains mostly fertile |

**LANDSCAPE-CRITICAL THRESHOLDS**

| | |
|---|---|
| Forest | Large remnants remain as seed sources |
| Seed dispersers | Common; both large and small species |
| Fire risk | Low to medium |

**RAPID SITE ASSESSMENT**

The total number of regenerants averages more than 25 per circle, with more than 30 tree species (or roughly 10% of the estimated number of tree species in the target forest, if known) commonly represented across 10 circles, including several climax species.
Saplings taller than 50 cm are common in all circles, with larger trees found in most.
Small tree seedlings are common amongst the ground flora.
Herbs and grasses cover less than 50% of circles and their average height is usually lower than that of the regenerants.

**RECOMMENDED RESTORATION STRATEGY:**

• Protection from encroachment, cattle, fire, and any other further disturbances and prevention of the hunting of seed-dispersing animals
• Reintroduction of locally extirpated species

---

**SITE-CRITICAL THRESHOLDS**

| | |
|---|---|
| Vegetation | Mixed trees and herbaceous weeds |
| Sources of regeneration | Seeds and seedling banks depleted; live tree stumps common |
| Soil | Remains mostly fertile; erosion low |

**LANDSCAPE-CRITICAL THRESHOLDS**

| | |
|---|---|
| Forest | Remnants remain as seed sources |
| Seed dispersers | Large species becoming rare, but small species still common |
| Fire risk | Medium to high |

**RAPID SITE ASSESSMENT**

The total number of regenerants averages more than 25 per circle, with more than 30 tree species (or roughly 10% of the estimated number of tree species in the target forest, if known) represented across 10 circles, but pioneer tree species are more common than climax species.
Saplings taller than 50 cm are common in all circles, but larger trees are rare and the crown cover is insufficient to shade out weeds.
Herbs and grasses cover more than 50% of the circle areas on average, although small tree seedlings might still be represented amongst the ground flora.
Herbs and grasses overtop the tree seedlings and often also the saplings and sprouts from tree stumps.

**RECOMMENDED RESTORATION STRATEGY:**

• Protection + ANR
• Reintroduction of locally extirpated species

---

**SITE-CRITICAL THRESHOLDS**

| | |
|---|---|
| Vegetation | Herbaceous weeds dominate |
| Sources of regeneration | Mostly from incoming seed rain; a few saplings and live tree stumps might remain |
| Soil | Remains mostly fertile; erosion low |

**LANDSCAPE-CRITICAL THRESHOLDS**

| | |
|---|---|
| Forest | Remnants remain as seed sources |
| Seed dispersers | Mostly small species dispersing small seeds |
| Fire risk | High |

**RAPID SITE ASSESSMENT**

The total number of regenerants falls below 25 per circle, with fewer than 30 tree species represented across 10 circles (or roughly 10% of the estimated number of trees in the target forest, if known).
Climax tree species are absent or very rare.
Tree seedlings are rarely found amongst the ground flora. Herbs and grasses dominate, covering more than 70% of the circle areas on average, and they usually grow taller than the few natural regenerants that may survive.
Remnants of intact climax forest remain in the landscape within a few kilometers of the site and viable populations of seed-dispersing animals remain.

**RECOMMENDED RESTORATION STRATEGY:**

• Site protection + ANR + planting framework species

FIGURE 5.18 Five levels of degradation in restoration areas, their critical thresholds, characteristics in site assessments, and the recommended restoration strategy (adapted from Elliott *et al.* 2013). Drawings by Surat Plukam.

| SITE-CRITICAL THRESHOLDS | | LANDSCAPE-CRITICAL THRESHOLDS | | RAPID SITE ASSESSMENT |
|---|---|---|---|---|
| Vegetation | Herbaceous weeds dominate | Forest | Remnants too few or too distant to disperse tree seeds to site | The conditions recorded during the site assessment are similar to those of stage 3 degradation, but at the landscape level, intact forest no longer remains within 10 km of the site and/or seed-dispersing animals have become so rare that they are no longer able to bring the seeds of climax tree species into the site in sufficient quantities. Recolonization of the site by the vast majority of tree species is therefore impossible by natural means. |
| Sources of regeneration | Low | Seed dispersers | Mostly gone | |
| Soil | Erosion risk increasing | Fire risk | High | |

**RECOMMENDED RESTORATION STRATEGY:**

• Site protection + ANR + planting framework species + enrichment planting with climax species
• Maximum diversity methods (Goosem and Tucker, 1995) such as the Miyawaki method

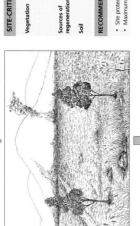

| SITE-CRITICAL THRESHOLDS | | LANDSCAPE-CRITICAL THRESHOLDS | | RAPID SITE ASSESSMENT |
|---|---|---|---|---|
| Vegetation | No tree cover. Poor soil might limit growth of herbaceous weeds | Forest | Usually absent within seed dispersal distances of the site | The total number of regenerants falls below 2 per circle (average spacing between regenerants > 6–7 m), with fewer than 3 tree species (or roughly 1% of the estimated number of tree species in the target forest, if known) represented across 10 circles. Climax tree species are absent. Bare earth is exposed over more than 30% of the circle areas on average and the soil is often compacted. Local people regard the soil conditions as exceedingly poor, and signs of erosion are recorded during the site assessment. Erosion gullies can be present, along with siltation of watercourses. The ground flora is limited by the poor soil conditions to less than 70% cover on average and is devoid of tree seedlings. |
| Sources of regeneration | Very few or none | Seed dispersers | Mostly gone | |
| Soil | Poor soil conditions limit tree establishment | Fire risk | Initially low (soil conditions limit plant growth); higher as the vegetation recovers | |

**RECOMMENDED RESTORATION STRATEGY:**

• Soil improvement by planting green mulches and the addition of compost, fertilizers, or soil micro-organisms
• ... followed by planting "nurse trees" — i.e. hardy nitrogen-fixing trees that will further improve the soil (also known as the "plantations as catalysts" method (Parrotta 2000))
• ... and then thinning of nurse trees and their gradual replacement by planting a wide range of native forest tree species

FIGURE 5.18 (cont.)

include both pioneer and climax species ideally different from those recorded during the site assessment. Under stages 4 and 5 degradation, extremely poor sources of natural regeneration (both internal and external) require the introduction of a large number of absent species, which is technically challenging and costly. In addition, degraded soil may require improvement by plowing, adding compost, green mulching, or planting "nurse" trees capable of improving the soil such as nitrogen-fixing trees of the family Leguminosae. Due to the very high costs involved, especially in comparison with the modest expected outcome in terms of the forest conservation value, the sites with stages 4 and 5 degradation are rather useless for conservation-oriented restoration.

### 5.3.10   Planting Designs in the Landscape Context

Until recently a target of restoration was "a characteristic assemblage of the species that occur in the reference ecosystem and that provide appropriate community structure" (SER 2004) but this definition seems to have been replaced in the latest literature by another one emphasizing a transition from a local to landscape scale in restoration planning – "a process that aims to regain ecological integrity and enhance human well-being in deforested or degraded landscapes" (Maginnis and Jackson 2007). Leaving aside the enhancement of human well-being, which is not a target for conservation-oriented restoration, this change in an operational definition of restoration goals is symptomatic. Because landscapes nowadays are mosaics of sites that vary greatly in size and configuration, land use, or degradation, restoration projects need to be planned at large spatial scales and based on an understanding that the areas available for restoration are not necessarily those parts of the landscape that are most important to conserve biodiversity (Lamb et al. 2012) and, in those available, the barriers to forest recovery are both site specific and context dependent (Tambosi et al. 2014). Such projects must consider the natural resilience of a given site, intrinsic and extrinsic barriers to its recovery, and human activities in the surrounding areas.

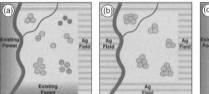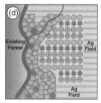

FIGURE 5.19 Different planting designs in the landscape context (from Stanturf *et al.* 2014). Scattered (nucleation) plantings involve planting either small clumps of late-successional tree species of limited dispersal ability near existing forests (a) or pioneer nurse tree species within an agricultural matrix (b) to promote natural colonization and succession. An analogous role can be played by small scattered plantations of species mixtures (c). The above two methods (nucleation plantings and mixed-species plantations) can be combined: supporting natural regeneration scattered planting in areas adjacent to remaining upland or riparian forests and within an effective seed dispersal distance, and extensive plantations elsewhere (d).

Based on these principles, conservation-oriented restoration planning can identify the target areas most appropriate to complement conservation efforts in protected areas. These restored sites with enhanced habitat quality will connect the protected areas and act as buffer zones between them and managed forest areas.

As for the choice of a suitable restoration method, natural regeneration should be practiced in areas that have a high capacity for natural colonization, with areas that have a low capacity being assigned to active restoration involving plantings (Murcia 1997; Elliot *et al.* 2013). Chazdon (2017) suggested natural regeneration as a preferred option for the buffer zones of protected areas, large forest fragments, riparian zones, and steep slopes, given that seed sources for colonization are available in the adjacent areas. Low-intensity active restoration, such as nucleation or group planting, can be a vital option for clear-cut forests or abandoned field areas close to the remaining forest fragments, and higher-intensity planting designs can be applied in areas far from seed sources (Figure 5.19).

# 6    Conservation-Oriented Restoration of Particular Systems

6.1 INCORPORATION OF RESTORATION OF PARTICULAR SYSTEMS INTO REGIONAL CONSERVATION PLANNING

The focus of this book is the restoration of damaged ecosystems that have considerable conservation value or potential, but not all degraded or destroyed ecosystems that are in need of restoration will be targeted or even considered as candidates. Some of the preconditions for a particular area to be included in a list of targets for conservation-oriented restoration are: full protection, at least relatively intact soil, and a lack of highly persistent exotic species.

Some areas can be *a priori* excluded from consideration, e.g., those highly devastated by mining, severely eroded by deforestation, or colonized by invasive species that are impossible to eradicate. These areas should be dealt with using traditional ecological restoration. However, less severely degraded areas, including those that underwent clear-cutting and agricultural use for certain periods of time, can be targeted for conservation-oriented restoration.

In many countries, previously heavily logged or poorly managed agricultural lands eventually become economically unprofitable and are abandoned. Because they have no – or very little – economic value, such areas are the first to be designated by local governments as protected "natural" areas. Although these areas, in reality, have very little in common with what the word "nature" means for conservation biologists, nevertheless they can be turned into something useful for conservation. For example, the abandoned pastures, croplands, or degraded forest plantations, after being incorporated into existing nature reserves and national parks, can become the buffer zones. This will decrease the detrimental edge effect and anthropogenic pressure of the outside human population. Another important role these areas can serve is as corridors allowing migration of fauna and dispersal of propagules. From a long-term perspective, if properly restored and managed, these areas can closely approach the once existing pre-disturbance habitats. Areas of different levels of degradation constitute a significant proportion of PAs; for example in Mexico 6% of the identified priority restoration sites were in PAs (Tobon *et al.* 2017).

This perception is gaining popularity and, in Europe, the current habitat conservation policies (e.g., European Union's Habitats Directive 92/43/CEE) state that management in abandoned and degraded habitats should aim to reestablish the structure and function of native plant communities (SER 2002).

Thus, degraded habitats can be targets for conservation-oriented restoration and, to maximize the conservation value of these areas, they should be a part of a bioregional planning approach, involving the protection of landscapes composed of archipelagos of fragments, connected by corridors of original or restored vegetation (da Silva and Tabarelli 2000; Tobon *et al.* 2017).

## 6.2 RESTORATION OF SELECTIVELY LOGGED FORESTS

There is no doubt that primary forests, i.e., those with no record of clearing, commercial timber harvesting, or other large-scale anthropogenic disturbance in recent centuries, have the highest conservation value and must be the number one target in conservation planning. Unfortunately, we are witnessing the rapid disappearance of these worldwide, and especially in the tropics and subtropics. One of the major causes of their disappearance is logging. Due to human's insatiable appetite for wood fiber from 3.5 billion to 7 billion trees are cut down each year either through clear-cutting or selective logging.

The drivers and impacts of logging, especially in the tropical forests are well known. Logging is no exception to the common economic model of natural resources utilization practiced by humans whenever possible (another example is the world's ocean fishery). The most valuable species are selectively harvested first, then, upon their depletion the next-most-valuable species are taken, and so on until no more timber is left and the only uses for the land devoid of forest vegetation are agriculture or ranching (Asner *et al.* 2006; Laporte *et al.* 2007; Hall 2008; Schulze *et al.* 2008; Zimmerman and Kormos 2012). One-third of the remaining tropical forest has already been degraded by selective harvesting of high-value timber, harvesting which is spreading rampantly through the remainder (Zimmerman and Kormos 2012 and references therein), leaving virtually no hope that any wilderness forest patch of the world will remain untouched. Because of rising market demands, the pursuit of high-value timber will lead to further expansion into remaining unlogged forests of Latin America, Central Africa, and East Asia (Laporte *et al.* 2007; Hall 2008; Asner *et al.* 2009) with no or little incentive for sustainable forest management or conservation (Zimmerman and Kormos 2012). Witnessing this relentless extraction of timber through selective logging, we should nevertheless realize that forest clear-cutting, which eliminates about 1 million $km^2$ every 5–10 years (Pimm and Raven 2000), has even worse consequences for biodiversity. Thus, although selective logging, defined as felling of selected trees based on species preference and/or cut-size limitations, has numerous negative impacts on forest structure, function, and species richness, as discussed below, these impacts are not comparable with the effects of clear-cutting (Zeng *et al.* 2007; Zimmerman and Kormos 2012; Xu *et al.* 2015) (Table 6.1); the conversion of land to agriculture has a stronger impact still. Therefore, among anthropogenically impacted systems, previously selectively logged forests have the highest conservation value with respect to different biodiversity measures (Gardner *et al.* 2010; Gibson *et al.* 2011; Edwards and Laurance 2013; Mori and Kitagawa 2014), and protection and restoration of these forests is likely to offer the greatest conservation benefits, especially when the logged forests are within national parks and nature reserves (Cannon *et al.* 1998; Berry *et al.* 2010; Gibson *et al.* 2011; Baraloto *et al.* 2012; Edwards and Laurance 2013; Wilcove *et al.* 2013; Crouzeilles *et al.*

Table 6.1 *Structural characteristics of forest ecosystems as a function of age and ecological legacies (complete clearance, selective logging, and pristine conditions) (from Clark 1996 with changes)*

| Characteristic | Young secondary forest | Old secondary forest | Selectively logged forest | Old-growth forest |
|---|---|---|---|---|
| Stand basal area | Lowest | Intermediate | Intermediate | Highest |
| Distribution of tree stem diameters | Lowest coefficient of variation | Intermediate coefficient of variation | Intermediate coefficient of variation | Highest coefficient of variation |
| Canopy structure | Even canopy, few gaps | Even canopy, gaps more frequent | Variable canopy height, small and large gaps common | Variable canopy height, small gaps common but large gaps infrequent |
| Large lianas and/or epiphytes | Absent | Rare | Variable depending on history | Common |
| Large logs | Present or absent | Usually absent | Usually present | Always present |
| Total quantity of coarse woody debris | Usually small | Small | Large | Large |
| Very large trees | Usually absent except as obvious remnants | Usually absent | Usually absent or scarce | Always present |
| Number of tree species with large animal-dispersed seeds with little or no dormancy | Few | Some | Many | Many |

2016). In protected areas, restored secondary forests will increase the percentage of high-quality habitat and connectivity among the remaining patches of pristine forest.

Superficially, selective logging is akin to naturally occurring treefalls creating small canopy gaps that constitute about 1–2% of the forest area (Denslow 1987). However, in natural treefalls elimination of trees occurs randomly, while in selective logging either particular species or stem diameters are deliberately targeted. Second, natural and felling gaps differ in disturbance conditions to the understory and soil. After felling, the logs are attached to winch cables and dragged from the gap by skidders, damaging understory vegetation and disturbing the litter and soil. This affects large areas, resulting in increased soil bulk density, decreased porosity, and loss of soil organic matter, which can strongly inhibit seedling establishment and growth (Guariguata and Ostertag 2001). In contrast, in natural gaps, disturbance to the soil and surrounding vegetation is primarily caused by uprooting, is very localized, and usually stimulates seed germination and seedling growth. Third, logging operations create much more fragmented canopy structure with larger and more numerous patches of felling gaps. Recovery of natural treefalls is highly predictable and quick because the gaps are usually small and saplings and seedlings from the understory shoot up in response to the improved light conditions, while in logged forests regeneration is longer and can take more than one pathway (Chapman and Chapman 1997; Brown and Gurevitch 2004; Piiroinen *et al.* 2017). In natural gaps, due to smaller gap sizes and less understory and substrate disturbance, advanced growth of saplings and large seedlings of shade-tolerant species is favored, while in felling gaps, larger canopy openings and higher levels of disturbance favor colonization by shade-intolerant species from seeds or small seedlings (Dickinson *et al.* 2000; Turner 2001).

Clearly, damage to the residual forest and soil impacts caused by skid trails and log landings all increase with increasing logging intensity, measured by the number of stems or cubic meters extracted per unit of area, but even moderate intensity logging can have very negative consequences. In some systems, moderate intensity logging can cause impeded or arrested succession due to the rapid growth of outcompeting tree seedlings, shrubs and herbs, and the grazing and trampling of seedlings by large ungulates (Chapman and Chapman 1997; Babaasa *et al.* 2004; Paul *et al.* 2004; Bonnell *et al.* 2011; Piiroinen *et al.* 2017). In others, arrested succession can result from initial colonization of gaps by invasive species that are able to maintain long-term viable populations after colonization and prevent recolonization by native species (Brown and Gurevitch 2004), or from rapid growth of vines, when instead of returning to its original tall forest state, logged forest might develop into low, impenetrable vine tangles (Uhl and Vieira 1989). *Psidium cattleianum* is an example of an invasive tree species that, after initial colonization of logged tropical forests, alters the forest recovery trajectory dramatically. This aggressive invader of tropical areas prevents the establishment of other plant species (Huenneke and Vitousek 1990) and predominates in stands that were clear-cut even 150 years ago, whereas it is absent in never-cut stands (Brown and Gurevitch 2004).

Alternatively, after canopy closure, the logged forest may have a brighter forest floor light environment in comparison with the primary forest, due to dominance in the canopy of trees of moderate size with smaller, less heterogeneous (monotonic) crowns growing at a high density (Okuda *et al.* 2003). This secondary forest will not evolve toward the pre-logged state unless the complexity of canopy structure and the heterogeneity of the forest floor light environment that existed in the primary forest is recreated. At the landscape level, probably the most undesirable impact of logging of even low intensity is increased access to previously inaccessible locations, always followed by accelerated visitation and exploitation of the forest, its fragmentation, and degradation (Schulze *et al.* 2008; Laurance *et al.* 2009).

When the intensity of logging passes a forest resilience threshold (which can be as small as 5 stems/ha), the new forest conditions no longer support the growth of nonpioneer vegetation (Sist and Nguyen-The 2002). Excessive logging will lead to suppressed regeneration of late-successional trees due to (1) high mortality of residual trees including seed trees; (2) creation of favorable conditions for vines and pioneer vegetation (i.e., large canopy gaps) allowing them to easily outcompete the slower-growing, nonpioneer tree species; and (3) making residual stands more susceptible to fire through reduced tree density and the presence of large amounts of woody debris (Uhl and Kauffman 1990; Kasenene and Murphy 1991; Kuusipalo et al. 1997; Sist and Nguyen-The 2002; Schulze et al. 2008; Ashton and Hall 2011; Zimmerman and Kormos 2012). In addition, elimination of reproductive individuals of low-density species may prevent pollen and seed production leading to local extinction of species (Guariguata and Pinard 1998; Degen et al. 2006; Schwartz et al. 2017). Analysis of species-specific tree performance and population demography using 10-year demographic data in selectively logged Malaysian rainforest, 50 years after the last logging, revealed lower recruitment in comparison with the primary forest (Yamada et al. 2013). Similarly, in selectively logged Indonesian forest, 8 years after harvest, seedlings and saplings were less dense in logged than in unlogged sites, which was especially evident for dipterocarp species (Cannon et al. 1994).

At the community level, logging not only inevitably changes proportions of successional stages in forest stands in favor of early-successional species but also causes other structural changes of forest stands. Besides the obvious impact of logging on abundance and demographic structure (i.e., age/size distribution) of the harvested species, other structural effects of selective logging include homogenization of canopy structure, loss of large trees (with a shift toward small- and medium-sized trees in the residual stand), higher stem density in general, and higher densities of small size-class stems of late-successional species in particular (Okuda et al. 2003; Kariuki et al. 2006; Bonnell et al. 2011; Rutten et al. 2015). The genetic impacts include reductions in effective population size and interruptions of gene flow. Importantly, the selective effect of the deliberate removal of the largest and usually most successful genotypes in terms of growth, competition, and seed production negatively affects the genetic composition of the remaining population (Degen *et al.* 2006).

Selective logging has a strong negative impact on biodiversity and we should not be misled by equally high or even higher species richness for certain groups of organisms in post-logged forests, as compared with primeval forests. Disturbed areas often have higher richness than analogous undisturbed ones due to colonization by ruderal species, which have no chance of survival in undisturbed habitats. For example, a meta-analysis of the relationship between logging intensity and bird species richness revealed a positive correlation, but this is the result of an influx of habitat generalists into heavily logged areas, while forest specialist species in these areas always decline (Burivalova *et al.* 2014). Comparisons of species richness in disturbed versus undisturbed habitats should always take into account the functional roles of comparable species (e.g., Berry *et al.* 2010; Whitworth *et al.* 2016), and comparisons should be done in a functionally pairwise manner, without combining in the same species list, for example, late-successional trees with early-successional graminoids and herbs. But even without accounting for the contribution of ruderal and exotic species after logging, loss of biodiversity in regenerating secondary forests can be substantial. For example, Brown and Gurevitch (2004) found that loss of species richness due to logging was irreversible and still evident 150 years after logging in the tropical forests of Madagascar, and a meta-analysis found a significant negative effect on tree species richness following logging in both tropical and temperate forests (Clark and Covey 2012). Selective logging inevitably shifts the composition and relative abundance of species (Baraloto *et al.* 2012), and the least abundant and rare species, which are more prone to extinction after habitat disturbance and fragmentation (e.g., Laurance *et al.* 1999), are usually the most vulnerable to these logging impacts and disappear first.

The resilience threshold for most animal taxonomic groups can be as small as 10 m³/ha of extracted timber (Burivalova *et al.* 2014).

Mammals and amphibians were found to be particularly sensitive to logging intensity and to suffer halving of species richness at logging intensities of just 38 m³/ha and 63 m³/ha, respectively (Burivalova *et al.* 2014). When the forests are logged the large frugivorous animals (e.g., primates) suffer the most and their food resources usually show the slowest rate of recovery (Bonnell *et al.* 2011).

Summarizing the numerous impacts of selective logging presented above, the logged forests usually have a disturbed structure with an increased proportion of pioneer species and poorer floristic composition with respect to the old-growth specialists. When logging has ceased under full protection, formerly logged forests as a rule slowly but steadily approach the forest structure and floristic composition of undisturbed old-growth forests (Guariguata and Ostertag 2001; DeWalt *et al.* 2003; Liebsch *et al.* 2008; Bruelheide *et al.* 2011), but the naturally occurring secondary succession can take a very long time (Finegan 1996; Plumptre 1996; Manokaran 1998; Fashing *et al.* 2004; Liebsch *et al.* 2008; Bonnell *et al.* 2011; Yamada *et al.* 2013; Xu *et al.* 2015) with no guarantee that a regenerated forest will closely resemble the pre-logged one. Restoration interventions should be aimed at speeding up this process and changing the species composition and proportion of late-successional, rare, and threatened species in a desirable direction. Exact convergence of the secondary forest species composition with that of the nearest undisturbed forests is less important than a convergence in structure and functional group composition. Given the heterogeneity of the undisturbed primeval forests over the landscape, close convergence in species composition should be a goal only at the regional, and not the local, scale.

Considering the experience of commercial forestry in testing different silvicultural methods to improve the growth of timber species (which almost always are late-successional species with high conservation value), the major silvicultural methods for conservation-oriented restoration of logged forests should be liberation cutting (Grauel and Putz 2004; Pena-Claros *et al.* 2008; Villegas *et al.*

2009; Schwartz *et al.* 2013; Lussetti *et al.* 2016), selective thinning (Swinfield *et al.* 2016), and seeding or enrichment planting with or without tending, depending on the situation (Adjers *et al.* 1995; Park *et al.* 2005; Schulze 2008; Doucet *et al.* 2009; Schwartz *et al.* 2013, 2017; Piiroinen *et al.* 2017).

Liberation cutting should be applied, first of all, to over-abundant lianas and bamboos that have strong effects on local vegetation, impeding forest regeneration and reducing local diversity and abundance of woody plants and/or altering the vegetation structure. In protected areas, bamboos and lianas can become a real concern (e.g., Lima *et al.* 2012). Lianas and bamboos can become so abundant in gaps that they can completely arrest reversion of the latter back to tall forest. These liana- or bamboo-dominated gaps can remain at a low canopy height for decades or longer (Veblen *et al.* 1979; Taylor and Qin 1988; Schnitzer *et al.* 2000; Tymen *et al.* 2016). Lianas can suppress recruitment of old-growth tree species with no such effect on pioneer trees, thus facilitating regeneration of the latter at the expense of slow-growing shade-tolerant trees competing with them (Putz 1984; Clark and Clark 1990; Schnitzer *et al.* 2000; Laurance *et al.* 2001). Analysis of 18 years of post-logging data following selective logging in Indonesian dipterocarp forest revealed that pre-harvest climber cutting increased ingrowth as well as decreased the mortality of dipterocarp species (Lussetti *et al.* 2016).

In addition to removing lianas, bamboos, and shrubs (liberation cutting), low-intensity thinning of small stems suppressing the target tree species (selective thinning) can also be useful. In general, removal of large trees is undesirable in the restoration of logged forests, as it leads to localized increases in light intensity, which stimulate the growth and recruitment of fast-growing pioneers. However, elimination of small trees only slightly elevates irradiance enhancing the growth of late-successional, but not pioneer, species (Swinfield *et al.* 2016). Besides improving growth conditions for late-successional tree species, such low-impact tree thinning also increases habitat structural complexity thereby creating conditions for a richer animal community (De la Montana *et al.* 2006; Edwards *et al.* 2009; Kalies *et al.* 2010; Verschuyl *et al.* 2011; Blakey *et al.* 2016). In a study by Swinfield and colleagues, testing the efficacy of selective pioneer removal on the recovery of a selectively logged Sumatran tropical forest, the cutting of pioneer stems of <10 cm DBH from the subcanopy and understory accelerated the growth of both large (>10 cm DBH) and small (2–10 cm DBH) late-successional stems (Swinfield *et al.* 2016). In contrast, after removal of bamboo in large forest gaps, the shrubs, herbs, ferns, and graminoids were the first to colonize the gap with impeded seed germination and seedling establishment of tree species (Carnpanello *et al.* 2007). Thus, removal of competing vegetation must be localized to the very proximity of the target tree species.

Another important silvicultural method for the restoration of selectively logged forests, as noted by Zimmerman and Kormos (2012) in their review of logging in tropical regions, is enrichment planting. Analyzing the available literature (e.g., Kuusipalo *et al.* 1997; Park *et al.* 2005; Grogan *et al.* 2008; Hall 2008; Schulze 2008; Doucet *et al.* 2009), these authors concluded that for the restoration of logged forests, enrichment planting of seedlings and their subsequent tending in small (single-tree size) logging gaps mimicking natural conditions is the most appropriate method for primary forest tree regeneration (see also d'Oliveira 2000; Lopes *et al.* 2008; Schwartz *et al.* 2017). To ensure that planted seedlings grow well in logging gaps, competing pioneer vegetation must undergo weeding and liberation cutting regularly during, at least, the first few years (Hall 2008; Schulze 2008; Zimmerman and Kormos 2012; Schwartz *et al.* 2013).

While I completely agree with the idea of using enrichment planting of seedlings in small logging gaps, I would add that the seedlings must represent not necessarily the overexploited and now underrepresented timber species, but first of all species of conservation concern and those vital for pre-disturbance ecosystem functioning. In a 20-year study of the effect of selective logging on dipterocarp rainforest in Indonesia, endangered *Agathis borneensis* was found to be abundant in virgin forest plots in the size class of 40 cm DBH and above, but completely absent in both moderate- and low-intensity logged forest plots (Verburg and van Eijk-Bos 2003). The lack of this species even in the smallest diameter class 20 years after logging means that its regeneration most probably will not occur without intervention. Moreover, the observed lack of *A. borneensis* in gaps of any size left by selective logging suggests that selective thinning in either the canopy or understory will not help to reestablish this species, and the only viable option is underplanting. There are case studies showing the feasibility of underplanting threatened species in the logging gaps. Seedlings of *Amburana cearensis* and *Cedrela fissilis*, two endangered tree species of Brazilian deciduous forests, showed equally high survival rates and good growth when planted in intact, moderately, and heavily logged forest fragments (Guarino and Scariot 2012). Although forest gaps are generally considered to be favorable for the regeneration of the Central African tree *Baillonella toxisperma*, which is threatened by logging, under natural regeneration this species demonstrates very low seedling density even in the gaps. The attempted assisted introduction of this species into logging gaps was successful. Both seeding and planting resulted in high survival and growth of seedlings during the first 30 months, even without additional site maintenance (clearing of debris and removal of competing vegetation) (Doucet *et al.* 2009).

Very few studies compared the demography of rare/threatened species in logged versus unlogged forests. The demography of five rare tree species, of which two are currently endangered, was studied in the Xuan Son National Park of Northern Vietnam in a fully protected forest without signs of disturbance, as well as in forest areas where these species were subject to either commercial or domestic logging of unknown intensity for decades, and were subsequently protected for 9 years before the study. In four out of five species, the largest DBH classes were absent or less frequent in logged compared with unlogged forest, but the smallest DBH classes in all five species were found at high frequency indicating their good regeneration in previously logged areas (The Long and Hoelscher 2014). Thus, protection of the formerly logged secondary forest, even without additional measures, will allow recovery of rare species impacted by logging.

All six known populations of the extremely endangered tree *Craigia yunnanensis* underwent logging for domestic use in the past. In addition, their habitat has been severely disturbed by cultivation of a cash crop *Amomum tsao-ko* in the forest understory, accompanied by near-total removal of herbs and shrubs plus thinning of the canopy. Nevertheless *C. yunnanensis*, even in such disturbed habitat, shows vigorous demographic structure, with all life-cycle stages and size categories present and the size–frequency distribution approaching an inverse J-shaped curve (Gao *et al.* 2010).

Critically endangered *Aquilaria crassna*, intensively harvested for agarwood in Laos, despite extensive felling of large trees, exhibits active regeneration in the form of seedling, sapling, and young tree growth, as well as resprouting of damaged old trees (Jensen and Meilby 2012).

Many conifers currently threatened due to overharvesting or disappearance of their habitat (e.g., *Araucaria angustifolia*, *Araucaria laubenfelsii*, *Araucaria hunsteinii*, *Agathis ovata*, *Agathis australis*) can survive in the understory, but grow extremely well under improved light conditions (Rigg *et al.* 1998; Enright *et al.* 1999, 2003; Souza *et al.* 2008). For these species, large gaps in the canopy created by logging provide better conditions for establishment than undisturbed primary forest, and these species are ideally suited to enrichment planting in selectively logged forests.

## 6.3 RESTORATION OF ABANDONED AGRICULTURAL LANDS

Currently, 25% of the world's land area is degraded (FAO 2011) and the majority of this land was previously forest. The estimated annual rate of forest conversion (deforestation) is 13 million hectares per year (FAO 2010), Agriculture is the major cause of, first, deforestation, and then land degradation following clearance, especially in the tropics and subtropics, where vast forest areas have been, and regretfully continue to be, cleared for conversion to agriculture. Agriculture practiced on fertile soils is usually more sustainable than on poor soils and these poor soils are at the highest risk of becoming degraded. Unfortunately, it is increasingly the poor soils that are being cleared (Dobson *et al.* 1997) and then abandoned for various ecological and socioeconomic reasons. These include mismanagement, low profitability, diversion of labor toward other sectors, and depopulation. Commonly, forest conversion to agriculture benefits only small section of rural communities, while it negatively impacts directly and/or indirectly the majority of others within the same region. Furthermore, economic benefits are short-lived in areas where the established land-use system does not sustain productivity, as is the case with the majority of tropical soils (Lamb *et al.* 2008).

In abandoned agricultural fields, early succession is dominated by past crops, followed by invasion of grasses and herbaceous species characteristic of open habitats, and then shrubs and vines from the adjacent forest, creating a patchy structure; later, trees colonize these patches and grow to form a canopy. Among the tree species, those capable of resprouting and that have a persistent soil seed bank or wind-dispersed seeds will be favored. The past crops can have long-lasting historical effects after abandonment. Persisting for a long time, crops with extensive root systems can affect litter, soil biota, and vegetation, and these effects will be crop specific (Myster and Pickett 1994; Rivera *et al.* 2000; Myster 2007). In the study of Myster (2007) conducted in Ecuador, natural forest regeneration was investigated in sugarcane and banana plantations, and pastures for the first 5 years after abandonment. The sugarcane fields had twice the number of dispersed seeds as banana fields and 20 times the number of dispersed seeds as pastures. Survival of planted seedlings of five locally common tree species showed a similar pattern (25%, 15%, and 0%, respectively). Some of the past crops remaining after abandonment can have a positive effect on forest regeneration. Fruit trees on abandoned farmlands (e.g., avocado and mango) provide perches and a favorable microclimate for germination and growth of seeds that arrive, and also food for seed-dispersing birds, greatly enhancing their visitation rate (Jacob *et al.* 2017).

The succession pathway in abandoned agricultural lands to a large extent will depend on how the land was cleared of forest vegetation and modified before agricultural use. The impact of past land transformation (e.g., whether a site was plowed or not) on vegetation patterns can be even stronger than the effects of topography, soil, and subsequent disturbances (Motzkin *et al.* 1996). In human history, agriculture was based on two general forest-clearing techniques, both of which decrease soil fertility and change soil physical properties, but substantially differ in rate and trajectory of recovery following cessation of land use (Figure 6.1). The first technique, used in slash-and-burn agriculture, involves cutting down the vegetation to clear the land and then burning it. This practice removes standing vegetation but does not completely eliminate the soil seed bank, roots, and stumps. Slash-and-burn agriculture, characterized by long fallow periods, allowing vegetation to regenerate after short periods of field utilization, still predominates in Latin America, Asia, and sub-Saharan Africa. After land abandonment, although both seed bank and seed rain contribute to succession, sprouts are the main source of recruits of primary forest species (De Rouw 1993; Guimaraes and Proctor 2007). Guimaraes and Proctor (2007), analyzing secondary succession after slash-and-burn agriculture in eastern Amazonia, found that, whereas pioneer species predominated among secondary forest seedlings, the sprouts included some old-growth species, many of which had no or poor seed rain and seed bank. Thus, sprouting may represent the only way for regeneration for many late-successional species in the secondary forests (Guimaraes and Proctor 2007). Although this type of agriculture is generally associated with the most rapid forest recovery compared with other types of agriculture involving mechanical land clearing (Guariguata and Ostertag 2001), even for this land use it is unlikely that species composition will reach pre-disturbance levels within a period of less than 200 years (Saldarriaga *et al.* 1986),

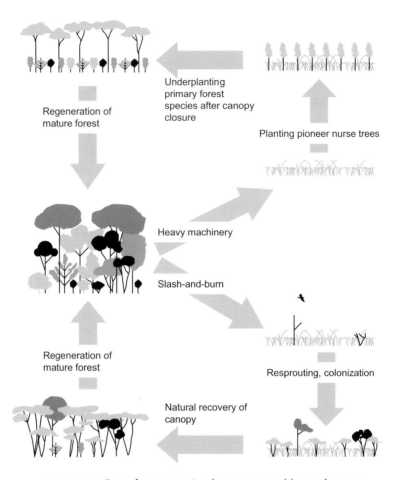

FIGURE 6.1 Secondary succession for two types of forest clearance (slash-and-burn and use of heavy machinery) followed by either cultivation of crops and, when soil fertility decreases, transformation into exotic grassland, or conversion into pasture right after forest clearance.

or even 500 years (Riswan *et al.* 1985). Especially sensitive to shifting cultivation practices are endemic species, and even after 50–60 years, the level of endemism is still significantly lower than in old-growth forest (Van Gemerden *et al.* 2003).

Mechanical clearing (e.g., use of a bulldozer) has replaced the above technique in many areas, especially in Europe and North

America. This method completely removes the natural vegetation leaving no stumps, seedlings, or woody remnants of any kind. In addition, mechanical clearing combined with plowing makes the soil surface flat and spatially homogeneous, reducing within-stand variability in soil moisture and pH (Flinn and Marks 2007). The forest regeneration potential of such lands is greatly limited in comparison with agricultural lands used at relatively low intensity for a short time (Holl 2007; Zermeno-Hernandez *et al.* 2015).

Due to a lack of resprouters and greatly depleted soil seed bank, bulldozed lands have significantly reduced species diversity and a higher percentage of exotic species than non-bulldozed sites (Chinea 2002). In bulldozed sites, and especially in those cultivated for many years under tillage or plowing before abandonment (i.e., old fields), for a long time after field abandonment colonizing woody vegetation will be limited to early successional species.

In general, abandoned agricultural lands, and especially those that in the past underwent mechanical clearing, have low potential conservation value as discussed in Section 5.2.2. The impact of agricultural use is so strong that forests developed on former agricultural fields differ in soil properties and other attributes from uncleared forests, even 2000 years after reforestation (Dupouey *et al.* 2002). Soils in such forests generally have a higher pH and nutrient concentrations and lower C content and C/N ratios (Koerner *et al.* 1997; Compton and Boone 2000), which, at least partly, is responsible for an increase in nitrogen-demanding ruderal species (Kopecký and Vojta 2009).

However, use of this kind of forest in conservation-oriented restoration is justified in cases when for some reason they become protected territories, because every scrap of protected land counts (e.g., Rivera *et al.* 2000). Conservation-oriented restoration of such secondary forests requires enrichment planting, which can serve two purposes. One is to introduce missing functionally important native or threatened species that cannot colonize the restoration site by themselves due to dispersal or establishment limitations. This type of enrichment planting was described and discussed in preceding chapters. Post-agricultural forests are known to take more than a century to recover their original species composition (Matlack 1994; Bossuyt *et al.* 1999; Flinn and Marks 2007) because the colonization ability of many old-growth forest species is extremely slow (from <0.05–1 m/year) (Bossuyt *et al.* 1999). Therefore, some slow-growing tree species of this kind may need to be reintroduced or translocated at a later stage of secondary forest development when they are less susceptible to desiccation and competitive exclusion (e.g., shade-tolerant, late-successional, animal-dispersed species) (Janzen 2002; Martinez-Garza and Howe 2003; Griscom and Ashton 2011).

The second purpose of enrichment planting is to accelerate or remove a barrier to succession created by aggressive colonizers: native or exotic invasive grasses which preclude native forest regeneration. Removal of grasses or application of herbicides in general are ineffective (Hooper *et al.* 2005b). Shading is a more efficient way to eliminate these grasses. Planting species that can act as nurse trees can catalyze natural regeneration by shading grasses, but also improving microclimate and soil conditions, and releasing dispersal limitations for vertebrate-dispersed tree species (Hooper *et al.* 2002, 2005b; Martinez-Garza and Howe 2003; Gomez-Aparicio 2009; Griscom and Ashton 2011). Tall tree species that are fast growing in open ground with broad but shallow crowns are the first choice, especially if they support frugivorous birds, as they promote seed dispersal (Martinez-Garza and Howe 2003; Jones *et al.* 2004). Some shrub species (Vieira *et al.* 1994) can have a similar catalyzing role in forest regeneration by attracting bats and birds, and facilitating the establishment of late-successional species.

While there is no issue of origin for the species of the first category (late-successional species), it is a very important one for the second category of nurse plants. It is not surprising that the fastest growing species, and the most tolerant to diseases, pests, and abiotic stresses, pioneers are extremely invasive. These species can rapidly establish and create a canopy. But many of these species will be impossible to eradicate later, when native species start to colonize the secondary forest. Moreover, due to high invasiveness, these species can rapidly spread into adjacent areas. Thus, use of exotic species requires extreme caution (see also Section 5.1).

Plantations of native pioneer tree species demonstrate higher abundance of understory woody regeneration than plantations of exotics (Keenan *et al.* 1997), and native tree species used as nurse trees often show a similar performance to most popular exotics in terms of growth and survival (Haggar *et al.* 1998; Carpenter *et al.* 2004; Wishnie *et al.* 2007; Roman-Danobeytia *et al.* 2012). Therefore, native tree species should always be strongly preferred to exotics, and planted in mixtures rather than as monospecific plantations (Carnevale and Montagnini 2002). Exceptions can be made for threatened species with very localized and rapidly shrinking current distributions, but poorly known historical ranges. For example, some threatened Mexican and Central American pines (such as *Pinus tecunumanii, P. chiapensis, P. greggii,* and *P. maximinoi*) were recognized as potential replacements for temperate *P. taeda* and *P. elliottii,* which are widely planted in tropical and subtropical regions (Lambeth and McCullough 1997). *Pinus tecunumanii,* non-native to Costa Rica but native to Central America, showed outstanding performance (growth and survival), especially on the most degraded sites, among two exotic and five native tree species planted in a highly degraded Costa Rican pasture (Carpenter *et al.* 2004).

## 6.4   RESTORATION OF PASTURES

In East Asia, cultivation of rubber, tea, and especially palm oil was, and continues to be, the major cause of deforestation. Similarly, in the Neotropics, vast forest areas became banana, sugar cane, coffee, and cotton plantations. But throughout the tropics, conversion to pastures for livestock grazing is by far the most widespread land use profoundly affecting tropical ecosystems (Anderson 1990; Hecht 1993; Garrity *et al.* 1996; Lowry *et al.* 1997; Aide *et al.* 2000). The reason for this is that, as well as deliberately created pastures after land clearance, grasslands can also develop on abandoned crop plantations. Because of the low nutrient levels of tropical soils, crop

plantations after being exploited for some time rapidly lose productivity, becoming less and less profitable, and are eventually abandoned
(Aide *et al.* 1995). Use of these lands as pasture begins when livestock
are let into the areas and grazing leads to dominance of grasses. Grass
species can be either seeded into the pasture or colonize naturally,
and can form dense monodominant stands that prevent early tree
seedling establishment. Grasses, due to their soil-level meristems,
are especially favored under regularly applied fires. In addition, in
abandoned pastures, vines frequently form dense blankets over the
few shrubs and trees smothering them (Uhl *et al.* 1988).

Even if pastures have had crops growing in them previously, they
have a distinct recovery due to the presence of livestock. Livestock acts
as a natural selection agent, favoring species that are unpalatable and
can withstand browsing and trampling. This selection leads to simplified ecosystems in which the remaining tree species are (1) stump,
stem, or root sprouters; (2) unpalatable to livestock or dispersed by
it; (3) resistant to fire; (4) desiccation tolerant; (5) wind dispersed; and
(6) have some human use and have therefore been selectively kept
in the landscape either as isolated trees or live fences (Griscom
and Ashton 2011). The presence of grasses and the rapid colonization and growth of ferns and vines inhibiting tree establishment are
major obstacles to the development of secondary forest in abandoned
pastures (Holl *et al.* 2000; Ortega-Pieck *et al.* 2011).

Seed rain appears to be a stronger regeneration filter in degraded
lands dominated by grasses than the soil seed bank (Myster 2007), but
the former is usually very low at distances more than 10 m or even 5 m
away from forest edges (Aide and Cavelier 1994; Holl 1999; Cubina
and Aide 2001; Charles *et al.* 2017). Even when tree seeds disperse
into the abandoned pastures, they have poor germination and development because of the competitive dominance of the thick vegetation
comprising grasses and ferns (Zahawi and Augspurger 1999; Myster
2004; Ortega-Pieck *et al.* 2011). This is why a general recovery, and
especially recovery of primary forest species, in pastures is greatly
delayed in comparison with lands where vegetation clearance was

not followed by use as pasture before abandonment (Aide *et al.* 1995; Steininger 2000; Mesquita *et al.* 2001; Wandelli and Fearnside 2015).

If forest clearing is not followed by agricultural use, woody vegetation recovers quickly, in less than 5 years, with the colonizing woody species being pioneer tree species found in natural disturbances in the nearby forest (Swaine and Hall 1983; Uhl and Jordan 1984; Guariguata and Ostertag 2001). In contrast, the first woody colonizers of abandoned pastures are shrubs and treelets that are able to compete with well-established grasses (Aide *et al.* 1995, 1996). During the first 10–15 years after abandonment, pastures are dominated by grasses and herbaceous vegetation (herbs, ferns, and vines). If the ecosystem resilience threshold is crossed or no sources of propagules exist in the surrounding matrix, natural regeneration to the previous state of a forest will not occur (Holl *et al.* 2000; Griscom *et al.* 2009; Ortega-Pieck *et al.* 2011; Mesquita *et al.* 2015). However, if the pastures are small, nearby forest patches exist, there is no fire, and soil is relatively intact, then the herbaceous vegetation will be gradually replaced by shrubs and small trees, followed by colonization by other woody species (Aide *et al.* 1995, 1996).

Pasture-use history (duration and intensity) might greatly influence succession following abandonment and the restoration strategy. The longer the pasture is grazed, the more pronounced are the effects of trampling of vegetation and deposition of dung. Prolonged pasture use leads to devastation of the soil seed bank, poor infiltration into the compacted soil, greater runoff, and increased erosion, and makes development of a secondary forest a much longer process (Uhl *et al.* 1988; Aide *et al.* 1995, 2000). Light use can be attributed to cases where the area is never weeded, grass establishment is poor, grazing intensity is low, and abandonment occurs shortly after pasture formation. Thirteen abandoned pastures grazed for <8 years in eastern Amazonia belonging to this category were described by Uhl *et al.* (1988). Regeneration in these pastures occurred within a few years of abandonment and sprouting was an important means of revegetation; more than half of the 171 tree species identified in these sites had the ability to resprout. Moderate use is characterized by good establishment of seeded grasses, regular application of weeding and burning, prolonged use (>10 years), and moderate grazing intensity. Heavy use differs from moderate use in that after several cutting and burning episodes to control weeds, all vegetation (including logs and standing trunks) is mechanically cleared, and the area is disked and leveled.

The pastures that underwent light use before abandonment can be restored relatively easily under natural regeneration. This restoration strategy is most effective if soils have not been severely degraded, grazing and fires are prevented, and remnant forests can efficiently act as seed sources for colonization. Available literature suggests that the impact of invasive species is not as strong as the effect of grazing and, in many cases, after removal of the invasive species, native tree species will invade grass-dominated ecosystems and become established (e.g., Cabin *et al.* 2002; Gilman *et al.* 2016). But if the success of a restoration project is measured not by basal area, forest structure, or species richness, but the similarity of species composition to the original forest, additional interventions are often necessary. Many old forest species exist in the landscape but are not colonizing the secondary forest derived from abandoned pastures either because grasses and ferns prevent their germination and development into seedlings or the distance separating them is too large (Aide and Cavelier 1994; Zahawi and Augspurger 1999; Charles *et al.* 2017). Given poor colonization of these species after more than 60 years of natural regeneration (Aide *et al.* 1996, 2000), enrichment planting may be a necessary active intervention required to enhance the natural regeneration (Sampaio *et al.* 2007; Powers *et al.* 2009), performed at least in a less costly form (i.e., through nucleation).

For moderately and heavily used pastures the role of barriers for forest recovery, such as grass cover preventing seed germination and seedling establishment, drought-induced seedling mortality, low seed rain, high seed predation, lack of nutrients, high light intensity, and herbivory, will be much more limiting than for lightly used ones (Nepstad *et al.* 1996). On severely disturbed sites, the seed bank and sprouting avenues of regeneration no longer exist. Because intensively used pastures are highly degraded ecosystems in which forest recovery is strongly limited by a suite of factors, to be successful, pasture restoration must address these multiple factors simultaneously (Holl *et al.* 2000; Cabin *et al.* 2002; Gunaratne *et al.* 2014). The most promising strategy is seeding or planting seedlings of pioneer species to shade out grasses and enhance seed dispersal, accompanied by prevention of grazing and fire (Holl *et al.* 2000; Cabin *et al.* 2002; Griscom and Ashton 2011) (Figure 6.1). In certain cases, planting must be preceded (Cohen *et al.* 1995) or accompanied by surface-soil disturbance (Gunaratne *et al.* 2010, 2014). Gunaratne *et al.* (2010, 2014) proposed tillage in strips up to 5 m wide within 10 m of the forest patches adjacent to the pasture as the means of removing entire grass plants, including their root systems, and thus enhancing seedling emergence of forest species. Passive regeneration, even with bird perching structures and protection from fire and grazing, as well as assisted natural regeneration, does not suit these systems. Especially important is enrichment early in succession as a means to alter the trajectory of the pasture succession (e.g., using a framework species method, Goosem and Tucker 1995; Tucker and Murphy 1997), with increase in functional diversity of the outplants leading to greater suppression of invasive grasses (Ammondt and Litton 2012). Enrichment planting must be done at sufficiently high density to overcome the establishment limitation imposed by grass and fern cover. This planting should use, whenever possible, native species (Celis and Jose 2011), some of which can be threatened ones. For example, individuals of critically endangered *Araucaria angustifolia* in grasslands of south Brazil act as nurse trees, facilitating the establishment of many vertebrate-dispersed tree species under their crown, and thus promoting the regeneration of degraded Atlantic Forest sites (Duarte *et al.* 2006). The feasibility of using threatened and even critically endangered species in restoration planting of abandoned pastures has been shown by several recent studies. Among 60 tree species used in reforestation trials of degraded grassland in the Philippines, 44 were native tree species, of which 23 were threatened including 9 critically endangered. Three critically endangered dipterocarp species, *Shorea guiso*, *S. contorta*, and *Parashorea malaanonan*, were among the best performing species (Schneider *et al.* 2014). In a reforestation planting trial on degraded lands (grasslands and shrublands established on abandoned farmlands) in Singapore, among 45 native tree species, four were threatened, of which three were evaluated as showing "excellent" and "good" performance (Shono *et al.* 2007). Two tested threatened Indonesian *Shorea* species, *S. leprosula* and *S. selanica*, demonstrated good survival and growth 15 and 17 years after planting in abandoned farmlands under a variety of spacing distances on mineral and swamp soil (Subiakto *et al.* 2016).

## 6.5    RESTORATION OF FOREST PLANTATIONS

Plantation forests are the main strategy against the trend of globally diminishing forest cover. While natural forest cover rampantly decreases, forest plantation cover only increases, with a gain of about 2.5 million hectares (*c.* 2%) annually during 2010–2015 (FAO 2015), and planted forests currently occupy 278 million hectares, which is 2% of the Earth's land surface (Keenan *et al.* 2015; Payn *et al.* 2015). These plantations are usually established to reduce erosion, produce timber, or both, through planting or seeding indigenous or introduced tree species. The creation and management of plantations often involves intensive site preparation (e.g., plowing), selective thinning, and clear-cut harvesting, often with short rotations.

In contrast to naturally regenerated forests which are compositionally, structurally, and functionally complex systems, plantations are simple systems generally composed of one artificially established tree species, with even-aged and regularly spaced individuals. Usually these species are fast-growing timber species with low requirements, such as those most commonly used in plantations

(30% of plantation forests; Carnus *et al.* 2006), species of the genera *Pinus* and *Eucalyptus*.

Nowadays, it is common to find exotic or indigenous mono-culture timber plantations where the management objective has subsequently changed from timber production to production plus con-servation or conservation alone (Ashton *et al.* 1997; Parrotta 2000; Parker *et al.* 2001; Zerbe 2002; Millet *et al.* 2013). The presence of plantations in high conservation priority ecosystems, where intact vegetation communities have disappeared or have been greatly dimin-ished in size, sparked an interest in converting these plantations to the preexisting vegetation type. Under these circumstances, nonharvest plantations provide a recognized global opportunity for restoring indi-genous forest communities (Lugo 1997; Lamb 1998; Martinez-Garza and Howe 2003; Lamb *et al.* 2005; Carnus *et al.* 2006; Ashton *et al.* 2014; Brown *et al.* 2015). The challenge is to use this opportunity in an effi-cient way in terms of the goal formulation, time to achieve it, and cost.

There is a principal difference between "plantations that have replaced natural forests of shade-tolerant species able to recruit within plantations, compared to where plantations replaced prairies, savannas, or shrublands containing light-demanding species unable to recruit in plantation understories" (Abella *et al.* 2017). For the latter, heavy thinning or complete removal of the plantations appears to be the best option, because the shade-intolerant vegetation has no chance of establishment in the plantation understory (Abella and MacDonald 2002; Catling and Kostiuk 2010; Harrington 2011). For the former, the situation is different. Generally, in the absence of silvicultural management, the monospecific plantation system is gradually replaced by a mixed forest due to colonization by floristic elements from surrounding forest areas (if the latter exist). In this mixed forest a proportion of the plantation species in the species pool will either eventually stabilize or disappear entirely from the system. The outcome will depend on life duration, shade tolerance, and competitive ability of the plantation species. In any case, a floris-tically rich secondary forest will be the result of the slowly occurring

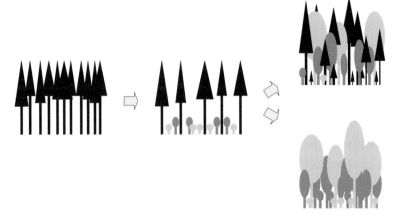

FIGURE 6.2 Two conservation-oriented restoration pathways for monospecific plantations composed of historically dominant local species (top) and exotic or native but never historically dominant or subdominant species (bottom).

succession. This process can be made much quicker by gradual removal of the planted trees and support for the establishment of alternative canopy species.

Considering restoration of plantations that have replaced natural forests, we again must recognize an important difference between two types of plantations (Figure 6.2). The first type are plantations representing endemic or local historically dominant tree species. Examples are plantations of species endemic to the Canary Islands *Pinus canariensis* (Otto *et al.* 2012), native to the Pacific Northwest of the United States *Pseudotsuga menziesii* (Maas-Hebner *et al.* 2005), or native to Ontario *Pinus resinosa* (Parker *et al.* 2001) established in their natural range. The second type are plantations of exotic or native, but never historically dominant or subdominant, species. The restoration pathways and the ultimate goals for these two types of plantations will be very different. For the first, the goal is improving regeneration, seed production, and making a more natural demographic structure of the target species, as well as increasing structural complexity and old-successional species richness in the ecosystem. For this type of plantation, thinning the canopy with or without underplanting is usually enough to speed the otherwise lengthy process (Figure 6.3). For the second type, the goal is to either completely replace or greatly decrease in abundance the plantation species through a series of interventions. When a plantation is created with a species that cannot produce viable seed in this environment, underplanting after thinning may suffice with no need for eradication treatments afterwards. If a species is capable of reproduction, removal of the seed trees or all the individuals of the undesirable species may be necessary after advancement of the underplanted late-successional species into the canopy. And, if a plantation species is an aggressive and persistent invader, like *Acacia* or *Leucaena*, eradication may simply not be possible even with the most intensive treatments, thus making this plantation useless for conservation-oriented restoration.

The aforementioned *Pinus canariensis*, endemic to the Canary Islands, is a good example of conservation-oriented restoration applied to a plantation of the first type, with the goal of promoting growth and complexity of the existing forest without changing the canopy dominance (Otto *et al.* 2012). The dense 60-year-old *P. canariensis* plantations (>1500 stems/ha) showed almost no seed production and, as a consequence, very scarce natural pine regeneration. Thinning of 60% of the basal area through single-tree selection applied to these plantations increased pine seed rain to levels similar to semi-natural old-growth stands. It also improved natural pine regeneration evident in increased seedling, sapling, and juvenile densities, and promoted understory plant diversity, including typical endemic pine forest species, without triggering plant invasion. The authors recommended a thinning intensity of 60% of the basal area as sufficient to enhance tree growth, facilitate functional regeneration of this conifer, and promote forest structural complexity, because lower thinning intensities could not provide these effects.

Unfortunately, plantations of the first type are rare exceptions among existing plantations. The vast majority of them are composed

FIGURE 6.3 Example of the transformation of a forest plantation that became part of a national park (a), into a natural boreal forest (b) via promoting complexity of its structure without change of the canopy dominance (c) (from Halme *et al.* 2013).

of exotic, or, if local, nondominant, or even uncommon species. Some of these plantations can convert to a natural forest composed of indigenous biota with no or minimal intervention, mostly through naturally occurring colonization and gradual replacement of the plantation species. But for the majority, conversion requires underplanting with indigenous species either without altering the forest canopy or after canopy thinning.

An example of plantations requiring minimal, if any, interventions are the planted forests of *Picea abies* in Western and Central Europe. These forests were widely planted in the past outside of their native range and, frequently, on unsuitable sites. Fortunately, conversion of many such plantations into deciduous woods occurs naturally. Herault *et al.* (2004) found intensive regeneration of a suite of broad-leaved species (*Fraxinus excelsior*, *Acer pseudoplatanus*, and *Sambucus nigra*) in plantations with low *P. abies* densities on base-rich soils, located in the proximity of riverine deciduous forests. Nevertheless, minor silvicultural treatments would be desirable even in these stands to foster the natural regeneration process, e.g., sowing, enrichment planting, and single-tree selection thinning.

An example of plantations for which underplanting is necessary are planted forests of *Pinus taeda* that replaced historically dominant *Pinus palustris* in the southeastern United States. Reestablishing *P. palustris* as a canopy-dominant species is currently required on sites occupied by other species within the former *P. palustris* range because this species provides a vital habitat for many animal species, including threatened gopher tortoise and red-cockaded woodpecker. In this system, retention of plantation canopy trees during stand conversion is desirable due to their facilitation effect on *P. palustris* seedling survival, though not after advancement of *P. palustris* into the canopy. Conversion of this system can use a variety of thinning treatments including single-tree selection to reduce basal area to 5–9 m²/ha or group selection with gaps of 0.1 ha followed by underplanting (Knapp *et al.* 2013).

Can we make any general recommendations for plantation restoration silvicultural treatments to recreate late-successional

characteristics, i.e., thinning and underplanting? These recommendations should be based on two things. The first is the importance of
creating different sized gaps for species coexistence. For example,
in boreal forests, coexistence of shade-tolerant conifers (spruce, fir,
and pine) regenerating in small gaps, with shade-intolerant birch is
mediated by the predominant regeneration of the latter in much larger
(up to 1 ha) gaps (Leemans 1991; Taylor *et al.* 1996; Kuuluvainen and
Juntunen 1998). The second is demonstrated in several studies of differential responses in mid- and late-successional species to canopy
gap size due to different requirements for light. Measuring growth
rates of 21 Dipterocarpaceae species from Malaysian Borneo grown
in shade houses for 2 years in three light treatments (0.3%, 3%, and
18% full sunlight) revealed no single growth rate hierarchy common
to all light treatments (Philipson *et al.* 2012). Differences in light-
dependent growth have been shown for transplanted seedlings of
four *Shorea* species that occur together as canopy emergents in Sri
Lanka (Ashton 1995), and in growth and survival for saplings of six
nonpioneer Costa Rican tree species (Clark and Clark 1992). In New
Zealand, two indigenous canopy tree species performed differently
in two circular canopy gaps of 4.6 and 11.2 m in diameter created
in a 18-year-old plantation of *Pinus radiata*. In that study, light-
demanding *Podocarpus totara* showed an increased rate of growth in
height in large gaps, whereas in *Beilschmiedia tawa* height growth
increased in gaps of any size (Forbes *et al.* 2016). Coates and Burton
(1999) reported variability in growth of five boreal conifers measured
5 years after planting at intermediate light levels 30–70% of full light
and suggested careful matching of planted tree species to light environment to maximize their growth rates. Differential light-dependent
mortality and growth was demonstrated for four related tropical
tree species by Kobe (1999) who transplanted seedlings into gaps of
varying size with light intensity ranging from 1% to 85% full sun.
In that study, all species decreased in mortality with increases in
light to 20% full sun. Above 20% full sun, the mortality of *Trophis
racemosa* and *Castilla elastica* continued to decrease, while that of
*Cecropia obtusifolia* remained constant and that of *Pourouma aspera*

increased. At 10% full sun, there was a fivefold difference in mortality probability among the species. Each species showed the highest growth relative to other surviving species at distinct light levels.

Determination of the optimal gap size for a species based only on seedling emergence and early growth may be unreliable. Whitmore and Brown, after measuring seedling growth of three dipterocarp species in canopy gaps of different sizes for three additional years, revealed a growth pattern completely different from that found after the initial 40 months period (Whitmore and Brown 1996). In Bolivian tropical rainforest, saplings of the canopy emergent tree *Bertholletia excelsa* only occurred in gaps >95 m² or >10.4% of the total above canopy irradiance that is received below the canopy, whereas seedlings occurred throughout a range of gap sizes and in the forest understory (Myers *et al.* 2000a). Similarly, in plantations of *Pinus thunbergii* in Japan, 1-year-old seedlings were present in gaps of all sizes, but seedlings older than 2 years were absent both under the canopy and in small gaps (less than 30% full sun) (Zhu *et al.* 2003).

Some authors have recognized the importance of variable-density thinning and the creation of multiple size canopy gaps in plantations to increase their structural complexity, and improve wildlife habitat and the aesthetic value of residual structures (Zhu *et al.* 2003; Parker *et al.* 2008). To achieve the desired vegetation structure with a variety of light levels interspersed within a matrix of more aggregated canopy tree retention, Parker and colleagues recommended "where possible, deviate from traditional row and selection removal of the overstory in initial and subsequent thinning" (Parker *et al.* 2008). What seems to be the future of plantation transformation into natural forests is to incorporate a diversity of species with different responses to canopy opening into artificially created canopy gaps. The latter requires determination of the gap size most suitable for a species according to the species-specific response to light availability (Figure 6.4). The next step, after determining the species-specific gap sizes, will be working out a site-specific gap size distribution to support late-successional multi-species establishment (Figure 6.5). This approach can be used for both passive restoration

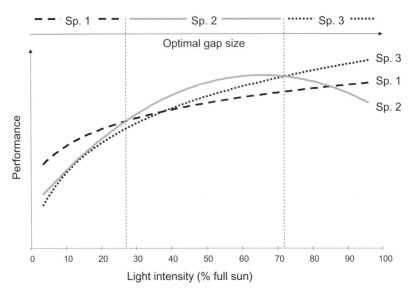

FIGURE 6.4 Seedling performance (survival and growth) of the three hypothetical late-successional species along the gradient of light intensity in the canopy, and the most appropriate canopy gap size for each species.

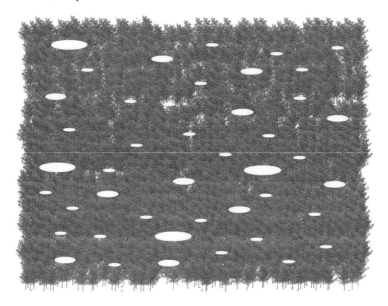

FIGURE 6.5 Group selection applied to a plantation with a variety of gap sizes based on species-specific light requirements.

Table 6.2 *Summary statistics (range values; median values in parentheses) from published literature for gap characteristics of gap-disturbed forests (from McCarthy 2001)*

| Forest type | Gap fraction (%) | Average gap size (m²) | Gap range (m²) |
|---|---|---|---|
| Boreal and subalpine | 6–36 (21) | 41–141 (78) | 15–1245 |
| Temperate hardwoods | 2–10 (10) | 28–239 (79) | 8–2009 |
| Temperate coniferous | 11–18 (14) | 77–131 (85) | 5–734 |
| Tropical | 0.8–8 (4) | 10–120 (50) | 4–700 |
| Southern hemisphere | 3–35 (8) | 40–143 (93) | 24–1476 |

and active restoration through enrichment planting. Periodic regulation of vegetation within the gaps created is required to exclude undesirable tree species and prevent high levels of competition among the desirable ones.

In natural forests, gaps have a wide range of sizes. Among different forest types, median values of average gap size are lowest for tropical forests and highest for the temperate coniferous forests and southern hemisphere forests (Table 6.2). However, despite these differences, a gap size–frequency distribution of many small single-treefall and a few large multi-treefall gaps is common for various forest types (Foster and Reiners 1986; Wagner *et al.* 2010). Such frequency distribution approximated by a log-normal or negative exponential distribution (Runkle 1982; McCarthy 2001) has an important implication for the creation of multiple size canopy gaps in plantations, as suggested above: the proportion of gaps created should decrease from small-sized to medium- and large-sized (Figure 6.6).

To increase the positive effect of thinning in transforming plantations into natural forests, gap creation can be accompanied by burning or creating deadwood. In northern Europe, natural regeneration in cultivated Scots pine *Pinus sylvestris* and Norway spruce *Picea abies* stands is often scarce because seed germination

FIGURE 6.6  Thinned Norway spruce forest plantation in Sweden enriched with broad-leaved species. Photo by David Lamb. *A black and white version of this figure will appear in some formats. For the color version, please refer to the plate section.*

and seedling development are retarded by the thick litter and raw humus layers. Thinning and burning treatments can create favorable conditions for their regeneration by exposing mineral soil and increasing light availability, and also increase the forest microtopographic and climatic heterogeneity (Vanha-Majamaa *et al.* 2007; Lilja-Rothsten *et al.* 2008). In Estonia, conversion of old (30–60 years), even-aged, cultivated coniferous forests designated for protection, toward more structurally complex and more natural forest stands was investigated, applying gap creation either alone or in combination with the addition of deadwood or burning branches and needles. The highest Scots pine regeneration was observed in gaps with the burning treatment, while Norway spruce regeneration was highest in gaps alone. Understory vegetation diversity increased the most in gaps with burning, lichens in gaps without burning, and bryophytes with the addition of deadwood (Laarmann *et al.* 2013). The results of this as well as several other studies (Vanha-Majamaa *et al.* 2007; Lilja-Rothsten *et al.* 2008) suggest that after thinning, the maximum positive effect in terms of microsite diversity can be achieved if in some gaps the debris is burned while in others it is left to decay.

In many cases, canopy manipulations cannot improve the establishment of late-successional species and should be accompanied by underplanting (Figure 6.6). In a study by Sciwa *et al.* (2012), 5 years after two thinning treatments (67% and 33%) in a 20-year-old *Cryptomeria japonica* plantation in Japan the number of late-successional species, in contrast to early- and mid-successional species, did not increase in comparison with the control. On the other hand, planting three indigenous Podocarpaceae conifers into a degraded plantation of *Pinus ponderosa* in New Zealand resulted in the development of a podocarp-dominated forest structure within *c.* 50 years of underplanting, but without a woody understory. This was due to the isolation of the plantation from indigenous forest and the absence of fruit and nectar sources to attract bird dispersers (Forbes *et al.* 2015). These and other studies (Parrotta 1993; Wunderle 1997; Cubina and Aide 2001; Aubin *et al.* 2008; Ashton *et al.* 2014) suggest that if plantations are not located in close proximity to natural forests, underplanting is necessary and should not be limited to the late-successional dominants only. Short geophytes and zoochorous perennials, identified as the two functional groups characteristic of mature deciduous forests, are the plant groups most severely underrepresented in plantations because of their low dispersal abilities (Herault *et al.* 2005).

What is the potential utility of different plantations, for example deciduous versus coniferous in conservation-oriented restoration? On the one hand, restoring deciduous plantations is easier than coniferous ones, because the acidic and nutrient-limited needle litter layer inhibits development of the understory. The extent to which deciduous and coniferous plantations develop understory attributes comparable to those of naturally regenerated stands was found to be markedly different in a study by Aubin *et al.* (2008). The understory functional groups and environmental conditions of deciduous plantations converged toward those of old naturally regenerated forests, while conifer plantations, a predominant plantation type in the northern hemisphere, showed a completely different pathway of understory development. Similarly, in the southern hemisphere, an assessment of understory colonization in plantation monocultures of species native to North Queensland, Australia, *Araucaria cunninghamii*, *Flindersia brayleyana*, and *Toona ciliata*, ranging in age from 5 to 63 years revealed a greater diversity of tree species under the broad-leaved *F. brayleyana* and *T. ciliata* than under the conifer *A. cunninghamii* (Keenan *et al.* 1997).

On the other hand, many deciduous plantation species are not only extremely aggressive invaders but also notoriously difficult to eradicate even from a very localized area due to their early reproductive age, production of numerous persistent seeds, and vigorous resprouting from stumps, stems, and roots (e.g., *Acacia*, *Prosopis*, *Leucaena*, *Populus*, *Eucalyptus*). In this respect, among plantations of exotic trees, coniferous plantations, and pine plantations in particular, seem to be more suitable for conversion with conservation goals than the other widespread plantations. Although quite a few conifers are known to be invasive (e.g., *Pseudotsuga menziesii*, *Larix occidentalis*, *Pinus contorta*, *P. radiata*, *P. ponderosa*, *P. strobus*, *P. sylvestris*, *P. nigra*), they can only reproduce through seeds. In contrast, those conifers that can resprout (e.g., *Sequoiadendron giganteum*, *Taxodium distichum*, *Sequoia sempervirens*, *Pinus rigida*) are not invasive. The seeds of the majority of conifers used for the creation of plantations are not persistent. Therefore, as long as seed trees are removed, regular removal of seedlings will do the work of eradication.

In temperate deciduous regions of Europe and North America, plantations until recently were almost exclusively composed of monospecific conifer stands (Aubin *et al.* 2008), and their conversion into mixed or deciduous forests appears to be feasible (e.g.,

Herault *et al.* 2004; Jonášová *et al.* 2006; Onaindia *et al.* 2013). In Europe, conversion of pine plantations into deciduous woodlands has become a major objective for forest management (Herault *et al.* 2005). There are examples of successful underplanting with native species in different parts of the globe to assist natural succession in pine plantations (e.g., United States, Knapp *et al.* 2013; Canada, Parker *et al.* 2001; Spain, Rodriguez-Calcerrada *et al.* 2008; Germany, Noack 2011; Sri Lanka, Ashton *et al.* 2001). Some of these and other studies examined the potential of *Pinus* plantations to assist the restoration of indigenous forest communities through canopy gap creation. In a study by Rodriguez-Calcerrada *et al.* (2008), two indigenous oak species were planted in the understory of even-aged plantations of *Pinus sylvestris* under two thinning intensities (33% and 50% stand density) and a control. Thinning of both intensities had no effect on *Quercus petraea*, but a very positive effect on the survival of *Quercus pyrenaica* after 2 years. In Japan, circular canopy gaps helped to promote forest succession from the understory of a 40-year-old *Pinus thunbergii* plantation, with growth and establishment being greatest in large gaps (Zhu *et al.* 2003). Removal of consecutive rows in an even-aged plantation of *Pinus caribaea* in Sri Lanka created different widths of canopy opening strips (4 and 8 m) and all five late-successional canopy tree species planted showed maximum growth in the 8 m wide strips (Ashton *et al.* 1997). These studies suggest that the creation of large canopy gaps in pine plantations, especially if followed by the planting of desirable species in gaps, and gradual removal of pine trees as the planted trees start to dominate the canopy, can effectively restore the preceding forest communities.

# 7 From Theory to Practice

As discussed in Section 4.2, ecosystems, and macro- and microhabitats differ in their conservation value and priority for conservation-oriented restoration, with the highest priority being the least altered habitats containing threatened species or having suitable conditions for them, and those areas in which restoration may provide a disproportionate increase in conservation benefits. Some such priority habitats are presented in Table 4.2. This section will go deeper into identifying those areas that should be the primary targets for protection and active interventions, starting at the global scale and proceeding to regional, and then to local scale. Many are emergency areas and require immediate action through strategically targeted site conservation programs that will stop the loss of natural habitat and of the species that they shelter (Eken *et al.* 2004; Langhammer *et al.* 2007). The importance of area prioritization has been recognized, and maps of priority areas are increasingly being used to set targets not only for protection but also for restoration (Tobon *et al.* 2017).

How can the top priority areas be identified? Seemingly straightforward criteria such as total species richness or number of threatened species do not suit, as for example species-poor sites can represent unique habitats or play a critical role for one or just a few critically endangered species, while species-rich areas can represent ecotones and widely dispersed and/or wide-ranging species (Eken *et al.* 2004). There is a suite of other more appropriate approaches for delineation of key and emergency areas for conservation, as will be discussed below.

## 7.1.1 Biodiversity Hotspots

Intuitively, areas with high species diversity, many endemics, or large numbers of rare/threatened species have always been seen as the most relevant targets for conservation. Norman Myers was the first to formally define large-scale conservation priority areas, proposing the term "biodiversity hotspot," a geographic unit containing large numbers of endemic species found in relatively small areas and facing significant threats from habitat loss (Myers 1989, 1990). Defined in this manner, hotspots are areas in which protection might prevent the extinction of a larger number of species than in similarly sized areas elsewhere (Mittermeier *et al.* 1998; Myers *et al.* 2000b; Brooks *et al.* 2006). Given that the cost of land protection generally scales by area, the main idea of creating lists of biodiversity hotspots is to focus the limited amount of resources on a limited number of areas that have high concentrations of species not found anywhere else, and by virtue of this, to minimize extinction rates for as many threatened species as possible. Myers' definition of a hotspot continues to be used today, because, in addition to the above reasons, endemic species are particularly vulnerable to habitat destruction and, for a given territory, endemic species are usually better studied than the complete flora or the threats (Myers *et al.* 2000b; Orme *et al.* 2005; Lamoreux *et al.* 2006; Cañadas *et al.* 2014). In the last decade, the definition of this term has been widened to include the total number of species, number of threatened species, and evolutionary distinctiveness (Rodrigues *et al.* 2005; Rosauer *et al.* 2009; Cadotte and Davies 2010). The importance of other criteria, in addition to the level of endemism for defining priority conservation regions, stems from the fact that some regions have extremely high levels of endemism but a relatively small number of species (e.g., Madagascar and Australia), while others have extremely high species richness (e.g., Brazil, Colombia, and Indonesia) or a large number of threatened species including "living fossils" (e.g., southern China) independent of the overall endemism of the region (Reid 1998). In addition,

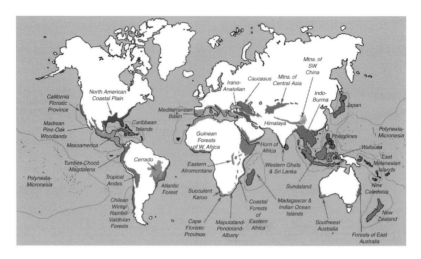

FIGURE 7.1 Currently recognized global hotspots of plant endemism, which are defined as having >1500 endemic plant species and >70% habitat conversion (from Harrison and Noss 2017).

distantly related threatened taxa are worth more in terms of phylogenetic diversity than are numerous closely related species and, for this reason, all other things being equal, areas with higher phylogenetic diversity of threatened species should have priority (Rodrigues and Gaston 2002; Brummitt and Lughadha 2003). Nevertheless, despite a low correlation between global richness and endemism, with a scarcity of other information, endemism is a useful criterion for defining broad conservation priority areas because regions selected for high levels of endemism effectively capture both narrow- and broad-ranging species (Lamoreux *et al.* 2006), and have high phylogenetic diversity (Sechrest *et al.* 2002).

The biodiversity hotspot concept is adopted by nongovernmental conservation organizations, such as Conservation International. They identify biodiversity hotspots as each having >1500 endemic plant species as well as >70% loss of primary vegetation. The number of recognized global hotspots from the 18 originally proposed (Myers *et al.* 2000b) was later expanded to 35 (Mittermeier *et al.* 1998) and then 36 (Critical Ecosystem Partnership Fund 2016; Figure 7.1). These

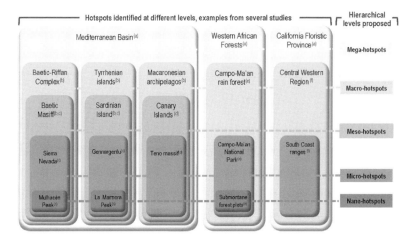

FIGURE 7.2 Examples of hotspots hierarchically organized into several levels (from Cañadas *et al.* 2014). Superscript letters show the specific studies in which hotspots were identified: (a) Myers *et al.* (2000b), (b) Médail and Quézel (1997), (c) Cañadas *et al.* (2014), (d) Reyes-Betancort *et al.* (2008), (e) Tchouto *et al.* (2006), (f) Kraft *et al.* (2010).

hotspots contain over 50% of the world's plant species and close to two-thirds of the IUCN red-listed plants and vertebrates.

However, although defined at a large geographic scale, global biodiversity hotspots are important for large-scale conservation planning, for finer scale planning their value is limited as they represent areas too large to be wholly protected (Reid 1998; Brummitt and Lughadha 2003). Besides, the huge size of some hotspots (e.g., Mediterranean Basin or Indo-Burma, each more than 2 million km$^2$) requires the close coordination of many national governments, which often is problematic.

A possible solution is to use a hierarchy of sub-hotspots within global biodiversity hotspots (Cañadas *et al.* 2014). Using the same criteria (i.e., endemicity and plant richness), smaller hotspots can be identified within larger hotspots at a variety of scales (e.g., Médail and Quézel 1999; Laffan and Crisp 2003; Murray-Smith *et al.* 2009; Raes *et al.* 2009; Kraft *et al.* 2010; Grant and Samways 2011) (Figure 7.2).

Fenu *et al.* (2010) even proposed the specific terms "micro-hotspots" and "nano-hotspots," the former corresponding to biogeographic units and the latter being areas smaller than 3 km$^2$ with an exceptional concentration of endemic species; Cañadas *et al.* (2014) proposed several analogous terms for the higher hierarchical levels (Figure 7.2).

Among the hotspots, some are "superhot" due to the extremely high rate of land conversion and habitat destruction. Predicted future extinctions based on actual current rates of deforestation in hotspots identified the Caribbean, Tropical Andes, Philippines, Mesoamerica, Sundaland, Indo-Burma, and Madagascar as areas likely to lose most of their endemic plants and vertebrates if their current intensity of deforestation continues (Brooks *et al.* 2002). Another approach to identify priority hotspots was used by Schmitt and colleagues. They found that 20 (58.8%) of the 34 forest hotspots analyzed had less than 10% of their forest area within IUCN I–IV protected areas (Schmitt *et al.* 2009) (Table 7.1). These least protected hotspots deserve first priority in conservation and restoration planning.

Originally, biodiversity hotspots were defined as remaining primary habitats, but in reality many of these "primary habitats" represent either partly degraded or secondary vegetation, which cannot be a substitute for the biodiversity of intact primary cover (Dunn 2004; Gibson *et al.* 2011). An assessment of vegetation cover in hotspots using satellite imagery by Sloan *et al.* (2014) revealed that they retain only 14.9% of their total area as natural intact vegetation, with half having ≤10% of intact vegetation. The most affected are hotspots representing dry forests, open woodlands, and grasslands, reflecting their historic conversion for agricultural use (Sloan *et al.* 2014). This means that conservation of hotspots cannot be limited to protection alone, but must include restoration as well.

### 7.1.2   *Priority Habitats*

There is a general consensus among conservation biologists that, because we cannot save everything, our goal is to at least ensure that

Table 7.1 *Forest area (>10% tree cover) within Conservation International biodiversity hotspots and its percentage protected under IUCN management categories I–IV and I–VI (from Schmitt et al. 2009)*

| Biodiversity hotspot | Forest cover (× 1000 km²) | % Protected (IUCN I–IV) | % Protected (IUCN I–VI) |
|---|---|---|---|
| Mountains of Southwest China | 125 | 0.0 | 13.8 |
| East Melanesian Islands | 72 | 0.0 | 0.7 |
| Succulent Karoo | 0.1 | 1.9 | 1.9 |
| Madrean Pine-Oak Woodlands | 281 | 2.1 | 6.4 |
| Coastal Forests of Eastern Africa | 188 | 2.2 | 5.7 |
| Irano-Anatolia | 2 | 2.6 | 8.0 |
| Japan | 244 | 3.3 | 15.9 |
| Mediterranean Basin | 265 | 4.2 | 11.8 |
| New Caledonia | 6 | 4.4 | 4.4 |
| Maputaland-Pondoland-Albany | 124 | 4.7 | 4.8 |
| Cerrado | 366 | 5.6 | 8.7 |
| Guinean Forests of West Africa | 223 | 7.0 | 7.5 |
| Mesoamerica | 595 | 7.3 | 16.6 |
| Wallacea | 195 | 7.4 | 8.6 |
| Polynesia-Micronesia | 6 | 7.5 | 8.2 |
| Madagascar and the Indian Ocean Islands | 129 | 7.6 | 9.9 |
| Atlantic Forest | 246 | 7.7 | 15.9 |
| Sundaland | 766 | 9.0 | 12.7 |
| Tumbes-Choco-Magdalena | 77 | 9.8 | 12.0 |
| Eastern Afromontane | 295 | 9.8 | 13.8 |
| Himalaya | 211 | 10.5 | 14.8 |
| Cape Floristic Region | 15 | 11.1 | 11.1 |
| California Floristic Province | 155 | 11.7 | 50.8 |
| Caucasus | 90 | 12.1 | 13.8 |
| Philippines | 83 | 12.6 | 17.6 |
| Indo-Burma | 742 | 14.2 | 19.2 |
| Horn of Africa | 2 | 15.5 | 18.4 |
| Caribbean Islands | 45 | 15.6 | 28.4 |

(continued)

Table 7.1 (*cont.*)

| Biodiversity hotspot | Forest cover (× 1000 km²) | % Protected (IUCN I–IV) | % Protected (IUCN I–VI) |
|---|---|---|---|
| Chilean Winter Rainfall and Valdivian Forests | 134 | 17.6 | 19.6 |
| Mountains of Central Asia | 11 | 17.7 | 18.4 |
| Western Ghats and Sri Lanka | 97 | 17.8 | 17.8 |
| Tropical Andes | 426 | 18.3 | 24.0 |
| Southwest Australia | 73 | 26.0 | 26.1 |
| New Zealand | 76 | 40.7 | 54.5 |

all ecosystem and habitat types are represented in regional conservation strategies. The conservation approach based on a "representation" principle adds to the goal of maintaining species diversity – the traditional focus of biodiversity conservation – another level of conservation action focusing on distinct ecosystems and ecological processes rather than on particular species and their habitats. Some ecosystems (e.g., tundra, mangroves, or savannas) are not as species rich as tropical rainforests, but they contain species assemblages adapted to distinct environmental conditions with unique evolutionary histories. The idea of preventing the loss of world ecosystem diversity and associated species diversity motivated the creation of the Global 200 (Olson and Dinerstein 1998), an ambitious project to create a representative list of habitat types and ecosystems which harbor globally important biodiversity and ecological processes. The Global 200, using ecoregions as the units of analysis, identified the major habitat types (MHTs), the areas of the world that share similar environmental conditions and habitat structure, with community members that represent similar functional types and adaptations. The MHTs identified were further subdivided by biogeographic realm (e.g., Neotropical, Indian Ocean) in order to represent unique faunas and floras of different continents or ocean basins, and within each realm, ecoregions were carefully chosen to represent the most distinctive examples of biodiversity for a given MHT (Table 7.2). The criteria for choosing the set of ecoregions with the greatest biological distinctiveness were: species richness, endemism, taxonomic uniqueness (e.g., unique genera or families, relict taxa or communities, primitive lineages), unusual ecological or evolutionary phenomena (e.g., intact large vertebrate faunas or migrations, extraordinary adaptive radiations), and global rarity of MHT.

From the 233 identified ecoregions whose biodiversity and representation values are outstanding on a global scale, 136 are terrestrial ecosystems in 12 MHTs. These ecoregions vary greatly in terms of current and future ability to maintain their integrity and viable populations of species inhabiting them (i.e., in their conservation status), and therefore were classified into one of three broad conservation status categories: critical/endangered, vulnerable, or relatively stable/relatively intact. As a result of the conservation status assessment based on total habitat loss, degree of fragmentation, future threat, and protection, 67 terrestrial ecoregions were classified as critical/endangered. Among them, ecoregions representing temperate grasslands, Mediterranean shrublands, temperate broad-leaved forests, and tropical dry forests (especially those located on oceanic islands) are the most threatened.

The majority of terrestrial ecosystems with critical/endangered status listed in the Global 200 require restoration. Table 7.2 lists these emergency areas for both protection and restoration. Currently, only a few of the 67 regions listed have ongoing restoration projects (Atlantic forests of Brazil, Seychelles and Mascarene Islands forests, Hawaii moist and dry forests).

The priority ecoregions recognized by the World Wildlife Fund (WWF) (Olson and Dinerstein 1998) are a valuable supplement to 36 biodiversity hotspots identified by Conservation International. Together they represent some of the most significant remaining regions for conserving the world's biological diversity at both ecosystem and species levels, and should be the first targets for conservation-oriented restoration.

Table 7.2 *The 67 terrestrial ecoregions from the Global 200 list that have critical/endangered conservation status (from Olson and Dinerstein 1998)*

| Major habitat type and biogeographic realm | Ecoregion |
| --- | --- |

**Tropical and subtropical moist broad-leaved forests**

| | |
| --- | --- |
| Neotropical | Atlantic forests – Brazil, Paraguay, Argentina |
| | Northern Andean montane forests – Ecuador, Colombia, Venezuela, Peru |
| | Coastal Venezuela montane forests – Venezuela |
| | Greater Antillean moist forests – Haiti, Cuba, Dominican Republic, Jamaica, Puerto Rico |
| | Varzea flooded forests – Peru, Brazil, Venezuela |
| Afrotropical | Madagascar moist forests – Madagascar |
| | Guinean moist forests – Ghana, Guinea, Côte d'Ivoire, Liberia, Sierra Leone, Togo |
| | Eastern Arc montane forests – Tanzania, Kenya |
| | East African coastal forests – Tanzania, Kenya, Mozambique, Somalia |
| | Albertine Rift highland forests – DR Congo, Rwanda, Uganda, Burundi, Tanzania |
| | East African highland forests – Kenya, Tanzania, Uganda |
| | Seychelles and Mascarene Islands forests (e.g., Mauritius, Seychelles, Comoros, Réunion, Rodrigues) |
| | Gulf of Guinea Islands forests – São Tomé and Príncipe, Equatorial Guinea |
| | Macaronesian forests (Azores, Madeira, Canary, Cape Verde Islands) |
| | Congolian coastal forests – Cameroon, Gabon, R. Congo, Nigeria, Equatorial Guinea, Benin |
| Indo-Malayan | Western Ghats moist forests – India |
| | Sri Lankan moist forests – Sri Lanka |

Table 7.2 (cont.)

| Major habitat type and biogeographic realm | Ecoregion |
| --- | --- |
| | Peninsular Malaysian lowland and montane forests – Malaysia, Thailand |
| | Sumatran-Nicobar Islands lowland forests – Indonesia, India |
| | Northern Borneo-Palawan moist forests – Malaysia, Indonesia, Philippines, Brunei |
| | Philippines moist forests – Philippines |
| | Southeast China subtropical forests – China |
| | Hainan Island forests – China |
| | Nansei Shoto Archipelago forests – Japan |
| Australasian | New Caledonia moist forests – New Caledonia, France |
| | New Zealand tropical forests – New Zealand |
| | Lord Howe and Norfolk Island forests – Australia |
| Oceanian | Hawaii moist forests – United States |
| | South Pacific Islands forests – Fiji, Tonga, Samoa, I Sisito, American Samoa |
| **Tropical, subtropical dry, and monsoon broad-leaved forests** | |
| Neotropical | Bolivian lowland dry forests – Bolivia, Brazil |
| | Tumbesian and North Inter-Andean Valleys dry forests – Ecuador, Peru, Colombia |
| | Southern Mexican dry forests – Mexico |
| Afrotropical | Madagascar dry forests – Madagascar |
| | Maputaland-Pondoland dry forests – South Africa, Swaziland, Mozambique |
| Australasia | New Caledonia dry forests – New Caledonia, France |
| Oceanian | Hawaii dry forests – United States |
| **Tropical and subtropical conifer forests** | |
| Neotropical | Mexican pine-oak forests – Mexico, United States |
| | Greater Antillean pine forests – Haiti, Cuba, Dominican Republic |

(continued)

Table 7.2 (*cont.*)

| Major habitat type and biogeographic realm | Ecoregion |
| --- | --- |
| **Temperate conifer and broad-leaved forests** | |
| Nearctic | Klamath-Siskiyou coniferous forests – United States |
| | Appalachian and mixed mesophytic forests – United States |
| | Pacific temperate rainforests – United States, Canada |
| | Sierra Nevada conifer forests – United States |
| | Southeastern conifer and broad-leaved forests – United States |
| Neotropical | Valdivian temperate rainforests – Chile, Argentina |
| Palearctic | Western Himalayan temperate forests – Pakistan, India, Nepal |
| | Southern European montane forests – Bulgaria, Greece, Spain, Italy, France, Andorra, Switzerland, Austria, Slovenia, Poland, Slovakia, Hungary, Czech Republic, Germany, Romania, Ukraine, Yugoslavia |
| | Central China temperate forests – China |
| Australasian | Tasmanian temperate rainforests – Australia |
| **Temperate grasslands, savannas, and shrublands** | |
| Nearctic | Tallgrass prairies – United States |
| **Tropical and subtropical grasslands, savannas, and shrublands** | |
| Afrotropical | Angolan Escarpment woodlands – Angola |
| Indo-Malayan | Terai-Duar savannas and grasslands – Nepal, India, Bhutan, Bangladesh |
| Palearctic | Red Sea fog woodlands – Egypt, Sudan, Djibouti, Eritrea |
| **Flooded grasslands and savannas** | |
| Neotropical | Everglades flooded grasslands – United States |
| Afrotropical | Zambezian flooded savannas – Botswana, Namibia, Angola, Zambia, Malawi, Mozambique |

Table 7.2 (*cont.*)

| Major habitat type and biogeographic realm | Ecoregion |
|---|---|
| **Tropical montane grasslands and savannas** | |
| Afrotropical | Ethiopian Highlands – Ethiopia, Somalia, Eritrea, Sudan |
| | Zambezian montane savannas and woodlands – Zimbabwe, Mozambique, Malawi, Zambia, Tanzania |
| | South African montane grasslands and shrublands – South Africa, Lesotho, Swaziland |
| **Deserts and xeric shrublands** | |
| Neotropical | Atacama Desert – Chile |
| Afrotropical | Namib and Karoo deserts and shrublands – South Africa, Namibia |
| | Madagascar Spiny Desert – Madagascar |
| Palearctic | Central Asian deserts – Turkmenistan, Kazakstan, Uzbekistan, Tajikistan |
| **Mediterranean shrublands and woodlands** | |
| Neotropical | Chilean matorral – Chile |
| | California chaparral and woodlands – United States, Mexico |
| Afrotropical | Fynbos – South Africa |
| Palearctic | Mediterranean shrublands and woodlands – Portugal, Spain, France, Italy, Monaco, Greece, Yugoslavia, Bosnia and Herzegovina, Croatia, Albania, Turkey, Libya, Lebanon, Israel, Morocco, Algeria, Tunisia, Malta, Cyprus, Macedonia, Bulgaria, Egypt, Syria, Jordan, Slovenia, Gibraltar |
| Australasian | Southwest Australian shrublands |

### 7.1.3   Ecoregions with the Highest Deforestation-to-Protection Ratio

The world's priority ecoregions described in the previous section that have been given critical/endangered conservation status stand out because of their conservation value and risk of being lost. Their status was decided upon using a large set of parameters, including current and predicted habitat loss, degree of fragmentation, present and future threats, and level of protection. However, since the publication of the Global 200 (Olson and Dinerstein 1998), several other approaches have been proposed to formally define the world's priority ecoregions.

Hoekstra et al. (2005) proposed the identification of terrestrial biomes and ecoregions with the greatest risk of losing biodiversity and ecosystem integrity by using a combination of two criteria: extensive habitat conversion and limited habitat protection. To classify the ecoregions into categories analogous to those established for IUCN Red List species (IUCN 2001), they calculated the Conservation Risk Index (CRI), an index of relative risk of biome-wide biodiversity loss, which is the ratio of percentage area converted to percentage protected area, for 13 terrestrial biomes and each of the world's 810 ecoregions (Olson et al. 2001). Ecoregions in which habitat conversion >20% with CRI >2 were classified as vulnerable ($n = 161$); those in which conversion >40% and CRI >10 were classified as endangered ($n = 80$); and those with conversion >50% and CRI >25 were classified as critically endangered ($n = 64$) (Figure 7.3). The highest CRI was exhibited by temperate grasslands and savannas, and Mediterranean forests, woodlands, and scrub biomes (more than 8:1) (Figure 7.4), and in more than 140 ecoregions this ratio was more than 10:1.

Another approach to identifying priority areas among the world's ecoregions that have forests within their area was proposed by Gillespie et al. (2012). In this approach, estimates of native vegetation cover and extent of protected areas are used to identify the ecoregions with the least amount of forest in protected areas.

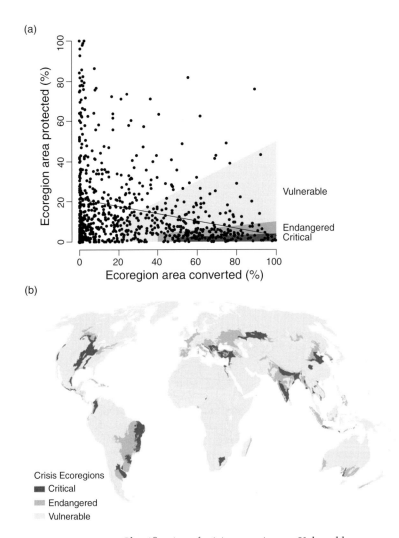

FIGURE 7.3 Classification of crisis ecoregions as Vulnerable,
Endangered, and Critically endangered (a), and the world map of
crisis ecoregions (b) (from Hoekstra *et al.* 2005). Scatterplot shows
the relationship between percentage area converted and percentage
area protected in each of 810 terrestrial ecoregions (slope = –0.18).
Ecoregions with >50% habitat conversion and Conservation Risk
Index (CRI) >25 are classified as Critically endangered (red); ecoregions
with >40% conversion and CRI >10 are classified as Endangered
(orange); and those with >20% conversion and CRI >2 are classified as
Vulnerable (yellow). CRI for each ecoregion was calculated as the ratio
of percentage area converted to percentage area protected. *A black and
white version of this figure will appear in some formats. For the color
version, please refer to the plate section.*

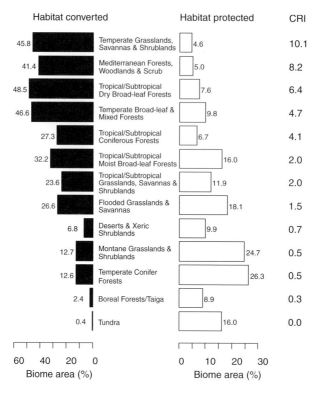

FIGURE 7.4 Habitat conversion and protection in the world's 13 terrestrial biomes ordered by their Conservation Risk Index (CRI) calculated as the ratio of percentage area converted to percentage area protected (from Hoekstra *et al.* 2005).

Applying this method to the ecoregions within the biodiversity hotspots, Gillespie *et al.* (2012) identified 47 forest ecoregions that contained less than 10% forest cover in protected areas, the minimum extent of forest cover to be protected to prevent ecosystem collapse according to the Convention on Biological Diversity (CBD 2009) (Table 7.3). These heavily deforested habitats should be the conservation priority areas.

## 7.1.4 Past and Future Refugia

As discussed in Section 2.3, refugia, locations where species have persisted during long-term climatic changes due to their eco-climatic

Table 7.3 *Rare forest ecoregions within biodiversity hotspots having less than 10% of forest cover in protected areas (from Gillespie et al. 2012)*

| Biodiversity hotspot and WWF forest ecoregion | Total forest cover (km²) | Forest in protected areas (%) |
|---|---|---|
| Atlantic Forest | | |
|   Pernambuco Coastal Forests | 802.9 | 5 |
| California Floristic Province | | |
|   East Cascades–Modoc Plateau Forests | 1 828.9 | 1 |
| Caribbean Islands | | |
|   Bahamian Pine Mosaic | 1 336.8 | 9 |
| Caucasus | | |
|   Northern Anatolian Conifer and Deciduous Forests | 2 965 | 4 |
| Chilean Winter Rainfall and Valdivian Forests | | |
|   Juan Fernández Islands Temperate Forests | 2.4 | 0 |
|   San Félix-San Ambrosio Islands Temperate Forests | NA | 0 |
| East Melanesian Islands | | |
|   Admiralty Islands Lowland Rain Forests | 1 111.6 | <1 |
|   New Britain-New Ireland Montane Rain Forests | 6 936.8 | <1 |
|   Solomon Islands Rain Forests | 14 307.0 | 1 |
| Guinean Forests of West Africa | | |
|   Cross-Niger Transition Forests | 13 824.0 | 1 |
| Indo-Burma | | |
|   Andaman Islands Rain Forests | 2 791.8 | 2 |
|   Irrawaddy Dry Forests | 1 576.9 | 8 |
|   Meghalaya Subtropical Forests | 10 452.1 | 1 |
|   Northeast India-Myanmar Pine Forests | 5 315.5 | <1 |
|   Northern Khorat Plateau Moist Deciduous Forests | 1 852.5 | 4 |

*(continued)*

Table 7.3 (*cont.*)

| Biodiversity hotspot and WWF forest ecoregion | Total forest cover (km²) | Forest in protected areas (%) |
|---|---|---|
| Irano-Anatolian | | |
| Eastern Anatolian Deciduous Forests | 278.1 | 2 |
| Kopet Dag Woodlands and Forest Steppe | 0 | 0 |
| Zagros Mountains Forest Steppe | 21.0 | 3 |
| Mediterranean Basin | | |
| Cape Verde Islands Dry Forests | 70.4 | 7 |
| Eastern Med. Conifer-Sclerophyllous Broad-Leaved Forests | 3 210.9 | 2 |
| Mesoamerica | | |
| Chiapas Depression Dry Forests | 2 425.7 | 6 |
| Chiapas Montane Forests | 4 875.2 | 5 |
| Chimalapas Montane Forests | 1 664.8 | 5 |
| Cocos Island Moist Forests | 0 | 0 |
| Oaxacan Montane Forests | 6 746.9 | 3 |
| Veracruz Dry Forests | 371.8 | 2 |
| Veracruz Montane Forests | 4 069.4 | 4 |
| New Zealand | | |
| Chatham Island Temperate Forests | NA | 0 |
| Kermadec Islands Subtropical Moist Forests | 13.4 | 1 |
| Philippines | | |
| Sulu Archipelago Rain Forests | 764.9 | <1 |
| Polynesia-Micronesia | | |
| Carolines Tropical Moist Forests | 2.4 | 4 |
| Central Polynesian Tropical Moist Forests | NA | 0 |
| Cook Islands Tropical Moist Forests | NA | 0 |
| Eastern Micronesia Tropical Moist Forests | 0 | 0 |
| Fiji Tropical Dry Forests | 948.1 | 1 |
| Fiji Tropical Moist Forests | 3 739.9 | 2 |
| Marquesas Tropical Moist Forests | NA | 0 |
| Palau Tropical Moist Forests | 208.2 | 7 |

Table 7.3 (*cont.*)

| Biodiversity hotspot and WWF forest ecoregion | Total forest cover (km²) | Forest in protected areas (%) |
|---|---|---|
| Society Tropical Moist Forests | NA | 0 |
| Tuamotu Tropical Moist Forests | NA | 0 |
| Tubuai Tropical Moist Forests | NA | 0 |
| Western Polynesian Tropical Moist Forests | 0.6 | 0 |
| Yap Tropical Dry Forests | 41.3 | 0 |
| Tropical Andes | | |
| Cauca Valley Dry Forests | 3 603 | 0 |
| Magdalena Valley Dry Forests | 10 324.5 | <1 |
| Patía Valley Dry Forests | 948.2 | 0 |
| Western Ghats and Sri Lanka | | |
| Maldives-Lakshadweep-Chagos Tropical Moist Forests | 24.9 | 0 |

stability, have high conservation value. Because these were locations where species that disappeared in surrounding areas were preserved during past climate fluctuations, refugia are more likely than other areas to preserve species living in locations that are predicted to become unsuitable due to ongoing anthropogenic climate changes in the near and more distant future. These considerations motivated recent and rapidly growing interest in the identification and incorporation of potential climate refugia in conservation planning (Shoo *et al.* 2011; Ashcroft *et al.* 2012; Groves *et al.* 2012; Keppel *et al.* 2012, 2015; Reside *et al.* 2014; Morelli *et al.* 2016). Including refugia in protected area networks as a means to keep pace with the anthropogenically changing climate was first proposed by Noss (2001), and now they are among priority areas for conserving biodiversity (Loarie *et al.* 2008; Klein *et al.* 2009; Game *et al.* 2011; Shoo *et al.* 2011; Olson *et al.* 2012).

Refugia, by definition, are areas that have some features that maintain favorable climatic conditions, even if in the surrounding

Topographically complex terrain creates varied microclimates and increases the likelihood that current climates will continue to exist nearby.

Deep snow drifts provide insulation to the surface below and provide water later in the season.

Valleys that harbor cold air pools and inversions can decouple local climatic conditions from regional circulation patterns.

Canopy cover can buffer local temperature maximums and minimums throughout the year.

Poleward-facing slopes and aspects result in shaded areas that buffer solar heating, particularly during the low solar angles of winter and early spring.

Cold groundwater inputs produce local cold-water refuges in which stream temperature is decoupled from air temperature.

Areas near or in large deep lakes or oceans will warm more slowly due to the high heat capacity of water.

FIGURE 7.5 Examples of the physical basis for geographic locations likely to act as refugia (from Morelli *et al.* 2016).

landscape the conditions become more climatically extreme. The scale of these features allowing decoupling of local from regional climates and the creation of (micro)climate heterogeneity ranges from relatively large geographic scales (mountain range characteristics and proximity to coasts) to small-scale topography (mountain slope aspect, depression depth) (Dobrowski 2011; Ashcroft *et al.* 2012; Keppel *et al.* 2012; Reside *et al.* 2014; Morelli *et al.* 2016; Harrison and Noss 2017). Examples of these features are shown in Figure 7.5 and discussed below.

The most important environmental feature in creating refugia is mountainous topographic heterogeneity. The latter usually leads to cold-air pooling, the formation of a cooler and moister air layer near the ground in valleys and other topographic depressions in comparison with surrounding uplands. This local climatic buffering of regionally hot and dry climate is common in sheltered mountain valleys. A similar cooling role allows inland penetration of coastal fog and low stratus clouds caused by offshore upwelling. Sheltered gorges and karst depressions provide protection from solar radiation and hot drying winds. Mountain tops can also act as refugia, as was found by Ohlemueller *et al.* (2008), for many cold-adapted small-ranged European plants and butterflies that found refuge there during unusually warm periods of the last 10 000 years. In the tropics, cool mountain environments are major areas of endemism and are very likely to act as refugia under climate warming.

Another type of topographic heterogeneity creating refugia is one redistributing runoff and soil moisture. The riparian zones, rock glaciers, ravines, brooks, and groundwater-fed seeps and springs can act as climate change refugia supporting many plant and animal taxa, especially in arid regions.

Both large areas that served as refugia for many species during the Pleistocene glacial/interglacial temperature fluctuations, and isolated enclaves that played the same role but for fewer species over a much smaller geographic extent, are important for conservation. The names of these two refugium types reflect the difference in geographic size: the former called macrorefugia, and the latter called microrefugia (Rull 2009; Hampe and Jump 2011; Mosblech *et al.* 2011; Mee and Moore 2014). Both macrorefugia (e.g., mountain ranges) and microrefugia (e.g., springs and gorges) are of conservation interest. Generally, the larger the refugium, the higher the chance that a larger number of species can survive in it when changing climate causes shrinking of their ranges. For animal species with large home ranges and plants with large natural breeding unit sizes (see Section 3.4.2), apparently only macrorefugia can provide a shelter. However, microrefugia have a greater resistance to the deterioration of regional climatic conditions than macrorefugia, and even if no macrorefugia remain in the former species range, many species with small home ranges and those with naturally low abundance can survive in very localized microrefugia that still have conditions suitable for the species. Thus, microrefugia can play crucial roles in hosting small populations of endemic and rare species, both within regional climatic refugia and in regions that are not refugia (Harrison and Noss 2017).

All of the above suggests that refugia should be included in networks of protected areas. Unfortunately, most of them are currently outside nature reserves and national parks, as, for example, was found by Deng and Volis for China (Tao Deng and Sergei Volis, unpublished results). In their investigation, they used published studies reporting precise locations of every haplotype detected with chloroplast markers for 152 species represented by 3364 populations to locate putative refugia in China. As a result of multi-species haplotype distribution mapping, they identified areas with high intraspecific diversity and/or those harboring high number of unique haplotypes not found elsewhere (Deng *et al.* 2019). The majority of these putative refugia were outside protected areas (Figure 7.6). These areas, like any other unprotected areas in China and elsewhere, suffer from anthropogenic disturbance and fragmentation, and can rapidly lose their capacity for buffering the effects of regional climate change if disturbed. Several studies have shown that the buffering of regional climate changes by refugia is due not only to their topography but also to such ecosystem parameters as canopy density and foliage cover (Shoo *et al.* 2011; Lenoir *et al.* 2017). Karst depressions can act as microclimatic refugia (Bátori *et al.* 2014), but deforestation of their landscape can significantly change their microclimate. Comparison of microclimate variability for two large depressions in the karst landscape of Hungary, one of which was covered with mesic forest and the other a clear-cut forest, revealed that their air humidity and temperature ranged from 95 to 100% versus 50 to100% and 18 to 25°C versus 10 to 32°C, respectively (Lehmann 1970). Similarly, logging can reduce or eliminate the buffering effect of canopy on the understory vegetation. Analysis of more than 1400 vegetation plots in temperate forests across Europe and North America revealed that the microclimatic effect of forest canopy closure buffers thermophilization of forest biodiversity caused by macroclimate warming (i.e., increase in abundance of species adapted to warmer conditions and/or decrease in abundance of cold-adapted organisms), while the microclimatic effect of canopy openness only accelerates thermophilization (De Frenne *et al.* 2013).

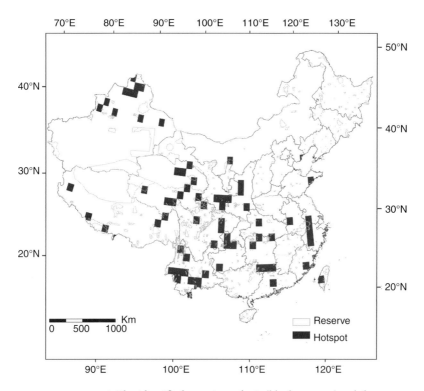

FIGURE 7.6 The identified putative refugia (black squares) and the network of Chinese protected areas (green contour lines). *A black and white version of this figure will appear in some formats. For the color version, please refer to the plate section.*

The aforementioned study of Deng and Volis revealed many previously unidentified putative refugia in China, which suggests that numerous very localized and small areas could act as microrefugia in this country's topographically diverse mountain chains, providing a variety of microclimates – including milder and moister conditions – throughout the glacial/interglacial cycles. These allowed the long-term survival of many species. This high climatic and topographic complexity made possible the survival of taxa that went extinct elsewhere (e.g., *Ginkgo biloba* and *Metasequoia glyptostroboides*) in very localized mountainous enclaves of central and southern China that apparently served as refugia throughout the Pleistocene.

Some specific microhabitats (e.g., riverbanks in deep gorges or closed mountain valleys) maintained moist conditions even during periods of drought, enabling some Tertiary relict plants that required a humid-warm climate to persist (e.g., *G. biloba*, *Davidia involucrata*, *Cathaya argyrophylla*, and *Taiwania cryptomerioides*) (Tang *et al.* 2014). Nowadays, many of these locations that acted as refugia in the past are mismanaged and require restoration (see also Sections 5.1.5 and 7.2.2). Other specific microhabitats located in low-altitude areas that acted as refugia in the past no longer exist in China because they were largely destroyed or significantly altered by humans in historic times. The moist and warm conditions of subtropical Chinese valley bottoms, coastal plains, and wetlands once provided a home for diverse relict warm-temperate, mesophyllous, and sclerophyllous vegetation, but these are now almost completely gone. A good example is *Metasequoia glyptostroboides*, which currently grows on wet lower slopes and along streams in a very few locations in the Yangtze River valley but apparently covered the whole valley floor before humans settled there, as indicated by numerous *M. glyptostroboides* logs up to 8 m in diameter that were discovered in the valley's rice paddies (Yang *et al.* 2004). At least some of these past refugia must be both protected and restored, or even recreated, as they are critical as habitats for many relict species.

Several authors have provided a similar perspective in recently published papers. In Southwestern Australia, which is a global biodiversity hotspot, rising water temperature under global warming is a conservation concern because it may exceed the thermal tolerances of local stenobiotic aquatic fauna. Modeling showed that 50–70% shade is needed in areas adjacent to the channel to maintain the rivers' and streams' water temperature within the tolerance limits of aquatic fauna, and that an increase in shade of about 10% is required to nullify every 1°C rise in water temperature. Therefore, replanting riparian zones, especially upland stream banks close to the channel, is recommended to keep the local freshwater ecosystem as a refugium for the local aquatic fauna (Davies 2010).

The Wet Tropics ecoregion of Northeastern Australia comprises a series of disjunct mountain chains running in parallel to the coast that support diverse rainforest and sclerophyll forest vegetation. These cool mountains have high endemism and are predicted to serve as important refugia under global warming. However, although a large part of this ecoregion is under protection as a World Heritage Site, about 21% of the pre-European distribution of rainforest in this area has been cleared of native vegetation. Shoo *et al.* (2011) found that foliage cover is an important factor ameliorating the effect of solar radiation in this area; the independent effect of dense foliage cover typical of rainforest was about 3°C during the hottest months. The authors identified the areas in the landscape where targeted restoration (i.e., reestablishment of foliage cover through reforestation) is likely to yield the greatest benefit in terms of increasing the availability of cool habitat (Figure 7.7). The areas proposed for restoration will fulfill several tasks. They will increase the local extent of cool habitat, consolidate extant patches of remnant rainforest within cool refugia, and facilitate dispersal between small and disconnected populations in this heavily fragmented landscape.

In light of the above, not only protection of potential refugia but also conservation-oriented restoration of those that have been altered or transformed must be a regional priority. As was noted by Harrison and Noss (2017), "rapid warming will make even relatively stable climatic refugia unsuitable for the long-term sustainability of biodiversity, while continuing loss of natural habitat will eliminate even the most climatically stable macro- and microrefugia."

## 7.1.5 *Important Plant Areas*

Within large regions of conservation importance, identified using the criteria described in the previous sections, identification of particular sites can use the concept of Important Plant Areas (IPA), a procedure that prioritizes sites of high botanical importance. IPA is defined as "a natural or semi-natural site exhibiting exceptional botanical richness and/or supporting an outstanding assemblage of rare, threatened

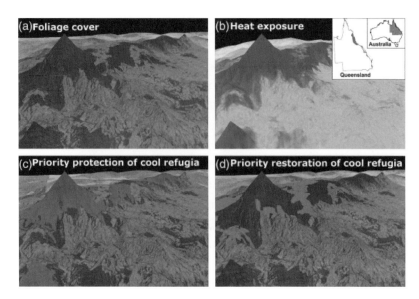

FIGURE 7.7  Cool rainforest refugia within a modified landscape (from Shoo *et al.* 2011). Patterns of foliage cover (green, a) have a strong influence on buffering heat exposure (i.e., maximum temperature of the warmest period, blue to red, b). To increase the local extent of cool rainforest refugia, it is proposed to supplement the extant protected areas (blue, c) with formerly supporting cool habitat but currently deforested areas after restoration of their foliage cover through reforestation (purple, d). *A black and white version of this figure will appear in some formats. For the color version, please refer to the plate section.*

and/or endemic plant species and/or vegetation of high botanic value" (Anderson 2002). The IPAs program, developed in the early 2000s by Plantlife International, uses a set of widely applicable criteria for this purpose. These criteria utilize national, regional, and global data to select national IPA networks. IPAs are identified at a national level on the basis of three consistent criteria: the presence of threatened species, exceptional botanical richness, and threatened habitats (Anderson 2002; Plantlife International 2004; Darbyshire *et al.* 2017). To qualify as an Important Plant Area, a site needs to satisfy one or more of these three criteria. Figure 7.8 describes the quantifiable thresholds and the sources of data for each criterion.

| CRITERION/SUBCRITERION | THRESHOLD |
|---|---|
| **(A) Threatened species** | |
| A(i) Site contains one or more globally threatened species | Site known, thought or inferred to contain ≥1% of the global population AND/OR ≥5% of the national population OR the 5 "best sites" for that species nationally, whichever is most appropriate |
| A(ii) Site contains one or more regionally threatened species | Site known, thought or inferred to contain ≥5% of the national population OR the 5 "best sites" for that species nationally, whichever is most appropriate |
| A(iii) Site contains one or more highly restricted endemic species that are potentially threatened | Site known, thought or inferred to contain ≥1% of the global population AND/OR ≥5% of the national population OR the 5 "best sites" for that species nationally, whichever is most appropriate |
| A(iv) Site contains one or more range restricted endemic species that are potentially threatened | |
| **(B) Botanical richness** | |
| B(i) Site contains a high number of species within defined habitat or vegetation types | For each habitat or vegetation type: up to 10% of the national resource can be selected within the whole national IPA network OR the 5 "best sites" nationally, whichever is the most appropriate |
| B(ii) Site contains an exceptional number of species of high conservation importance | Site known to contain ≥3% of the selected national list of species of conservation importance OR the 15 richest sites nationally, whichever is most appropriate |
| B(iii) Site contains an exceptional number of socially, economically or culturally valuable species | Site known to contain ≥3% of the selected national list of socially, economically or culturally valuable species OR the 15 richest sites nationally, whichever is most appropriate |
| **(C) Threatened habitat** | |
| C(i) Site contains globally threatened or restricted habitat/vegetation type | Site known, thought or inferred to contain ≥5% of the national resource (area) of the threatened habitat type OR site is among the best quality examples required to collectively prioritise 20–60% of the national resource OR the 5 "best sites" for that habitat nationally, whichever is the most appropriate |
| C(ii) Site contains regionally threatened or restricted habitat/vegetation type | |
| C(iii) Site contains nationally threatened or restricted habitat/vegetation type, AND/OR habitats that have severely declined in extent nationally | Site known, thought or inferred to contain ≥10% of the national resource (area) of the threatened habitat type OR site is among the best quality examples required to collectively prioritise up to 20% of the national resource OR the 5 "best sites" for that habitat nationally, whichever is most appropriate |

FIGURE 7.8 IPA selection criteria/subcriteria with their threshold values (from Darbyshire *et al.* 2017).

Identification of IPAs is important for highlighting gaps in the national protected area networks and for localizing sites for which conservation action is more urgent and essential. IPAs, because they use a standard, robust, and transparent selection procedure, provide clear evidence of the biodiversity value of a site, which is important for influencing decision-making in regional development projects. Identification of sites of botanical importance through the application of IPA criteria has been successful in Europe, North Africa, and the Middle East (Darbyshire *et al.* 2017), but has been less popular in other parts of the world. One reason is that it suits countries well with relatively poor endemism or species richness, producing IPA lists that have a manageable number of sites. In megadiverse countries, the number of IPA sites can be so huge that such a list will have no practical use. Regardless, the IPA concept is a very useful tool for identifying site-based plant conservation priorities, and therefore for choosing sites for conservation-oriented restoration. The IPA guidance documentation produced by Plantlife International repeatedly states that the identified IPAs are not necessarily sites to be designated for strict protection. Rather they are sites for appropriate management by protected area managers, land-owners, and other interested parties (Palmer and Smart 2001; Anderson 2002; Plantlife International 2004; Darbyshire *et al.* 2017). IPAs and PMRs have much in common (see Section 2.10).

The IPA concept explicitly targets botanical diversity, and therefore IPAs identified will not necessarily be habitats that are important for maintaining animal biodiversity. Thus, IPAs are either a subset or a complement to the areas identified using criteria accounting for both plants and animals, such as key biodiversity areas described in the next section.

### 7.1.6   *Key Biodiversity Areas*

The key biodiversity areas (KBA) approach, first systematically presented in Eken *et al.* (2004), is a rapid assessment methodology for identifying local-scale priority conservation areas based on global-scale data. KBAs are sites that contain globally important populations of threatened species or threatened ecosystems and, thus, contribute significantly to the global persistence of biodiversity in terrestrial and aquatic environments. Identification of a KBA is done through site assessment against all relevant quantitative criteria listed by IUCN (2016) (Figure 7.9) for which data are available, but meeting the thresholds under any one of the criteria or subcriteria is sufficient for a site to be recognized as a KBA. The criteria (and subcriteria) thresholds are designed to be applied nationally, while allowing

| CRITERION/SUBCRITERION | THRESHOLD |
| --- | --- |
| **A. Threatened biodiversity**<br>A1. Threatened species | Site regularly holds one or more of the following:<br>(a) ≥0.5% of the global population size AND ≥5 reproductive units of a CR or EN species;<br>(b) ≥1% of the global population size AND ≥10 reproductive units of a VU species;<br>(c) ≥0.1% of the global population size AND ≥5 reproductive units of a species assessed as CR or EN due only to population size reduction in the past or present;<br>(d) ≥0.2% of the global population size AND ≥10 reproductive units of a species assessed as VU due only to population size reduction in the past or present;<br>(e) Effectively the entire global population size of a CR or EN species |
| A2. Threatened ecosystem types | Site holds one or more of the following:<br>(a) ≥5% of the global extent of a globally CR or EN ecosystem type;<br>(b) ≥10% of the global extent of a globally VU ecosystem type |
| **B. Geographically restricted biodiversity**<br>B1. Individual geographically restricted species | Site regularly holds ≥10% of the global population size AND ≥10 reproductive units of a species |
| B2. Co-occurring geographically restricted species | Site regularly holds ≥1% of the global population size of each of a number of restricted-range species in a taxonomic group, determined as either ≥2 species OR 0.02% of the global number of species in the taxonomic group, whichever is larger |
| B3. Geographically restricted assemblages | Site regularly holds one or more of the following:<br>(a) ≥0.5% of the global population size of each of a number of ecoregion-restricted species within a taxonomic group, determined as either ≥5 species OR 10% of the species restricted to the ecoregion, whichever is larger;<br>(b) ≥5 reproductive units of ≥5 bioregion-restricted species OR 30% of the bioregion-restricted species known from the country, whichever is larger, within a taxonomic group;<br>(c) Part of the globally most important 5% of occupied habitat for each of ≥5 species within a taxonomic group |
| B4. Geographically restricted ecosystem types | Site holds ≥20% of the global extent of an ecosystem type |
| **C. Ecological integrity** | Site is one of ≤2 per ecoregion characterized by wholly intact ecological communities, comprising the composition and abundance of native species and their interactions |
| **D. Biological processes**<br>D1. Demographic aggregations | Site predictably holds one or more of the following:<br>(a) An aggregation representing ≥1% of the global population size of a species, over a season, and during one or more key stages of its life cycle;<br>(b) A number of mature individuals that ranks the site among the largest 10 aggregations known for the species. |
| D2. Ecological refugia | Site supports ≥10% of the global population size of one or more species during periods of environmental stress, for which historical evidence shows that it has served as a refugium in the past and for which there is evidence to suggest it would continue to do so in the foreseeable futures |
| D3. Recruitment sources | Site predictably produces propagules, larvae, or juveniles that maintain ≥10% of the global population size of a species |

FIGURE 7.9  KBA selection criteria/subcriteria with their threshold values (IUCN 2016).

identification of KBAs across all biogeographic regions and taxonomic groups globally. Areas that satisfy these criteria are priority targets for conservation management and represent current or potential protected areas (Langhammer *et al.* 2007).

The KBAs identified should support the strategic expansion of protected area networks in a country and influence decision-making in national development plans (Dudley *et al.* 2014). Although Eken *et al.* (2004) recommended that "KBAs should not be identified for potentially important sites requiring restoration until such sites firmly meet the criteria," the later literature lists prioritization of sites for restoration as another important role of KBAs (Dudley *et al.* 2014). Indeed, many sites with the highest conservation importance will not be able to maintain viable populations of threatened species or fully functional ecosystems without some form of restoration activity. Islands, discussed in Section 7.1.8, are clear examples.

KBAs have several advantages in comparison with IPAs. First, as already stated above, KBA criteria apply to the whole spectrum of regional biota, while IPA identification ignores its animal component. Second, KBA selection has a wider range of criteria and considers attributes neglected in IPA target identification, such as their potential to act as a refugium during periods of environmental stress, ecosystem degradation level, population demographic structure, and recruitment. As has been shown in the preceding sections, these attributes are very important for choosing target areas if the goal is either species-targeted conservation or conservation-oriented restoration.

### 7.1.7 Areas of Imminent Species Extinction

As was discussed in Sections 7.1.1–7.1.3, conservation planning should focus first of all on a limited number of the most important ecoregions to protect biodiversity, especially those regions that have experienced extensive habitat loss and are underprotected, with the aim of protecting entire ecosystems. Within ecoregions, as discussed in Sections 7.1.4–7.1.6, some environmental features crucial for ecosystem functioning under current or future climate, and key for preserving biodiversity sites should be the priority targets. However, among small-scale conservation priority areas, there exists a particularly important subset of sites: those that harbor the last remaining populations of threatened species. These sites represent the most urgent site-scale priorities.

Among those species that are critically endangered, the most vulnerable, other things being equal, are species occurring at a single location. Species with an extremely localized range and a single population face imminent extinction in the absence of urgent conservation action. Therefore, areas representing the last known species location are irreplaceable (i.e., cannot be compensated by protecting another site) and must have top priority for conservation through strict protection. However, such sites are also especially important targets for conservation-oriented restoration, because, although a variety of conservation approaches may be helpful, the first and foremost will be to do what is necessary to the habitat to sustain a viable population of a species in its single remaining site.

In an attempt to identify such conservation emergency sites, the Alliance for Zero Extinction (AZE; Alliance for Zero Extinction 2017) program (Ricketts *et al.* 2005) examined global distribution data for five taxonomic groups (mammals, birds, reptiles, amphibians, and conifers) in searching for sites that represent the sole area of occurrence for at least one highly threatened species. Methodologically, the AZE sites are a subset of KBAs. From the 595 irreplaceable areas identified (Figure 7.10), 103 sites had >1 single population species and many had numerous less range-restricted species of conservation concern. The sites were concentrated in tropical forests, on islands, and in mountainous areas. Despite the extreme extinction threat faced by the species they located, 257 (43%) of the identified sites currently lack any legal protection and only 203 (34%) sites are fully contained within a PA. Most sites are surrounded by intense human development, while the intensity of human activity in protected and unprotected sites is nearly identical. All this suggests that the protection of a site's habitat is an essential but insufficient measure, and what is needed, as Ricketts *et al.* (2005) note, "is a combination of site-level activities and broader-scale efforts to conserve and restore habitats, address

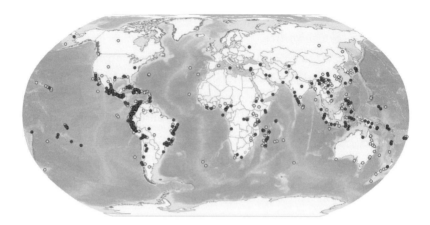

FIGURE 7.10 Map of 595 sites of imminent species extinction (from Ricketts *et al.* 2005). Light gray circles are sites that are either fully or partially contained within declared protected areas and black circles are completely unprotected or with unknown protection status sites. In areas of overlap, unprotected (black) sites are mapped above protected (gray) sites to highlight the more urgent conservation priorities. Copyright (2005) National Academy of Sciences, USA.

regional threats, and maintain ecological processes." It is impossible to disagree with their further suggestion that "the primary response to avoid these impending extinctions will be to safeguard their sites through land purchase, conservation easements, community management, or protected area enforcement, and to monitor their condition over time. In some cases, such measures will need to be complemented with control of invasive species or disease, translocation, or ex situ breeding, or cultivation." In other words many such locations will need species-targeted conservation and narrowly focused restoration of the habitat.

The species identified in Ricketts *et al.* (2005) represent only five taxonomic groups and therefore are only a subset of those species that are at high risk of extinction; the list would be much longer if it also included plants, fishes, terrestrial invertebrates, etc.

The approach of Ricketts *et al.* (2005) can be applied at a national level, using national Red Lists of threatened taxa. Diniz *et al.* (2017) used this methodology to identify conservation priority areas for Brazilian flora, and the number of emergency sites to prevent plant extinctions was found to be 140, while the global-scale analysis identified only 27 conservation emergency sites in Brazil (Alliance for Zero Extinction 2017). Of these 140 irreplaceable targets for conservation only 21 are located within protected areas.

## 7.1.8   Oceanic Islands

At all geographical scales, from global to regional, and then to local, oceanic islands have a special position with respect to biodiversity, endemism, and vulnerability.

Island biota usually exhibit a high degree of endemism because of their geographic isolation and the limited interchange with neighboring mainland or island biota. As a result of their long-term isolation oceanic islands and archipelagos harbor many unique lineages, both old (e.g., New Caledonia, Madagascar, the Seychelles, or New Zealand) and new (e.g., the Canary Islands or Hawaii). On the other hand, having a high proportion of relict taxa, oceanic islands show remarkable stability of their ecosystems during long geographic periods allowing the persistence of species while close relatives on the mainland went extinct. This stability, making oceanic islands climate refugia, is apparently due to the buffering of climatic fluctuations by the relatively thermally unresponsive ocean mass.

High levels of insular endemism and the presence of diverse adaptively radiated paleo- or neoendemic lineages gives oceanic islands unique conservation value (Kier *et al.* 2009; Caujapé-Castells *et al.* 2010; Whittaker *et al.* 2017) (Table 7.4). Unfortunately, islands are major sites of past and ongoing species extinction, and of some 80 documented plant extinctions in the last 400 years, about 50 have happened on islands (Sax and Gaines 2008). Islands, as was noted by Florens (2013), are among the first victims of the relentless human impact, and a similar daunting future awaits the rest of the tropical world as human overpopulation, habitat destruction, and alien species invasion show no signs of slowing down. The

Table 7.4 *Size, indicators of protection, and botanical richness of the nine oceanic archipelagos (from Caujapé-Castells et al. 2010)*

| | Land area (km²) | No. of PAs | Protected land area (%) | Population density per km²ᵃ | Endemic genera (species) | Endemic species | Tᵇ (%) |
|---|---|---|---|---|---|---|---|
| Azores | 2 332 | ~22 | ~20 | 105 | 1 (1) | 72 | na |
| Canaries | 7 545 | ~15 | ~40 | 269 | 23 (49) | ~607 | ~30 |
| Cape Verde | 4 033 | 2 | 0.2 | 102 | 1 (7) | 82 | ~43 |
| Galápagos | 7 900 | 2 | 96 | 3 | 7 (35) | 180 | ~60 |
| Hawaii | 16 636 | ~27 | ~5 | 77 | 32 (288) | 929 | ~53 |
| Juan Fernández | 100 | 1 | ~90 | 6 | 12 (33) | 133 | ~75 |
| Madeira and Selvagens | 794 | 5 | ~67 | 307 | 5 (11) | 136 | 49 |
| Mascarenes | 4 528 | 21 | 25 | 455 | 35 (81) | ~688 | ~50 |
| Seychelles | 235 | 7 | 16 | 345 | 12 (13) | ~70 | 71 |

ᵃ Approximate numbers.

ᵇ T proportion of endemic taxa that are in the Red List under some kind of threat (CR/EN/VU).

situation is particularly exacerbated on tropical oceanic islands, where between 3500 and 6800 of the estimated 70 000 insular endemic plant species might be highly threatened (CR and EN categories) and, of these, more than 2000 in critical danger of extinction (Caujapé-Castells *et al.* 2010) (Table 7.4). These threats result mainly from isolated and small population sizes, and simplified food webs and species interaction networks, making them prone to invasion by alien species and highly vulnerable to human-induced pressures (Sax and Gaines 2008; Kaiser-Bunbury *et al.* 2010).

An index-based assessment using the combined metric of endemism and species richness ("endemism richness"), which gives the most reliable and unbiased estimate of the contribution of an area to global biodiversity, clearly identified islands as globally important for conservation areas (Kier *et al.* 2009) (Figure 7.11). Half of the top 20 regions in terms of endemism richness per standard area were island regions (Figure 7.11). All but one island region ranked among the top third of the 90 terrestrial biogeographic regions and all had above average ranks. Endemism richness values for plants and vertebrates were strongly related, and values on islands exceeded those of mainland regions by a factor of 9.5 and 8.1 for plants and vertebrates, respectively. These results emphasize the outstanding importance of oceanic islands for global conservation. But their properties (small size and high susceptibility to human impact) make them especially vulnerable. Kier *et al.* (2009) quantified the spatial relationship between endemism richness and various measures of past and projected future human impact. They found that island and mainland regions were similarly affected by habitat loss in the past, but significantly differ in the level of current threat and its major cause. The "human impact index" quantifying the worldwide presence of humans per 1 $km^2$ was significantly higher for islands than for mainlands (Figure 7.11) and, while for mainland regions the major cause of their original land-cover future loss is climate change, for islands it is intensifying land use (agriculture, deforestation, urbanization). The islands where further vegetation loss does not happen

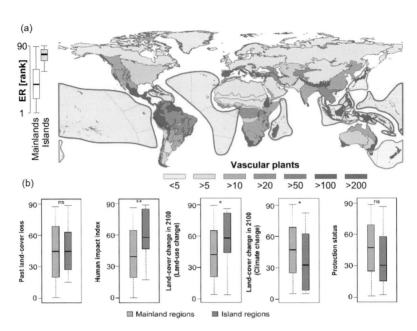

FIGURE 7.11 A global assessment of endemism and species richness across island and mainland regions (extracted from Kier *et al.* 2009). (a) Global patterns of endemism richness (ER; range equivalents per 10 000 km²) for vascular plants across 90 biogeographic regions. Map legends were classified using quantiles, i.e., each color class contains a comparable number of regions. Box-and-whisker plots illustrate rank-based differences in ER between mainland ($n$ = 76; white boxes) and island regions ($n$ = 14; gray boxes). Boxes mark second and third quartiles; whiskers mark the range of the data. (b) Comparison of key conservation features (expressed as ranks) for mainland (green) and island regions (orange), including past land-cover loss (year 2000), human impact (year 2000), future land-cover loss (projections for the year 2100) because of land-use change and climate change, and current protection status (proportion covered by protected areas). Significance codes indicate differences (Mann–Whitney U-test) between mainlands and islands: **, $P < 0.01$; *, $P < 0.1$; ns, not significant. *A black and white version of this figure will appear in some formats. For the color version, please refer to the plate section.*

are those where the native vegetation has already been completely eradicated. For example, the once-forested Easter Island currently has no remaining native forest cover and the only survivor of the island's endemic tree flora *Sophora toromiro* is maintained *ex situ*. On other

Table 7.5 *Conservation-related features of the Western Indian Ocean Islands (from Baret et al. 2013)*

| Island(s) | Land area (km²) | Land area with remaining native remnant vegetation cover (%)[a] | Protected land area (IUCN categories I–VI) (%) |
|---|---|---|---|
| Comoros | 1862 | 18.0 | 0.9 |
| Réunion | 2512 | 39.8 | 77.1 |
| Mauritius | 1865 | 1.6 | 4.7 |
| Mayotte | 363 | 5.0 | 11.5 |
| Rodrigues | 109 | 0.5 | 0.6 |
| Seychelles | 455 | 1.0 | 59.9 |

[a] Forest with at least 50% canopy cover of natives including lightly or moderately invaded.

islands, further land-cover loss is inevitable under local population growth due to islands' small sizes giving easy access to remote parts with remaining primary vegetation. Ironically, this imminent threat of destruction is accompanied by lower levels of protection than in the mainland regions (8.3% versus 10.6% of area protected; Kier *et al.* 2009), and for some islands, the percentage of protected area is negligible (Tables 7.4 and 7.5), especially in view of rapidly disappearing native vegetation cover (Table 7.5).

From the 47 rare forest ecoregions located within biodiversity hotspots and having less than 10% of forest cover in protected areas identified by Gillespie *et al.* (2012) (Table 7.3), the majority of the areas with the heaviest deforestation and lowest protection levels are located on islands. Thus, it is impossible to disagree with Kier *et al.* (2009) when they suggest "that conservation of island biodiversity requires the expansion of existing protected area networks to counteract the threats that island diversity faces as a result of the extraordinarily high levels of human impact," and that "furthermore, effective measures need to be taken to address threats that cannot be mitigated by protected areas alone, including threats arising from invasive species."

The impacts of invasive aliens combined with deforestation and overgrazing have turned many islands into novel ecosystems; this new trajectory is irreversible. However, in those islands that still have noneroded soil and at least some remnants of the indigenous vegetation, even if many former island inhabitants are extinct and the original vegetation has been replaced, recovery is still possible. Unfortunately, in comparison with a great deal of research on the negative impacts that tropical islands have undergone in the past, very limited attention has been given to their restoration. Nevertheless, although still limited, there is some experience of how to restore the island ecosystems (Burney and Burney 2007, 2016; Samways *et al.* 2010b; Griffin-Noyes 2012; Kueffer *et al.* 2013) and a few attempts are being made around the world to restore a whole island (Samways *et al.* 2010a, b; Zender 2014). Because islands are physically defined units, their limited territorial extent and isolation from other territories allow easier eradication and control of invasive species, in comparison with mainland areas. If given focused attention, with halted habitat destruction, they can be efficiently restored, in some cases surprisingly rapidly (Samways *et al.* 2010b; Baret *et al.* 2013).

It would be incorrect to say that oceanic islands are not seen as a very serious conservation concern by international conservation organizations. The problem is that the majority of funds are used for case-by-case projects of threatened vertebrate species recovery. However, the available evidence suggests that such a species-centric conservation approach focusing on a few charismatic vertebrates: (1) has in many cases very limited success and low impact, and (2) is expensive and takes away money that could be used for much higher impact and cost efficiency conservation work, such as larger scale habitat restoration. Florens (2013), in his review of the conservation activities of the last four decades on Mauritius and Rodrigues islands of the Mascarene Archipelago, provided several examples supporting this view and concluded that "over the course of some four decades of sustained and often intense and costly conservation efforts, only a handful of species seem to have been saved from virtually certain extinction." One of several species rescued from extinction, the pink pigeon *Nesoenas mayeri* on Mauritius, however, is under continuous artificial management through the provision of supplementary food, the control of alien predators, and the management of disease in "wild" populations. The use of feeding hoppers is necessary for this species because the native forest – its only suitable natural habitat – has been reduced to less than 1% of its original cover, which is too little to support the recovering bird population. The heavy dependence of restored populations of pink pigeon and other species (e.g., echo parakeet *Psittacula eques*) on intensive management calls into question the viability of their "restored" populations. True restoration of their populations will require restoration of their former habitat to the extent that allows their existence without supply of food and other management.

As a rule, restoration efforts require both unprotected and protected island areas because even within PAs invasive plants constitute a substantial threat (Baret *et al.* 2013; Florens and Baider 2013; Florens *et al.* 2016; Bellingham *et al.* 2018), and many PAs in part or as a whole represent highly degraded or alien-dominated ecosystems. In these systems, some native plant species continue to persist, but sometimes as only a few individuals per protected area. Several restoration strategies exist for highly invaded forests. In some cases, a solution is the maintenance of small stands of native-dominated vegetation patches in a matrix dominated by invaders with gradual enlargement of alien-eradicated areas. This can be done by periodic weeding of remnant stands of native trees (Baret *et al.* 2007; Baider and Florens 2011). In other cases, a solution is the creation of small gaps through the felling of one or a few invasive trees and replanting them with native plants (Kueffer *et al.* 2010; Griffin-Noyes 2012). The restored stands can serve as sources of native seed and promote spreading of native species into the matrix of invaders, and after expansion become corridors connecting fragmented native habitats. In highly degraded dry forests dominated by non-native graminoids, a promising strategy is non-native grass removal, ungulate exclusion, and reestablishment of native woody understory through nurse planting. Established nurse vegetation creates conditions necessary for both unassisted native seedling recruitment and the establishment of other planted tree and shrub species (Medeiros *et al.* 2014).

To conclude, having high conservation risks, small areas, and high levels of endemism richness, oceanic islands are areas of particularly high returns for protection and conservation-oriented restoration, and therefore must have a high priority in global conservation planning.

## 7.2    EXAMPLES OF CONSERVATION-ORIENTED RESTORATION PROJECTS

This section will present detailed examples of conservation-oriented restoration, envisioned, planned, or partly implemented. These examples are arranged by the scale (from single species to entire ecosystems) and intensity of the necessary interventions.

### 7.2.1   *Restoration of Critically Endangered* Wollemia nobilis

Restoration projects involving highly threatened species that have intact indigenous habitat (either occupied or potentially suitable) require well-planned and carefully performed interventions of very limited extent. These interventions, beside planting or seeding of the target species, may include minor liberation cutting or selective thinning. Such restoration is exemplified by the restoration of *Wollemia nobilis*, a critically endangered shade-tolerant rainforest conifer confined to a single wild population of 83 mature trees and 100–200 juveniles, and an area of less than 10 km$^2$ (Figure 7.12).

The *W. nobilis* habitat is a warm-temperate rainforest community in the sandstone canyon landscape of Wollemi National Park in Southeastern Australia. The viability of *W. nobilis* seed is low (~10%), and seed dispersal appears to be limited as no seedlings have been found farther than 50 m from mature trees. Establishment of seedlings is also limited, and growth of juveniles is very slow (<2 cm/year in height), although established juveniles can survive in the deep

FIGURE 7.12 Experimental translocation of critically endangered
*Wollemia nobilis*. The photos show: (a) the outplants produced at the
Australian Botanic Garden Mount Annan, (b) the translocation site, (c) a
small introduced outplant, and (d) a larger outplant. Photos by Haidi
Zimmer. *A black and white version of this figure will appear in some
formats. For the color version, please refer to the plate section.*

shade for 16 (and probably more) years. The understory of *W. nobilis*
habitat is characterized by ~3% light, but greenhouse experiments
have shown that stem growth of *W. nobilis* juveniles increases with
light (up to 50% light).

To find out the optimal natural conditions that would serve as
a basis for future translocations for *W. nobilis*, experimental colonization through underplanting was done in rainforest canopy gaps
spanning a range of different light conditions from deeply shaded
to high light gaps, as described in Zimmer *et al.* (2016). The introduction site was approximately 120 m long, following a creek, and
extending 60 m upslope from the creek, in another canyon with
similar vegetation, precipitation, and soil conditions. A total of 191

plants, grown for 2–3 years from cuttings from 20 different wild parents were planted in late winter, before the start of the *W. nobilis* growing season. Planting of six to seven plants was undertaken in plots of approximately 4 × 4 m within each plot, with at least 1 m between plants. Clearing of ground-level vegetation in area 1 m in diameter around planted individuals was done when required. The plants received supplementary watering of 5 l/plant twice. Two years after planting, 85% of plants survived. Despite previous studies showing long-term survival of wild *W. nobilis* juveniles in deep shade, light intensity had a strong positive effect on the survival of introduced plants. Subsequent interventions to stimulate good *W. nobilis* growth appear to be liberation cutting of the surrounding vegetation.

This study shows that the introduction of a highly threatened species into seemingly suitable locations outside the current range can overcome the apparent dispersal limitations, and that necessary interventions can be limited to planting in suitable microsites, and some minimal care after planting such as supplementary watering and caging.

### 7.2.2 Restoration of Threatened Relics in Mount Jinfo Nature Reserve

The next level of complexity in restoration is represented by projects focusing on one or several threatened species whose persistence requires alteration of their common habitat. The interventions applied to the habitat usually involve creating safe sites needed for these species recruitment. A good example is Mount Jinfo Nature Reserve in China which is a World Heritage Site and a home for several endangered Tertiary relict trees including *Cathaya argyrophylla*, *Davidia involucrata*, *Tetracentron sinense*, *Tapiscia sinensis*, and *Pterostyrax psilophyllus* (Figure 7.13).

All the above relict species are now represented by small and highly fragmented populations with regeneration problems. Being inferior competitors to broad-leaved evergreen *Castanopsis*, *Machilus*, and *Cyclobalanopsis*, and deciduous *Styrax*, *Fagus*, and

FIGURE 7.13 Mount Jinfo Nature Reserve. (a, b) The reserve landscape, (c) depressed growth of two juvenile *Cathaya argyrophylla* trees (marked by arrows) under the canopy of a large evergreen tree, (d) *C. argyrophylla* sapling growing in the canopy gap, and (e) clearcutting of understory during the bamboo shoot season. Photos by Yongchuan Yang. *A black and white version of this figure will appear in some formats. For the color version, please refer to the plate section.*

*Prunus* in Mt. Jinfo Nature Reserve and elsewhere, these species survive in unstable habitats such as scree slopes and ridges with high solar radiation and reduced competition from other trees. For example, young individuals of *C. argyrophylla* are usually seen in forest gaps and open places rather than under dense canopies (Qian *et al.* 2016) (Figure 7.13). It was repeatedly stressed that these species are disturbance-dependent and survive in locations where frequent erosion or landslides maintain availability of open, vegetation-free forest gaps (Tang *et al.* 2013, 2014).

Thus, to improve regeneration in these species, the creation of forest gaps through canopy thinning of broad-leaved trees could improve understory light conditions promoting growth and establishment of their seedlings. The first choice habitats for thinning should be stream banks and adjacent slopes, and scree slopes. Additionally, liberation cutting of the surrounding young trees and seedlings vegetation, first of all bamboos, should also be applied. Unfortunately, current Chinese laws and policies strictly prohibit any kind of interventions in the nature reserves, making the above activities impossible.

What seems a real paradox, however, is that by not allowing any conservation activities in the reserves, the current Chinese nature reserve policy requires only that large trees of threatened species be kept alive. This indulges and even motivates the local residents and visitors to the reserve to extract virtually any reserve resources either legally or illegally. In Mount Jinfo Nature Reserve, contractors and local workers are not only allowed but even supported by the local government of Nanchuan District, where the reserve is located, in any actions lowering the cost of bamboo shoot production and promoting the growth and spread of bamboo stands. In practical terms, this means that emerging tree seedlings (including seedlings of threatened species) are removed from the forest floor, the young sprouts are intentionally cut, and adult trees are ring-barked to accelerate their death. The dead trees are then used as firewood for boiling and preprocessing bamboo shoots *in situ* before they are dried (Qian *et al.* 2018). These actions effectively maintain the forest floor free of any non-bamboo vegetation in the nature reserve (Figure 7.13). If this practice is not interrupted, the non-bamboo vegetation will simply disappear in the future as still present adult trees die. Thus, the current reserve policy, instead of being a major guarantee of preserving the habitat and its wild inhabitants in Mount Jinfo National Nature Reserve, directly interferes with these goals. As the reserve has unique importance for several highly endangered relicts, the situation must radically change before it is too late.

## 7.2.3   *Restoration of a Native Hawaiian Forest at Limahuli Preserve*

Some protected but severely degraded areas where the native seed bank has been lost due to the dominance of invasive species can be not only reclaimed but returned to the "natural" state, i.e., to structurally complex forests with diverse native flora. This requires substantial efforts in terms of money and human effort, but these costs can certainly pay off if efforts lead to the recreation of the once existing ecosystem and delisting of endangered species.

In 2007, the Limahuli Preserve staff started the Groundcover Restoration Project aimed at restoring 4 ha of its territory that had been taken over by invasive species back into diverse native lowland forest by using a bottom-up approach. The project is described in Griffin-Noyes (2012). Limahuli Preserve is a part of the National Tropical Botanical Garden and the project would be impossible without the intensive preceding work of the garden botanists to conduct botanical surveys to document the historic diversity in the ecoregion, collect propagation material in the context of appropriate gene pools, and create *ex situ* living collections of many native species that went extinct in the wild or have otherwise undergone catastrophic population collapse.

Restoration of the preserve required the reintroduction of native plants into a highly degraded habitat dominated by invasive species with only a few native plant individuals per hectare. At first, the restoration technique involved clearing non-native vegetation and then planting a high number of fast-growing native trees to reestablish a native canopy. While initially effective on a small scale (0.2 ha) (Figure 7.14), scaling up turned out to be ineffective. This was due to the damaging effect of direct sun and strong winds which resulted from the removal of too many canopy trees, thus creating environmental conditions that gave invasive grass species a competitive advantage over both outplanted and regenerating native species. A new approach was attempted whereby the existing non-native canopy was left intact preventing the influx of solar radiation and wind that would otherwise allow for the colonization of the site by invasive grasses and moisture loss in the soil. Intensive planting in the understory (nearly 9000 plants annually) was accompanied by gradual removal of the canopy trees to stimulate growth of the planted trees. Low-intensity thinning allowed the growth of ferns and sedges but not grasses in

FIGURE 7.14 Restoration of a native Hawaiian forest at Limahuli
Preserve. (a–c) A time series of photographs taken at 6-month
intervals: (a) was taken immediately after initial tree planting, but a few
days before outplanting of groundcover species. The next two photos
show good establishment and vigorous growth of both common and
endangered species 6 months (b) and 1 year after planting (c). (d–e) The
reestablished native plant community that resulted from restoration,
using two different methods, small (0.2 ha) and large (2.0 ha) sections
of forest that had been completely taken over by invasive species.
Thirteen native plant species including three endangered tree species
(0.2 ha, d) and 14 native plant species including four endangered
tree species (2.0 ha, e) are identifiable in these pictures. Photo
credit: National Tropical Botanical Garden. *A black and white version
of this figure will appear in some formats. For the color version, please
refer to the plate section.*

the understory and, after the latter covered the forest ground, more
thinning and increased light levels allowed the introduction of less
shade-tolerant shrubs and trees. With a gradual increase in light level
a more and more diverse array of native species could be introduced.
Thinning involved prioritizing the non-native trees for removal based

on the threat posed by each species, compared to the amount of shade it provided. Besides the canopy trees, coffee trees were among the major targets for removal. The reestablished vegetation community that resulted from the new restoration approach was indistinguishable from the one recreated on the 0.2 ha plot (Figure 7.14).

In 2012, when a relatively stable forest had been recreated, the project started to focus on the utilization of rare and endangered species. In that year, 9000 individuals of approximately 100 different species were planted, of which 24 were endangered.

This project is an excellent example of habitat restoration with the objective of transforming a habitat where non-native and weed species replaced natives to one in which native, including endangered, species dominate and are no longer threatened.

## 7.2.4 Restoration of Cousine Island of the Seychelles

Cousine Island, Seychelles, is a small (27 ha) and until recently highly degraded tropical island that has received intensive restoration to "what is believed to be a semblance of the natural state" (Samways *et al.* 2010b). Until the mid-1970s, when the first phase of Cousine restoration started, the island had only small remnants of the original woody vegetation (mostly among the granite outcrops), but these remnants served as a baseline for recovery (Figure 7.15). During the first phase (until the early 1990s), major activities included putting a stop to poaching of indigenous animals, such as shearwaters and turtles, removal of some of the lowland coconut trees, and eradication of feral cats. These activities continued during the second phase (from the early to late 1990s) but the focus of the latter changed to restoring the coastal plain to what was believed to be its former natural state (Figure 7.15). A reference for restoration was worked out from (1) surveys of the remnants of the coastal vegetation on neighboring islands; (2) identification of seeds washed up onto the drift line of the beach, as this is the major colonization route for Seychelles islands' plants; and (3) monitoring of seeds dropped by the visiting island fruit bats.

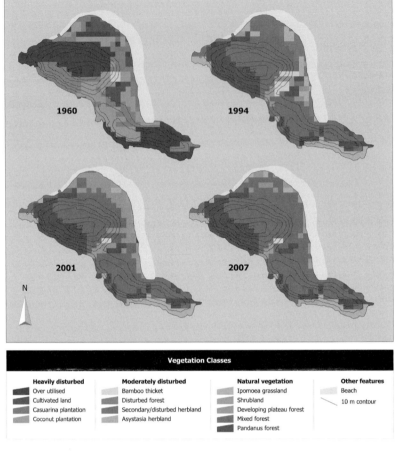

FIGURE 7.15 Changes in vegetation during Cousine Island restoration (from Samways *et al.* 2010a). *A black and white version of this figure will appear in some formats. For the color version, please refer to the plate section.*

The second phase included an intense removal program against alien plants and a propagation program to provide seedlings of native species for planting. The agricultural plants necessary for local residents' subsistence were confined to a designated 0.3 ha horticultural area, while all crops outside this area were removed. Newly established, from seeds dropped by birds, crop plants (e.g., papaya and soursop) were continuously removed upon detection.

FIGURE 7.16 (a–d) Recovered Cousine Island vegetation. Photos by
Michael Samways. *A black and white version of this figure will appear
in some formats. For the color version, please refer to the plate section.*

Chosen plant species were propagated in a plastic/shade cloth
covered tunnel from seeds or cuttings collected from both Cousine
and neighboring islands. Planting took place during the rainy season
(November–March) with temporary irrigation established for some
species that were more vulnerable to drought stress and temporary
shade cloth barriers erected for some more vulnerable to salt spray
species. Logs of felled alien trees were used to retain high soil moisture around the planted seedlings and protect them from grazing by
tortoises. A total of 2543 trees and shrubs of various species characteristic of lowland granitic Seychelles vegetation were planted from
1993 to 2001. Particularly important was planting and subsequent
establishment of Mapou trees (*Pisonia grandis*), the dominant component of a forest type that had almost completely disappeared in
the Seychelles and a favored nesting tree for the lesser noddy (*Anous
tenuirostris*).

Other activities of the second phase included enforced non-disturbance of the nesting seabirds and the reintroduction of highly threatened, Seychelles-endemic land birds, primarily the Seychelles magpie robin (*Copsychus sechellarum*) and the Seychelles white-eye (*Zosterops modestus*).

Demarcation of the third phase (year 2001 on) (Figure 7.16) coincided with canopy emergence in the planted stands. The 755 trees planted after 2001 served the goal of increasing species richness of the planted stands, and hence the structural and compositional diversity of the ecosystem.

Summarizing the Cousine restoration project, the coastal plain, once devoid of woody vegetation, is now covered with a forest carpet composed of native plant species, and of the 60 wild aliens, 25 have been eradicated, with the remainder in the process of being eradicated (17) or being controlled (26). The critically endangered Seychelles magpie robin has grown in number, as well as the endangered Seychelles warbler and Seychelles white-eye, allowing downlisting of the last two species from CR to EN. Other birds such as the shearwater, noddy, and land birds have also considerably increased in number; the sooty tern, *Onychoprion fuscatus*, reappeared in 2003, after its absence from the island for 31 years. Most important, the ecosystem trajectory appears to have been reverted and is now heading to the former island natural state. This project successfully demonstrates "that with determined effort, it is indeed possible to reverse the devastation so often seen on small tropical islands" (Samways *et al.* 2010b).

# Concluding Remarks and Prospects for the Proposed Strategy

This book presents a new concept of plant conservation that adopts as a cornerstone the idea of creating partly novel (i.e., having species compositions that differ from historical analogs) ecosystems for biodiversity conservation's sake. This concept has as crucial components interventions, assisted colonization, and experimentation, and utilization of threatened plant species in the restoration of degraded habitats. It has been conceived as a solution to numerous conservation challenges of the Anthropocene. If proved useful and of wide applicability, this concept may become a new conservation paradigm, the *habitat restoration paradigm*.

The concept includes the following basic elements:

1. conservation planning on a regional basis, which includes prioritization of species for *ex situ* and *in situ* actions in coordination;
2. assisting colonization of threatened species into as many locations as possible with conditions suitable in the near and more distant future;
3. prioritizing the restoration of least degraded areas that have extant populations of the threatened species;
4. targeting alternative states as reference ecosystems through adaptive learning, as a replicated over space experiment where outcomes are critically evaluated and used to inform the next introductions;
5. establishing conservation seed banks, i.e., germplasm storages of threatened and rare species designed to store, handle, and use a large number of seeds per species rather than just a large number of species, with the collected seeds being fully available for *in situ* actions;
6. obtaining a sufficient number of outplants through seed production in living *quasi in situ* collections, and then producing seedlings in specialized nurseries;

355

7. planting design based on experimentation with species assemblages composition and replicating these experimental assemblages over space;

8. reestablishing the integrity of disrupted interactions crucial for ecosystem functioning (seed dispersal, pollination, nutrient cycling, and food webs);

9. obtaining legal status of the restored site to prevent unauthorized anthropogenic disturbance but permitting pre- and postplanting management interventions;

10. planning and implementing habitat restoration in the landscape context, with assisted natural regeneration practiced in areas with high capacity for natural colonization, and areas with low capacity being assigned to active restoration involving plantings.

Some preliminary guidelines for applying this concept are provided, but more detailed and nuanced guidelines will be developed from the results of studies implementing this strategy. What is needed, however, to make the implementation of this strategy possible, are several changes in the current legislation of protected areas. The existing regulations do not allow active interventions in areas with the strictest protection (Categories I and II). The current definitions of these categories must be changed to permit management through active interventions while forbidding any unauthorized activities. Second, the introduction of critically endangered species based on predictions of SDM into these areas should be allowed even if there are no records of their past occurrence.

# References

Abeli T, Cauzzi P, Rossi G, *et al.* (2016) Restoring population structure and dynamics in translocated species: learning from wild populations. *Plant Ecology* 217, 183–192.

Abella SR, MacDonald NW (2002) Spatial and temporal patterns of eastern white pine regeneration in the northwestern Ohio oak stand. *The Michigan Botanist* 41, 115–123.

Abella SR, Schetter TA, Walters TL (2017) Restoring and conserving rare native ecosystems: a 14-year plantation removal experiment. *Biological Conservation* 212, 265–273.

Abrahamson IL, Nelson CR, Affleck DLR (2011) Assessing the performance of sampling designs for measuring the abundance of understory plants. *Ecological Applications* 21, 452–464.

Acacio V, Holmgren M, Jansen PA, Schrotter O (2007) Multiple recruitment limitation causes arrested succession in Mediterranean cork oak systems. *Ecosystems* 10, 1220–1230.

Acharya B, Bhattarai G, de Gier A, Stein A (2000) Systematic adaptive cluster sampling for the assessment of rare tree species in Nepal. *Forest Ecology and Management* 137, 65–73.

Adjers G, Hadengganan S, Kuusipalo J, Nuryanto K, Vesa L (1995) Enrichment planting of dipterocarps in logged-over secondary forests: effect of width, direction and maintenance method of planting line on selected *Shorea* species. *Forest Ecology and Management* 73, 259–270.

Adjers G, Otsamo A (1996) Seedling production methods of dipterocarps. In *Dipterocarp Forest Ecosystems Towards Sustainable Management*, Schulte A, Schone D (eds). Singapore: World Publishing, pp. 391–410.

Adler PB, HilleRisLambers J (2008) The influence of climate and species composition on the population dynamics of ten prairie forbs. *Ecology* 89, 3049–3060.

Agren J (1996) Population size, pollinator limitation, and seed set in the self-incompatible herb *Lythrum salicaria*. *Ecology* 77, 1779–1790.

Aide TM, Cavelier J (1994) Barriers to tropical lowland forest restoration in the Sierra Nevada de Santa Marta, Colombia. *Restoration Ecology* 33, 219–229.

Aide TM, Zimmerman JK, Herrera L, Rosario M, Serrano M (1995) Forest recovery in abandoned tropical pastures in Puerto Rico. *Forest Ecology and Management* 77, 77–86.

Aide TM, Zimmerman JK, Pascarella JB, Rivera L, Marcano-Vega H (2000) Forest regeneration in a chronosequence of tropical abandoned pastures: implications for restoration ecology. *Restoration Ecology* 8, 328–338.

Aide TM, Zimmerman JK, Rosario M, Marcano H (1996) Forest recovery in abandoned cattle pastures along an elevational gradient in northeastern Puerto Rico. *Biotropica* 28, 537–548.

Aitken M, Roberts DW, Shultz LM (2007) Modeling distributions of rare plants in the Great Basin, western North America. *Western North American Naturalist* 67, 26–38.

Aitken SN, Whitlock MC (2013) Assisted gene flow to facilitate local adaptation to climate change. *Annual Review of Ecology, Evolution, and Systematics* 44, 367–388.

Aitken SN, Yeaman S, Holliday JA, Wang T, Curtis-McLane S (2008) Adaptation, migration or extirpation: climate change outcomes for tree populations. *Evolutionary Applications* 1, 95–111.

Aizen MA, Feinsinger P (1994a) Forest fragmentation, pollination, and plant reproduction in a Chaco dry forest, Argentina. *Ecology* 75, 330–351.

Aizen MA, Feinsinger P (1994b) Habitat fragmentation, native insect pollinators, and feral honey bees in Argentine "Chaco Serrano". *Ecological Applications* 4, 378–392.

Aizen MA, Sabatino M, Tylianakis JM (2012) Specialization and rarity predict nonrandom loss of interactions from mutualist networks. *Science* 335, 1486–1489.

Akeroyd J, Wyse Jackson P (1995) *A Handbook for Botanic Gardens on the Reintroduction of Plants to the Wild.* Richmond, UK: Botanic Gardens Conservation International.

Akhalkatsi M, Abdaladze O, Nakhutsrishvili G, Smith WK (2006) Facilitation of seedling microsites by *Rhododendron caucaskum* extends the *Betula litwinowii* Alpine treeline, Caucasus Mountains, Republic of Georgia. *Arctic, Antarctic and Alpine Research* 38, 481–488.

Alagona PS, Sandlos J, Wiersma YF (2012) Past imperfect: using historical ecology and baseline data for conservation and restoration projects in North America. *Environmental Philosophy* 9, 49–70.

Albornoz FE, Gaxiola A, Seaman BJ, Pugnaire FI, Armesto JJ (2013) Nucleation-driven regeneration promotes post-fire recovery in a Chilean temperate forest. *Plant Ecology* 214, 765–776.

Albrecht MA, Guerrant Jr. EO, Maschinski J, Kennedy KL (2011) A long-term view of rare plant reintroduction. *Biological Conservation* 144, 2557–2558.

Albrecht MA, Maschinski J (2012) Influence of founder population size, propagule stages, and life history on the survival of reintroduced plant populations. In *Plant Reintroduction in a Changing Climate: Promises and Perils*, Maschinski J, Haskins KE (eds). Washington, DC: Island Press, pp. 171–188.

Albrecht MA, McCue KA (2010) Changes in demographic processes over long time scales reveal the challenge of restoring an endangered plant. *Restoration Ecology* 18, 235–243.

Alleaume-Benharira M, Pen IR, Ronce O (2006) Geographical patterns of adaptation within a species' range: interactions between drift and gene flow. *Journal of Evolutionary Biology* 19, 203–215.

Allee WC (1931) *Animal Aggregation: A Study in General Sociology.* Chicago, IL: University of Chicago Press.

Alley H, Affolter JM (2004) Experimental comparison of reintroduction methods for the endangered *Echinacea laevigata* (Boynton and Beadle) Blake. *Natural Areas Journal* 24, 345–350.

Alliance for Zero Extinction (2017) Website. www.zeroextinction.org/

Alliende MC, Hoffmann AJ (1985) Plants intruding *Laretia acaulis* (Umbelliferae), a high Andean cushion plant. *Vegetatio* 60, 151–156.

Allison SK (2007) You can't not choose: embracing the role of choice in ecological restoration. *Restoration Ecology* 15, 601–605.

Allison TD (1990) Pollen production and plant density affect pollination and seed production on *Taxus canadensis*. *Ecology* 71, 516–522.

Alpert H, Loik ME (2013) *Pinus jeffreyi* establishment along a forest-shrub ecotone in eastern California, USA. *Journal of Arid Environments* 90, 12–21.

Alvarez-Aquino C, Williams-Linera G (2012) Seedling survival and growth of tree species: site conditions and seasonality in tropical dry forest restoration. *Botanical Sciences* 90, 341–351.

Alvarez-Aquino C, Williams-Linera G, Newton AC (2004) Experimental native tree seedling establishment for the restoration of a Mexican cloud forest. *Restoration Ecology* 12, 412–418.

Alvarez-Yepiz CJ, Martinez-Yrizar A, Burquez A, Lindquist C (2008) Variation in vegetation structure and soil properties related to land use history of old-growth and secondary tropical dry forests in northwestern Mexico. *Forest Ecology and Management* 256, 355–366.

Ammondt SA, Litton CM (2012) Competition between native Hawaiian plants and the invasive grass *Megathyrsus maximus*: implications of functional diversity for ecological restoration. *Restoration Ecology* 20, 638–646.

Ammondt SA, Litton CM, Ellsworth LM, Leary JK (2013) Restoration of native plant communities in a Hawaiian dry lowland ecosystem dominated by the invasive grass *Megathyrsus maximus*. *Applied Vegetation Science* 16, 29–39.

Anacker BL, Gogol-Prokurat M, Leidholm K, Schoenig S (2013) Climate change vulnerability assessment of rare plants in California. *Madrono* 60, 193–210.

Andalo C, Beaulieu J, Bousquet J (2005) The impact of climate change on growth of local white spruce populations in Quebec, Canada. *Forest Ecology and Management* 205, 169–182.

Andelman SJ, Fagan WF (2000) Umbrellas and flagships: efficient conservation surrogates or expensive mistakes? *Proceedings of the National Academy of Sciences of the United States of America* 97, 5954–5959.

Anderson AB (1990) *Alternatives to Deforestation: Steps Toward Sustainable Use of the Amazon Rain Forest.* New York: Columbia University Press.

Anderson NJ (1995) Using the past to predict the future: lake sediments and the modeling of limnological disturbance. *Ecological Modelling* 78, 149–172.

Anderson S (2002) *Identifying Important Plant Areas: A Site Selection Manual for Europe.* London: Plantlife International.

Anderson TM, Schuetz M, Risch AC (2014) Endozoochorous seed dispersal and germination strategies of Serengeti plants. *Journal of Vegetation Science* 25, 636–647.

Angelstam P (1997) Landscape analysis as a tool for the scientific management of biodiversity. *Ecological Bulletins* 46, 140–170.

Antonsson H, Björk RG, Molau U (2009) Nurse plant effect of the cushion plant *Silene acaulis* (L.) Jacq. in an alpine environment in the subarctic Scandes, Sweden. *Plant Ecology and Diversity* 2, 17–25.

Araujo MB, Cabeza M, Thuiller W, Hannah L, Williams PH (2004) Would climate change drive species out of reserves? An assessment of existing reserve-selection methods. *Global Change Biology* 10, 1618–1626.

Arena G, Witkowski ETF, Symes CT (2015) Growing on rocky ground: microhabitat predictors for site-occupancy of *Aloe peglerae*, an endangered endemic species with a restricted range. *South African Journal of Botany* 100, 174–182.

Aronson J, Blignaut JN, Aronson TB (2017) Conceptual frameworks and references for landscape-scale restoration: reflecting back and looking forward. *Annals of the Missouri Botanical Garden* 102, 188–200.

Aronson J, Brancalion PHS, Durigan G, *et al.* (2011) What role should government regulation play in ecological restoration? Ongoing debate in Sao Paulo state, Brazil. *Restoration Ecology* 19, 690–695.

Aronson J, Floret C, Le Floc'h E, Ovalle C, Pontanier R (1993) Restoration and rehabilitation of degraded ecosystems in arid and semi-arid lands. I. A view from the south. *Restoration Ecology* 1, 8–17.

Arponen A, Heikkinen RK, Thomas CD, Moilanen A (2005) The value of bio-diversity in reserve selection: representation, species weighting, and benefit functions. *Conservation Biology* 19, 2009–2014.

Arroyo MTK, Cavieres LA, Penaloza A, Arroyo-Kalin MA (2003) Positive associ-ations between the cushion plant *Azorella monantha* (Apiaceae) and alpine plant species in the Chilean Patagonian Andes. *Plant Ecology* 169, 121–129.

Arroyo-Rodriguez V, Melo FPL, Martinez-Ramos M, *et al.* (2017) Multiple suc-cessional pathways in human-modified tropical landscapes: new insights from forest succession, forest fragmentation and landscape ecology research. *Biological Reviews* 92, 326–340.

Arroyo-Rodriguez V, Roes M, Escobar F, *et al.* (2013) Plant beta-diversity in fragmented rain forests: testing floristic homogenization and differentiation hypotheses. *Journal of Ecology* 101, 1449–1458.

Arthur JL, Haight RG, Montgomery CA, Polasky S (2002) Analysis of the threshold and expected coverage approaches to the probabilistic reserve site selection problem. *Environmental Modeling and Assessment* 7, 81–89.

Ashcroft MB, Gollan JR, Warton DI, Ramp D (2012) A novel approach to quantify and locate potential microrefugia using topoclimate, climate stability, and iso-lation from the matrix. *Global Change Biology* 18, 1866–1879.

Ashton MS (1995) Seedling growth of co-occurring *Shorea* species in the simulated light environments of a rain forest. *Forest Ecology and Management* 72, 1–12.

Ashton MS, Gunatilleke CVS, Gunatilleke IAUN, *et al.* (2014) Restoration of rain forest beneath pine plantations: a relay floristic model with special application to tropical South Asia. *Forest Ecology and Management* 329, 351–359.

Ashton MS, Gunatilleke CVS, Singhakumara BMP, Gunatilleke I (2001) Restoration pathways for rain forest in southwest Sri Lanka: a review of concepts and models. *Forest Ecology and Management* 154, 409–430.

Ashton MS, Hall JS (2011) The ecology, silviculture and use of tropical wet forests with special emphasis on timber rich types. In *Silviculture in the Tropics*, Gunter S, Weber M, Stimm B, Mosandl R (eds). New York: Springer, pp. 145–192.

Ashton PMS, Gamage S, Gunatilleke I, Gunatilleke CVS (1997) Restoration of a Sri Lankan rainforest: using Caribbean pine *Pinus caribaea* as a nurse for establishing late-successional tree species. *Journal of Applied Ecology* 34, 915–925.

Asner GP, Broadbent EN, Oliveira PJC, *et al.* (2006) Condition and fate of logged forests in the Brazilian Amazon. *Proceedings of the National Academy of Sciences of the United States of America* 103, 12947–12950.

Asner GP, Rudel TK, Aide TM, Defries R, Emerson R (2009) A contemporary assessment of change in humid tropical forests. *Conservation Biology* 23, 1386–1395.

Asquith NM, Wright SJ, Clauss MJ (1997) Does mammal community composition control recruitment in neotropical forests? Evidence from Panama. *Ecology* 78, 941–946.

Atondo-Bueno EJ, Bonilla-Moheno M, López-Barrera F (2018) Cost-efficiency analysis of seedling introduction vs. direct seeding of *Oreomunnea mexicana* for secondary forest enrichment. *Forest Ecology and Management* 409, 399–406.

Atondo-Bueno EJ, Lopez-Barrera F, Bonilla-Moheno M, Williams-Linera G, Ramirez-Marcial N (2016) Direct seeding of *Oreomunnea mexicana*, a threatened tree species from Southeastern Mexico. *New Forests* 47, 845–860.

Aubin I, Messier C, Bouchard A (2008) Can plantations develop understory biological and physical attributes of naturally regenerated forests? *Biological Conservation* 141, 2461–2476.

Augspurger CK (1983) Seed dispersal of tropical tree, *Platypodium elegans*, and the escape of its seedlings from fungal pathogens. *Journal of Ecology* 71, 759–771.

Avendano-Yanez ML, Sanchez-Velasquez LR, Meave JA, Pineda-Lopez MR (2016) Can *Pinus* plantations facilitate reintroduction of endangered cloud forest species? *Landscape and Ecological Engineering* 12, 99–104.

Babaasa D, Eilu G, Kasangaki A, Bitariho R, McNeilage A (2004) Gap characteristics and regeneration in Bwindi Impenetrable National Park, Uganda. *African Journal of Ecology* 42, 217–224.

Baider C, Florens FBV (2011) Control of invasive alien weeds averts imminent plant extinction. *Biological Invasions* 13, 2641–2646.

Bainbridge DA (2007) *A Guide for Desert and Dryland Restoration: New Hope for Arid Lands.* Washington, DC: Island Press.

Bakker ES, Olff H, Vandenberghe C, *et al.* (2004) Ecological anachronisms in the recruitment of temperate light-demanding tree species in wooded pastures. *Journal of Applied Ecology* 41, 571–582.

Bakker JD, Colasurdo LB, Evans JR (2012) Enhancing Garry oak seedling performance in a semiarid environment. *Northwest Science* 86, 300–309.

Balaguer L, Escudero A, Martin-Duque JF, Mola I, Aronson J (2014) The historical reference in restoration ecology: re-defining a cornerstone concept. *Biological Conservation* 176, 12–20.

Balcomb SR, Chapman CA (2003) Bridging the gap: influence of seed deposition on seedling recruitment in a primate–tree interaction. *Ecological Monographs* 73, 625–642.

Baldeck CA, Harms KE, Yavitt JB, *et al.* (2013) Habitat filtering across tree life stages in tropical forest communities. *Proceedings of the Royal Society B: Biological Sciences* 280, 20130548.

Ball IR, Possingham HP, Watts M (2009) Marxan and relatives: software for spatial conservation prioritisation. In *Spatial Conservation Prioritisation: Quantitative Methods and Computational Tools*, Moilanen A, Wilson KA, Possingham HP (eds). Oxford, UK: Oxford University Press, pp. 185–195.

Banks-Leite C, Pardini R, Tambosi LR, *et al.* (2014) Using ecological thresholds to evaluate the costs and benefits of set-asides in a biodiversity hotspot. *Science* 345, 1041–1045.

Bannister JR (2015) Recuperar bosques no es solo plantar árboles: lecciones aprendidas luego de 7 años restaurando bosques de *Pilgerodendron uviferum* (D. Don) Florin en Chiloé. *Anales del Instituto de la Patagonia* 43, 35–51.

Bannister JR, Wagner S, Donoso PJ, Bauhus J (2014) The importance of seed trees in the dioecious conifer *Pilgerodendron uviferum* for passive restoration of fire disturbed southern bog forests. *Austral Ecology* 39, 204–213.

Barajas-Guzmán MG, Campo J, Barradas VL (2006) Soil water, nutrient availability and sapling survival under organic and polyethylene mulch in a seasonally dry tropical forest. *Plant and Soil* 287, 347–357.

Barak RS, Hipp AL, Cavender-Bares J, *et al.* (2016) Taking the long view: integrating recorded, archeological, paleoecological, and evolutionary data into ecological restoration. *International Journal of Plant Sciences* 177, 90–102.

Baraloto C, Goldberg DE (2004) Microhabitat associations and seedling bank dynamics in a neotropical forest. *Oecologia* 141, 701–712.

Baraloto C, Herault B, Paine CET, *et al.* (2012) Contrasting taxonomic and functional responses of a tropical tree community to selective logging. *Journal of Applied Ecology* 49, 861–870.

Baraza E, Zamora R, Hodar JA (2006) Conditional outcomes in plant-herbivore interactions: neighbours matter. *Oikos* 113, 148–156.

Barbera GG, Martinez-Fernandez F, Alvarez-Rogel J, Albaladejo J, Castillo V (2005) Short- and intermediate-term effects of site and plant preparation techniques on reforestation of a Mediterranean semiarid ecosystem with *Pinus halepensis* Mill. *New Forests* 29, 177–198.

Barea J, Palenzuela J, Cornejo P, *et al.* (2011) Ecological and functional roles of mycorrhizas in semi-arid ecosystems of Southeast Spain. *Journal of Arid Environments* 75, 1292–1301.

Baret S, Baider C, Kueffer C, Foxcroft LC, Lagabrielle E (2013) Threats to paradise? Plant invasions in protected areas of the Western Indian ocean islands. In *Plant Invasions in Protected Areas: Patterns, Problems and Challenges*, Foxcroft LC, Pyšek P, Richardson DM, Genovesi P (eds). Dordrecht, The Netherlands: Springer, pp. 423–447.

Baret S, Le Bourgeois T, Riviere J-N, *et al.* (2007) Can species richness be maintained in logged endemic *Acacia heterophylla* forests (Reunion Island, Indian Ocean)? *Revue d'Ecologie (Terre Vie)* 62, 273–284.

Barnes AD, Chapman HM (2014) Dispersal traits determine passive restoration trajectory of a Nigerian montane forest. *Acta Oecologica* 56, 32–40.

Barnosky AD, Hadly EA, Gonzalez P, *et al.* (2017) Merging paleobiology with conservation biology to guide the future of terrestrial ecosystems. *Science* 355, 4787.

Barroetavena C, Gisler SD, Luoma DL, Meinke RJ (1998) Mycorrhizal status of the endangered species *Astragalus applegatei* Peck as determined from a soil bioassay. *Mycorrhiza* 8, 117–119.

Barton NH (2001) Adaptation at the edge of a species' range. In *Integrating Ecology and Evolution in a Spatial Context*, Silvertown J, Antonovics J (eds). Oxford, UK: Blackwell Science, pp. 365–392.

Baskin CC, Baskin JM (2003) When breaking seed dormancy is a problem try a move-along experiment. *Native Plants Journal* 4, 17–21.

Bátori Z, Farkas T, Erdős L, *et al.* (2014) A comparison of the vegetation of forested and non-forested solution dolines in Hungary: a preliminary study. *Biologia* 69, 1339–1348.

Batty AL, Brundrett MC, Dixon KW, Sivasithamparam K (2006) In situ symbiotic seed germination and propagation of terrestrial orchid seedlings for establishment at field sites. *Australian Journal of Botany* 54, 375–381.

Baumberger T, Croze T, Affre L, Mesleard F (2012) Co-occurring species indicate habitats of the rare *Limonium girardianum*. *Plant Ecology and Evolution* 145, 31–37.

Bazzaz FA (1975) Plant species diversity in old-field successional ecosystems in southern Illinois. *Ecology* 56, 485–488.

Beaune D (2015) What would happen to the trees and lianas if apes disappeared? *Oryx* 49, 442–446

Beaune D, Bretagnolle F, Bollache L, *et al.* (2013) Seed dispersal strategies and the threat of defaunation in a Congo forest. *Biodiversity and Conservation* 22, 225–238.

Bebber DP, Thomas SC, Cole WG, Balsillie D (2004) Diameter increment in mature eastern white pine *Pinus strobus* L. following partial harvest of old-growth stands in Ontario, Canada. *Trees* 18, 29–34.

Bechara FC, Dickens SJ, Farrer EC, *et al.* (2016) Neotropical rainforest restoration: comparing passive, plantation and nucleation approaches. *Biodiversity and Conservation* 25, 2021–2034.

Beckage B, Clark JS (2003) Seedling survival and growth of three forest tree species: the role of spatial heterogeneity. *Ecology* 84, 1849–1861.

Beckage B, Clark JS, Clinton BD, Haines BL (2000) A long-term study of tree seedling recruitment in southern Appalachian forests: the effects of canopy gaps and shrub understories. *Canadian Journal of Forest Research* 30, 1617–1631.

Beisner BE, Haydon DT, Cuddington K (2003) Alternative stable states in ecology. *Frontiers in Ecology and the Environment* 1, 376–382.

Beissinger SR (2002) Population viability analysis: past, present, future. In *Population Viability Analysis*, Beissinger SR, McCullough DR (eds). Chicago, IL: University of Chicago Press, pp. 5–17.

Bell G (2001) Ecology: neutral macroecology. *Science* 293, 2413–2418.

Bell G, Gonzalez A (2011) Adaptation and evolutionary rescue in metapopulations experiencing environmental deterioration. *Science* 332, 1327–1329.

Bell TJ, Bowles ML, McEachern AK (2003) Projecting the success of plant population restoration with viability analysis. In *Population Viability in Plants*, Brigham CA, Schwartz MW (eds). Berlin: Springer, pp. 313–348.

Bellingham PJ, Richardson J (2006) Tree seedling growth and survival over 6 years across different microsites in a temperate rain forest. *Canadian Journal of Forestry Research* 36, 910–918.

Bellingham PJ, Tanner EVJ, Martin PH, Healey JR, Burge OR (2018) Endemic trees in a tropical biodiversity hotspot imperilled by an invasive tree. *Biological Conservation* 217, 47–53.

Bellot J, De Urbina JMO, Bonet A, Sanchez JR (2002) The effects of treeshelters on the growth of *Quercus coccifera* L. seedlings in a semiarid environment. *Forestry* 75, 89–106.

Belsky AJ (1994) Influences of trees on savanna productivity: tests of shade, nutrients, and tree-grass competition. *Ecology* 75, 922–932.

Berry NJ, Phillips OL, Lewis SL, *et al.* (2010) The high value of logged tropical forests: lessons from northern Borneo. *Biodiversity and Conservation* 19, 985–997.

Bertacchi MIF, Amazonas NT, Brancalion PHS, *et al.* (2016) Establishment of tree seedlings in the understory of restoration plantations: natural regeneration and enrichment plantings. *Restoration Ecology* 24, 100–108.

Betz C, Scheuerer M, Reisch C (2013) Population reinforcement – a glimmer of hope for the conservation of the highly endangered Spring Pasque flower (*Pulsatilla vernalis*). *Biological Conservation* 168, 161–167.

Beyer HL, Merrill EH, Varley N, Boyce MS (2007) Willow on Yellowstone's northern range: evidence for a trophic cascade? *Ecological Applications* 17, 1563–1571.

BGCI (2012) *International Agenda for Botanic Gardens in Conservation*, 2nd edition. Richmond, UK: Botanic Gardens Conservation International.

Biesmeijer JC, Roberts SPM, Reemer M, *et al.* (2006) Parallel declines in pollinators and insect-pollinated plants in Britain and the Netherlands. *Science* 313, 351–354.

Birkinshaw C, Porter P. Lowry II, Raharimampionona J, Aronson J (2013) Supporting target 4 of the global strategy for plant conservation by integrating ecological restoration into the Missouri Botanical Garden's conservation program in Madagascar. *Annals of the Missouri Botanical Garden* 99, 139–146.

Birks HJB (1996) Contributions of Quaternary palaeoecology to nature conservation. *Journal of Vegetation Science* 7, 89–98.

Bissessur P, Baider C, Florens FBV (2017) Rapid population decline of an endemic oceanic island plant despite resilience to extensive habitat destruction and occurrence within protected areas. *Plant Ecology and Diversity* 10, 293–302.

Blackburn TM, Essl F, Evans T, *et al.* (2014) A unified classification of alien species based on the magnitude of their environmental impacts. *PLoS Biology* 12, e1001850.

Blackburn TM, Pyšek P, Bacher S, *et al.* (2011) A proposed unified framework for biological invasions. *Trends in Ecology and Evolution* 26, 333–339.

Bladow JM, Bohner T, Winn AA (2017) Comparisons of demography and inbreeding depression in introduced and wild populations of an endangered shrub. *Natural Areas Journal* 37, 294–308.

Blakey RV, Law BS, Kingsford RT, *et al.* (2016) Bat communities respond positively to large-scale thinning of forest regrowth. *Journal of Applied Ecology* 53, 1694–1703.

Blanquart F, Kaltz O, Nuismer SL, Gandon S (2013) A practical guide to measuring local adaptation. *Ecology Letters* 16, 1195–1205.

Blignaut A, Milton SJ (2005) Effects of multispecies clumping on survival of three succulent plant species translocated onto mine spoil in the Succulent Karoo Desert, South Africa. *Restoration Ecology* 13, 15–19.

Blois JL, Zarnetske PL, Fitzpatrick MC, Finnegan S (2013) Climate change and the past, present, and future of biotic interactions. *Science* 341, 499–504.

Bommarco R, Lindborg R, Marini L, Öckinger E (2014) Extinction debt for plants and flower-visiting insects in landscapes with contrasting land use history. *Diversity and Distributions* 20, 591–599.

Bond W, Slingsby P (1984) Collapse of an ant-plant mutualism: the Argentine ant (*Iridomyrmex humilis*) and myrmecochorous Proteaceae. *Ecology* 65, 1031–1037.

Bond WJ (1989) The tortoise and the hare: ecology of angiosperm dominance and gymnosperm persistence. *Biological Journal of the Linnean Society* 36, 227–249.

Bond WJ (1994) Do mutualisms matter? Assessing the impact of pollinator and disperser disruption on plant extinction. *Philosophical Transactions of the Royal Society of London Series B: Biological Sciences* 344, 83–90.

Bonnell TR, Reyna-Hurtado R, Chapman CA (2011) Post-logging recovery time is longer than expected in an East African tropical forest. *Forest Ecology and Management* 261, 855–864.

Bontrager M, Webster KL, Elvin M, Parker IM (2014) The effects of habitat and competitive/facilitative interactions on reintroduction success of the endangered wetland herb, *Arenaria paludicola*. *Plant Ecology* 215, 467–478.

Boojh R, Ramakrishnan PS (1982a) Growth strategy of trees related to successional status. I. Architecture and extension growth. *Forest Ecology and Management* 4, 359–374.

Boojh R, Ramakrishnan PS (1982b) Growth strategy of trees related to successional status. II. Leaf dynamics. *Forest Ecology and Management* 4, 375–386.

Bosch M, Waser NM (2001) Experimental manipulation of plant density and its effect on pollination and reproduction of two confamilial montane herbs. *Oecologia* 126, 76–83.

Boshier DH, Chase MR, Bawa KS (1995) Population genetics of *Cordia alliodora* (Boraginaceae), a neotropical tree. 2. Mating system. *American Journal of Botany* 82, 476–483.

Bossuyt B, Hermy M, Deckers J (1999) Migration of herbaceous plant species across ancient-recent forest ecotones in central Belgium. *Journal of Ecology* 87, 628–638.

Bottin L, Le Cadre S, Quilichini A, *et al.* (2007) Re-establishment trials in endangered plants: a review and the example of *Arenaria grandiflora*, a species on the brink of extinction in the Parisian region (France). *Ecoscience* 14, 410–419.

Bottrill MC, Mills MH, Pressey RL, Game ET, Groves C (2012) Evaluating perceived benefits of ecoregional assessments. *Conservation Biology* 26, 851–861.

Bower AD, St Clair B, Erickson V (2014) Generalized provisional seed zones for native plants. *Ecological Applications* 24, 913–919.

Bowles ML, McBride JL, Bell TJ (2015) Long-term processes affecting restoration and viability of the federal threatened Mead's milkweed (*Asclepias meadii*). *Ecosphere* 6, 1–22.

Bowles ML, McBride JL, Betz RF (1998) Management and restoration ecology of the federally threatened Mead's milkweed, *Asclepias meadii* (Asclepiadaceae). *Annals of the Missouri Botanical Garden* 85, 110–125.

Bowman G, Perret C, Hoehn S, Galeuchet DJ, Fischer M (2008) Habitat fragmentation and adaptation: a reciprocal replant-transplant experiment among 15 populations of *Lychnis flos-cuculi*. *Journal of Ecology* 96, 1056–1064.

Box EO (1996) Plant functional types and climate at the global scale. *Journal of Vegetation Science* 7, 309–320.

Boyce MS (2002) Reconciling the small-population and declining-population paradigms. In *Population Viability Analysis*, Beissinger SR, McCullough DR (eds). Chicago, IL: University of Chicago Press, pp. 41–49.

Boyle T, Liengsiri C, Piewluang C (1991) Genetic studies in a tropical pine – *Pinus kesiya*. III. The mating system in four populations from northern Thailand. *Journal of Tropical Forest Science* 4, 37–44.

Bradshaw AD (1984) Ecological significance of genetic variation between populations. In *Perspectives on Plant Population Ecology*, Dirzo R, Sarukhan J (eds). Sunderland, MA: Sinauer Associates, pp. 213–228.

Brambilla M, Caprio E, Assandri G, *et al.* (2017) A spatially explicit definition of conservation priorities according to population resistance and resilience, species importance and level of threat in a changing climate. *Diversity and Distributions* 23, 727–738.

Brancalion PHS, van Melis J (2017) On the need for innovation in ecological restoration. *Annals of the Missouri Botanical Garden* 102, 227–236.

Brashares JS, Arcese P, Sam MK (2001) Human demography and reserve size predict wildlife extinction in West Africa. *Proceedings of the Royal Society B: Biological Sciences* 268, 1–6.

Braun M, Schindler S, Essl F (2016) Distribution and management of invasive alien plant species in protected areas in Central Europe. *Journal for Nature Conservation* 33, 48–57.

Breed MF, Stead MG, Ottewell KM, Gardner MG, Lowe AJ (2013) Which provenance and where? Seed sourcing strategies for revegetation in a changing environment. *Conservation Genetics* 14, 1–10.

Bremer LL, Farley KA (2010) Does plantation forestry restore biodiversity or create green deserts? A synthesis of the effects of land-use transitions on plant species richness. *Biodiversity and Conservation* 19, 3893–3915.

Brewer JS, Menzel T (2009) A method for evaluating outcomes of restoration when no reference sites exist. *Restoration Ecology* 17, 4–11.

Bridle JR, Vines TH (2007) Limits to evolution at range margins: when and why does adaptation fail? *Trends in Ecology and Evolution* 22, 140–147.

Bried JT (2013) Adaptive cluster sampling in the context of restoration. *Restoration Ecology* 5, 585–591.

Briggs D, Walters SM (1997) *Plant Variation and Evolution*, 3rd edition. Cambridge: Cambridge University Press.

Briggs JD, Leigh JH (1996) *Rare or Threatened Australian Plants*. Collingwood, Australia: CSIRO.

Broadhurst LM, Lowe A, Coates DJ, *et al.* (2008) Seed supply for broadscale restoration: maximizing evolutionary potential. *Evolutionary Applications* 1, 587–597.

Brodie JF, Aslan CE (2012) Halting regime shifts in floristically intact tropical forests deprived of their frugivores. *Restoration Ecology* 20, 153–157.

Brodie JF, Helmy OE, Brockelman WY, Maron JL (2009) Bushmeat poaching reduces the seed dispersal and population growth rate of a mammal-dispersed tree. *Ecological Applications* 19, 854–863.

Brokaw N, Busing RT (2000) Niche versus chance and tree diversity in forest gaps. *Trends in Ecology and Evolution* 15, 183–188.

Brooks TM, Mittermeier RA, da Fonseca GAB, *et al.* (2006) Global biodiversity conservation priorities. *Science* 313, 58–61.

Brooks TM, Mittermeier RA, Mittermeier CG, *et al.* (2002) Habitat loss and extinction in the hotspots of biodiversity. *Conservation Biology* 16, 909–923.

Brown ADH, Briggs JD (1991) Sampling strategies for genetic variation in ex situ collections of endangered plant species. In *Genetics and Conservation of Rare Plants*, Falk DA, Holsinger KE (eds). Oxford, UK: Oxford University Press, pp. 99–122.

Brown AHD, Brubaker CL, Grace JP (1997) Regeneration of germplasm samples: wild versus cultivated plant species. *Crop Science* 37, 7–13.

Brown ADH, Marshall DR (1995) A basic sampling strategy: theory and practice. In *Collecting Plant Genetic Diversity: Technical Guidelines*, Guarino L, Ramantha Rao VR (eds). Wallington, UK: CAB International, pp. 75–111.

Brown JH, Heske EJ (1990) Control of a desert-grassland transition by a keystone rodent guild. *Science* 250, 1705–1707.

Brown KA, Gurevitch J (2004) Long-term impacts of logging on forest diversity in Madagascar. *Proceedings of the National Academy of Sciences of the United States of America* 101, 6045–6049.

Brown ND, Curtis T, Adams EC (2015) Effects of clear-felling versus gradual removal of conifer trees on the survival of understorey plants during the restoration of ancient woodlands. *Forest Ecology and Management* 348, 15–22.

Brudvig LA, Barak RS, Bauer JT, *et al.* (2017) Interpreting variation to advance predictive restoration science. *Journal of Applied Ecology* 54, 1018–1027.

Brudvig LA, Mabry CM (2008) Trait-based filtering of the regional species pool to guide understory plant reintroductions in Midwestern oak savannas, USA. *Restoration Ecology* 16, 290–304.

Bruelheide H, Boehnke M, Both S, *et al.* (2011) Community assembly during secondary forest succession in a Chinese subtropical forest. *Ecological Monographs* 81, 25–41.

Brummitt N, Lughadha EN (2003) Biodiversity: where's hot and where's not. *Conservation Biology* 17, 1442–1448.

Brunet J, von Oheimb G (1998) Migration of vascular plants to secondary woodlands in southern Sweden. *Journal of Ecology* 86, 429–438.

Bryce SA, Omernik JM, Larsen DP (1999) Ecoregions: a geographic framework to guide risk characterization and ecosystem management. *Environmental Practice* 1, 141–155.

Brzosko E, Jermakowicz E, Ostrowiecka B, *et al.* (2018) Rare plant translocation between mineral islands in Biebrza Valley (northeastern Poland): effectiveness and recipient site selection. *Restoration Ecology* 26, 56–62.

Bucking W (2003) Are there threshold numbers for protected forests? *Journal of Environmental Management* 67, 37–45.

Budde KB, Nielsen LR, Ravn HP, Kjær ED (2016) The natural evolutionary potential of tree populations to cope with newly introduced pests and pathogens – lessons learned from forest health catastrophes in recent decades. *Current Forestry Reports* 2, 18–29.

Bunnell FL, Houde I (2010) Down wood and biodiversity: implications to forest practices. *Environmental Reviews* 18, 397–421.

Burivalova Z, Sekercioglu CH, Koh LP (2014) Thresholds of logging intensity to maintain tropical forest biodiversity. *Current Biology* 24, 1893–1898.

Burkey TV (1997) Ecological principles for natural habitats management. *SUM Working Papers* 07/97, 1–26.

Burney DA, Burney LP (2007) Paleoecology and "inter-situ" restoration on Kauai, Hawaii. *Frontiers in Ecology and the Environment* 5, 483–490.

Burney DA, Burney LP (2016) Monitoring results from a decade of native plant translocations at Makauwahi Cave Reserve, Kaua'i. *Plant Ecology* 217, 139–153.

Büscher B, Dressler W (2012) Commodity conservation: the restructuring of community conservation in South Africa and the Philippines. *Geoforum* 43, 367–376.

Büscher B, Fletcher R, Brockington D, *et al.* (2017) Half-Earth or whole Earth? Radical ideas for conservation, and their implications. *Oryx* 51, 407–410.

Butterfield BJ, Cavieres LA, Callaway RM, *et al.* (2013) Alpine cushion plants inhibit the loss of phylogenetic diversity in severe environments. *Ecology Letters* 16, 478–486.

Butterfield BJ, Copeland SM, Munson SM, Roybal CM, Wood TE (2017) Prestoration: using species in restoration that will persist now and into the future. *Restoration Ecology* 25, S155–S163.

Butterfield RP (1996) Early species selection for tropical reforestation: a consideration of stability. *Forest Ecology and Management* 81, 161–168.

Byers DL, Meagher TR (1997) A comparison of demographic characteristics in a rare and a common species of *Eupatorium*. *Ecological Applications* 7, 519–530.

Cabin RJ, Weller SG, Lorence DH, *et al.* (2002) Effects of light, alien grass, and native species additions on Hawaiian dry forest restoration. *Ecological Applications* 12, 1595–1610.

Cadotte MW, Davies TJ (2010) Rarest of the rare: advances in combining evolutionary distinctiveness and scarcity to inform conservation at biogeographical scales. *Diversity and Distributions* 16, 376–385.

Callaway RM (1992) Effects of shrubs on recruitment of *Quercus douglasii* and *Quercus lobata* in California. *Ecology* 73, 2118–2128.

Callaway RM (1995) Positive interactions among plants. *Botanical Review* 61, 306–349.

Calvo-Alvarado JC, Arias D, Richter DD (2007) Early growth performance of native and introduced fast growing tree species in wet to sub-humid climates of the Southern region of Costa Rica. *Forest Ecology and Management* 242, 227–235.

Cameron AD (2002) Importance of early selective thinning in the development of long-term stand stability and improved log quality: a review. *Forestry* 75, 25–35.

Cañadas EM, Fenu G, Peñas J, *et al.* (2014) Hotspots within hotspots: endemic plant richness, environmental drivers, and implications for conservation. *Biological Conservation* 170, 282–291.

Cannon CH, Peart DR, Leighton M (1998) Tree species diversity in commercially logged Bornean rainforest. *Science* 281, 1366–1368.

Cannon CH, Peart DR, Leighton M, Kartawinata K (1994) The structure of lowland rainforest after selective logging in West Kalimantan, Indonesia. *Forest Ecology and Management* 67, 49–68.

Caravaca F, Barea JM, Figueroa D, Roldan A (2002) Assessing the effectiveness of mycorrhizal inoculation and soil compost addition for enhancing reafforestation with *Olea europaea* subsp. *sylvestris* through changes in soil biological and physical parameters. *Applied Soil Ecology* 20, 107–118.

Caravaca F, Barea JM, Palenzuela J, *et al.* (2003) Establishment of shrub species in a degraded semiarid site after inoculation with native or allochthonous arbuscular mycorrhizal fungi. *Applied Soil Ecology* 22, 103–111.

Carey AB (2003) Biocomplexity and restoration of biodiversity in temperate coniferous forest: inducing spatial heterogeneity with variable-density thinning. *Forestry* 76, 127–136.

Carlsen TM, Espeland EK, Pavlik BM (2002) Reproductive ecology and the persistence of an endangered plant. *Biodiversity and Conservation* 11, 1247–1268.

Carnevale NJ, Montagnini F (2002) Facilitating regeneration of secondary forests with the use of mixed and pure plantations of indigenous tree species. *Forest Ecology and Management* 163, 217–227.

Carnpanello PI, Gatti MG, Ares A, Montti L, Goldstein G (2007) Tree regeneration and microclimate in a liana and bamboo-dominated semideciduous Atlantic Forest. *Forest Ecology and Management* 252, 108–117.

Carnus JM, Parrotta J, Brockerhoff E, *et al.* (2006) Planted forests and biodiversity. *Journal of Forestry* 104, 65–77.

Caro TM, O'Doherty G (1999) On the use of surrogate species in conservation biology. *Conservation Biology* 13, 805–814.

Carpenter FL, Nichols JD, Sandi E (2004) Early growth of native and exotic trees planted on degraded tropical pasture. *Forest Ecology and Management* 196, 367–378.

Carriere SM, Andre M, Letourmy P, Olivier I, McKey DB (2002a) Seed rain beneath remnant trees in a slash-and-burn agricultural system in southern Cameroon. *Journal of Tropical Ecology* 18, 353–374.

Carriere SM, Letourmy P, McKey DB (2002b) Effects of remnant trees in fallows on diversity and structure of forest regrowth in a slash-and-burn agricultural system in southern Cameroon. *Journal of Tropical Ecology* 18, 375–396.

Carter KK (1996) Provenance tests as indicators of growth response to climate change in 10 north temperate tree species. *Canadian Journal of Forest Research* 26, 1089–1095.

Case TJ, Taper ML (2000) Interspecific competition, environmental gradients, gene flow, and the coevolution of species' borders. *American Naturalist* 155, 583–605.

Caspersen JP, Saprunoff M (2005) Seedling recruitment in a northern temperate forest: the relative importance of supply and establishment limitation. *Canadian Journal of Forest Research* 35, 978–989.

Castellanos-Acuna D, Lindig-Cisneros R, Sáenz-Romero C (2015) Altitudinal assisted migration of Mexican pines as an adaptation to climate change. *Ecosphere* 6, 1–16.

Castellanos-Castro C, Bonfil C (2013) Propagation of three *Bursera* species from cuttings. *Botanical Sciences* 91, 217–224.

Castro J, Gomez JM, Garcia D, Zamora R, Hodar JA (1999) Seed predation and dispersal in relict Scots pine forests in southern Spain. *Plant Ecology* 145, 115–123.

Castro J, Zamora R, Hodar JA (2006) Restoring *Quercus pyrenaica* forests using pioneer shrubs as nurse plants. *Applied Vegetation Science* 9, 137–142.

Castro J, Zamora R, Hodar JA, Gomez JM, Gomez-Aparicio L (2004) Benefits of using shrubs as nurse plants for reforestation in Mediterranean mountains: a 4-year study. *Restoration Ecology* 12, 352–358.

Catling PM, King B (2007) Natural recolonization of cultivated land by native prairie plants and its enhancement by removal of Scots Pine, *Pinus sylvestris*. *Canadian Field Naturalist* 121, 201–205.

Catling PM, Kostiuk B (2010) Successful re-establishment of a native savannah flora and fauna on the site of a former pine plantation at Constance Bay, Ottawa, Ontario. *Canadian Field Naturalist* 124, 169–178.

Catorci A, Cesaretti S, Luis Velasquez J, Zeballos H (2011) Plant–plant spatial interactions in the dry Puna (southern Peruvian Andes). *Alpine Botany* 121, 113–121.

Catterall CP, Kanowski J, Wardell-Johnson GW, *et al.* (2004) Quantifying the biodiversity values of reforestation: perspectives, design issues and outcomes in Australian rainforest landscapes. In *Conservation of Australia's Forest Fauna*, Lunney D (ed.). Sydney: Royal Zoological Society of New South Wales, pp. 359–393.

Caughley G (1994) Directions in conservation biology. *Journal of Animal Ecology* 63, 215–244.

Caujapé-Castells J, Tye A, Crawford DJ, *et al.* (2010) Conservation of oceanic island floras: present and future global challenges. *Perspectives in Plant Ecology, Evolution and Systematics* 12, 107–129.

Cavieres LA, Badano EI, Sierra-Almeida A, Gomez-Gonzalez S, Molina-Montenegro MA (2006) Positive interactions between alpine plant species and the nurse cushion plant *Laretia acaulis* do not increase with elevation in the Andes of central Chile. *New Phytologist* 169, 59–69.

Cavieres LA, Chacon P, Penaloza A, Molina-Montenegro MA, Arroyo MTK (2007) Leaf litter of *Kageneckia angustifolia* D. Don (Rosaceae) inhibits seed germination in sclerophyllous montane woodlands of central Chile. *Plant Ecology* 190, 13–22.

Celis G, Jose S (2011) Restoring abandoned pasture land with native tree species in Costa Rica: effects of exotic grass competition and light. *Forest Ecology and Management* 261, 1598–1604.

Center for Plant Conservation (1991) Genetic sampling guidelines for conservation collections of endangered plants. In *Genetics and Conservation of Rare Plants*, Falk DA, Holsinger KE (eds). Oxford, UK: Oxford University Press, pp. 225–238.

Chambers FM, Mauquoy D, Todd PA (1999) Recent rise to dominance of *Molinia caerulea* in environmentally sensitive areas: new perspectives from palaeoecological data. *Journal of Applied Ecology* 36, 719–733.

Chapin FS, Starfield AM (1997) Time lags and novel ecosystems in response to transient climatic change in arctic Alaska. *Climatic Change* 35, 449–461.

Chapin MH, Wood KR, Perlman SP, Maunder M (2004) A review of the conservation status of the endemic *Pritchardia* palms of Hawaii. *Oryx* 38, 273–281.

Chapman CA, Chapman LJ (1997) Forest regeneration in logged and unlogged forests of Kibale National Park, Uganda. *Biotropica* 29, 396–412.

Chapman CA, Chapman LJ, Jacob AL, *et al.* (2010) Tropical tree community shifts: implications for wildlife conservation. *Biological Conservation* 143, 366–374.

Chapman CA, Onderdonk DA (1998) Forests without primates: primate/plant codependency. *American Journal of Primatology* 45, 127–141.

Charles LS, Dwyer JM, Mayfield MM (2017) Rainforest seed rain into abandoned tropical Australian pasture is dependent on adjacent rainforest structure and extent. *Austral Ecology* 42, 238–249.

Charron L, Hermanutz L (2016) Prioritizing boreal forest restoration sites based on disturbance regime. *Forest Ecology and Management* 361, 90–98.

Chase JM (2003) Community assembly: when should history matter? *Oecologia* 136, 489–498.

Chase JM, Leibold MA (2003) *Ecological Niches: Linking Classical and Contemporary Approaches.* Chicago, IL: University of Chicago Press.

Chazdon RL (2003) Tropical forest recovery: legacies of human impact and natural disturbances. *Perspectives in Plant Ecology, Evolution and Systematics* 6, 51–71.

Chazdon RL (2008) Beyond deforestation: restoring forests and ecosystem services on degraded lands. *Science* 320, 1458–1460.

Chazdon RL (2017) Landscape restoration, natural regeneration, and the forests of the future. *Annals of the Missouri Botanical Garden* 102, 251–257.

Chazdon RL, Guariguata MR (2016) Natural regeneration as a tool for large-scale forest restoration in the tropics: prospects and challenges. *Biotropica* 48, 716–730.

Chazdon RL, Letcher SG, van Breugel M, *et al.* (2007) Rates of change in tree communities of secondary Neotropical forests following major disturbances. *Philosophical Transactions of the Royal Society of London Series B: Biological Sciences* 362, 273–289.

Chen Y, Chen G, Yang J, Sun W (2016) Reproductive biology of *Magnolia sinica* (Magnoliaecea), a threatened species with extremely small populations in Yunnan, China. *Plant Diversity* 38, 253–258.

Chien PD, Zuidema PA, Nghia NH (2008) Conservation prospects for threatened Vietnamese tree species: results from a demographic study. *Population Ecology* 50, 227–237.

Chinea JD (2002) Tropical forest succession on abandoned farms in the Humacao Municipality of eastern Puerto Rico. *Forest Ecology and Management* 167, 195–207.

Chirino E, Vilagrosa A, Vallejo VR (2011) Using hydrogel and clay to improve the water status of seedlings for dryland restoration. *Plant and Soil* 344, 99–110.

Chisholm RA, Muller-Landau HC, Abdul Rahman K, *et al.* (2013) Scale-dependent relationships between tree species richness and ecosystem function in forests. *Journal of Ecology* 101, 1214–1224.

Choi YD (2004) Theories for ecological restoration in changing environment: toward "futuristic" restoration. *Ecological Research* 19, 75–81.

Choi YD (2007) Restoration ecology to the future: a call for new paradigm. *Restoration Ecology* 15, 351–353.

Choi YD, Temperton VM, Allen EB, *et al.* (2008) Ecological restoration for future sustainability in a changing environment. *Ecoscience* 15, 53–64.

Christe C, Kozlowski G, Frey D, *et al.* (2014) Do living *ex situ* collections capture the genetic variation of wild populations? A molecular analysis of two relict tree species, *Zelkova abelica* and *Zelkova carpinifolia. Biodiversity and Conservation* 23, 2945–2959.

Christensen NL, Bartuska AM, Brown JH, *et al.* (1996) The report of the Ecological Society of America committee on the scientific basis for ecosystem management. *Ecological Applications* 6, 665–691.

Christie DA, Armesto JJ (2003) Regeneration microsites and tree species coexistence in temperate rain forests of Chiloé Island, Chile. *Journal of Ecology* 91, 776–784.

Christy EJ, Mack RN (1984) Variation in demography of juvenile *Tsuga heterophylla* across the substratum mosaic. *Journal of Ecology* 72, 75–91.

Cibrian-Jaramillo A, Hird A, Oleas N, *et al.* (2013) What is the conservation value of a plant in a botanic garden? Using indicators to improve management of ex situ collections. *Botanical Review* 79, 559–577.

Clark CJ, Poulsen JR, Levey DJ, Osenberg CW (2007) Are plant populations seed limited? A critique and meta-analysis of seed addition experiments. *American Naturalist* 170, 128–142.

Clark DA, Clark DB (1992) Life history diversity of canopy and emergent trees in a neotropical rain forest. *Ecological Monographs* 62, 315–344.

Clark DB (1996) Abolishing virginity. *Journal of Tropical Ecology* 12, 735–739.

Clark DB, Clark DA (1990) Distribution and effects on tree growth of lianas and woody hemiepiphytes in a Costa Rican tropical wet forest. *Journal of Tropical Ecology* 6, 321–331.

Clark JA, Covey KR (2012) Tree species richness and the logging of natural forests: a meta-analysis. *Forest Ecology and Management* 276, 146–153.

Clark JS, Beckage B, Camill P, *et al.* (1999) Interpreting recruitment limitation in forests. *American Journal of Botany* 86, 1–16.

Clark JS, Macklin E, Wood L (1998) Stages and spatial scales of recruitment limitation in southern Appalachian forests. *Ecological Monographs* 68, 213–235.

Clark NE, Boakes EH, McGowan PJ, Mace GM, Fuller RA (2013) Protected areas in South Asia have not prevented habitat loss: a study using historical models of land-use change. *PLoS ONE* 8, e65298.

Clarke PJ, Lawes MJ, Midgley JJ, *et al.* (2013) Resprouting as a key functional trait: how buds, protection and resources drive persistence after fire. *New Phytologist* 197, 19–35.

Clausen J, Keck DD, Hiesey WM (1940) Experimental studies on the nature of species. I. Effect of varied environments on western North American plants. In *Carnegie Institute Washington Publ. No. 520.*

Clausen J, Keck DD, Hiesey WM (1948) Experimental studies on the nature of species. III. Environmental responses of climatic races of *Achillea. Carnegie Institute Washington Publ. No. 581.*

Cleland DT, Avers PE, McNab WH, *et al.* (1997) National hierarchical framework of ecological units. In *Ecosystem Management Applications for Sustainable Forest and Wildlife Resources*, Boyce MS, Haney A (eds). New Haven, CT: Yale University Press, pp. 181–200.

Clewell A, Rieger JP (1997) What practitioners need from restoration ecologists. *Restoration Ecology* 5, 350–354.

Clewell AF, Aronson J (2013) *Ecological Restoration: Principles, Values, and Structure of an Emerging Profession*, 2nd edition. Washington, DC: Island Press.

Clout MN, Craig JL (1995) The conservation of critically endangered flightless birds in New Zealand. *Ibis* 137, S181–S190.

Coates KD, Burton PJ (1999) Growth of planted tree seedlings in response to ambient light levels in northwestern interior cedar–hemlock forests of British Columbia. *Canadian Journal of Forest Research* 29, 1374–1382.

Cochrane JA, Crawford AD, Monks LT (2007) The significance of ex situ seed conservation to reintroduction of threatened plants. *Australian Journal of Botany* 55, 356–361.

Cohen AL, Singhakumara BMP, Ashton PMS (1995) Releasing rain forest succession: a case study in the *Dicranopteris linearis* fernlands of Sri Lanka. *Restoration Ecology* 3, 261–270.

Cohrane JA, Crawford AD, Monks LT (2007) The significance of *ex situ* seed conservation to reintroduction of threatened plants. *Australian Journal of Botany* 55, 356–361.

Colas B, Kirchner F, Riba M, *et al.* (2008) Restoration demography: a 10-year demographic comparison between introduced and natural populations of endemic *Centaurea corymbosa* (Asteraceae). *Journal of Applied Ecology* 45, 1468–1476.

Colas B, Olivieri I, Riba M (2001) Spatio-temporal variation of reproductive success and conservation of the narrow-endemic *Centaurea corymbosa* (Asteraceae). *Biological Conservation* 99, 375–386.

Cole KL, Ironside K, Eischeid J, *et al.* (2011) Past and ongoing shifts in Joshua tree distribution support future modeled range contraction. *Ecological Applications* 21, 137–149.

Cole RJ, Holl KD, Zahawi RA (2010) Seed rain under tree islands planted to restore degraded lands in a tropical agricultural landscape. *Ecological Applications* 20, 1255–1269.

Collins SL, Good RE (1987) The seedling regeneration niche: habitat structure of tree seedlings in an oak–pine forest. *Oikos* 48, 89–98.

Comita LS, Hubbell SP (2009) Local neighborhood and species' shade tolerance influence survival in a diverse seedling bank. *Ecology* 90, 328–334.

Comita LS, Muller-Landau HC, Aguilar S, Hubbell SP (2010) Asymmetric density dependence shapes species abundances in a tropical tree community. *Science* 329, 330–332.

Compton JE, Boone RD (2000) Long-term impacts of agriculture on soil carbon and nitrogen in New England forests. *Ecology* 81, 2314–2330.

Condit R, Pitman N, Leigh EG, *et al.* (2002) Beta-diversity in tropical forest trees. *Science* 295, 666–669.

Condit R, Sukumar R, Hubbell SP, Foster RB (1998) Predicting population trends from size distributions: a direct test in a tropical tree community. *American Naturalist* 152, 495–509.

Connell JH, Slatyer RO (1977) Mechanisms of succession in natural communities and their role in community stability and organization. *American Naturalist* 111, 1119–1144.

Convention on Biological Diversity (CBD) (2009) *Convention on Biological Diversity*. Montreal, Canada: Secretariat of the Convention on Biological Diversity. www.cbd.int/

Coomes DA, Grubb PJ (2000) Impacts of root competition in forests and woodlands: a theoretical framework and review of experiments. *Ecological Monographs* 70, 171–207.

Coop JD, Givnish TJ (2008) Constraints on tree seedling establishment in montane grasslands of the Valles Caldera, New Mexico. *Ecology* 89, 1101–1111.

Corbin JD, Holl KD (2012) Applied nucleation as a forest restoration strategy. *Forest Ecology and Management* 265, 37–46.

Cordeiro NJ, Howe HF (2003) Forest fragmentation severs mutualism between seed dispersers and an endemic African tree. *Proceedings of the National Academy of Sciences of the United States of America* 100, 14052–14056.

Cordell S, McClellan M, Yarber Carter Y, Hadway LJ (2008) Towards restoration of Hawaiian tropical dry forests: the Kaupulehu outplanting programme. *Pacific Conservation Biology* 14, 279–284.

Corlett RT (1998) Frugivory and seed dispersal by vertebrates in the Oriental (Indomalayan) Region. *Biological Reviews* 73, 413–448.

Costa H, Aranda SC, Lourenço P, *et al.* (2012) Predicting successful replacement of forest invaders by native species using species distribution models: the case of *Pittosporum undulatum* and *Morella faya* in the Azores. *Forest Ecology and Management* 279, 90–96.

Cox PA (1983) Extinction of the Hawaiian aviafauna resulted in a change of pollinators for the ieie. *Freycinetia arborea. Oikos* 41, 195–199.

Crandall KA, Bininda-Emonds ORP, Mace GM, Wayne RK (2000) Considering evolutionary processes in conservation biology. *Trends in Ecology and Evolution* 15, 290–295.

Crawley MJ (2000) Seed predators and plant population dynamics. In *Seeds: The Ecology of Regeneration in Plant Population Dynamics*, Fenner M (ed.). Wallington, UK: CAB International, pp. 167–182.

Critical Ecosystem Partnership Fund (2016) Announcing the world's 36th biodiversity hotspot: the North American Coastal Plain. www.cepf.net/stories/north-american-coastal-plain-recognized-worlds-36th-biodiversity-hotspot

Crouzeilles R, Curran M (2016) Which landscape size best predicts the influence of forest cover on restoration success? A global meta-analysis on the scale of effect. *Journal of Applied Ecology* 53, 440–448.

Crouzeilles R, Curran M, Ferreira MS, *et al.* (2016) A global meta-analysis on the ecological drivers of forest restoration success. *Nature Communications* 7, 11666.

Crouzeilles R, Prevedello JA, Figueiredo MDSL, Lorini ML, Grelle CEV (2014) The effects of the number, size and isolation of patches along a gradient of native vegetation cover: how can we increment habitat availability? *Landscape Ecology* 29, 479–489.

Crowe KA, Parker WH (2008) Using portfolio theory to guide reforestation and restoration under climate change scenarios. *Climatic Change* 89, 355–370.

Crowly JM (1967) Biogeography. *Canadian Geographer* 11, 312–326.

Cubina A, Aide TM (2001) The effect of distance from forest edge on seed rain and soil seed bank in a tropical pasture. *Biotropica* 33, 260–267.

Cugnac Ad (1953) Le role des jardines botaniques pour la conservation des especes menacees de disparition ou d'alteration. *Annales de Biologie* 29, 361–367.

Culot L, Bello C, Batista JLF, Couto HTZ, Galetti M (2017) Synergistic effects of seed disperser and predator loss on recruitment success and long-term consequences for carbon stocks in tropical rainforests. *Scientific Reports* 7, 7662.

Cunningham SA (2000) Depressed pollination in habitat fragments causes low fruit set. *Proceedings of the Royal Society B: Biological Sciences* 267, 1149–1152.

Curran LM, Webb CO (2000) Experimental tests of the spatiotemporal scale of seed predation in mast-fruiting Dipterocarpaceae. *Ecological Monographs* 70, 129–148.

Cutter BE, Lowell KE, Dwyer JP (1991) Thinning effects on diameter growth in black and scarlet oak as shown by tree ring analyses. *Forest Ecology and Management* 43, 1–13.

Cwynar LC, Macdonald GM (1987) Geographical variation of lodgepole pine in relation to population history. *American Naturalist* 129, 463–469.

d'Oliveira MVN (2000) Artificial regeneration in gaps and skidding trails after mechanised forest exploitation in Acre, Brazil. *Forest Ecology and Management* 127, 67–76.

Da Silva JM, Donaldson JS, Reeves G, Hedderson TA (2012) Population genetics and conservation of critically small cycad populations: a case study of the Albany cycad, *Encephalartos latifrons* (Lehmann). *Biological Journal of the Linnean Society* 105, 293–308.

da Silva JMC, Tabarelli M (2000) Tree species impoverishment and the future flora of the Atlantic forest of northeast Brazil. *Nature* 404, 72–74.

Daily GC, Polasky S, Goldstein J, *et al.* (2009) Ecosystem services in decision making: time to deliver. *Frontiers in Ecology and the Environment* 7, 21–28.

Dalmacio MV (1987) *Assisted Natural Regeneration: A Strategy for Cheap, Fast, and Effective Regeneration of Denuded Forest Lands*. Tacloban City, Philippines: Philippines Department of Environment and Natural Resources Regional Office.

Dalrymple SE, Banks E, Stewart GB, Pullin AS (2012) A meta-analysis of threatened plant reintroductions from across the globe. In *Plant Reintroduction in a Changing Climate: Promises and Perils*, Maschinski J, Haskins EH (eds). Washington, DC: Island Press, pp. 31–50.

Dani Sanchez M, Manco BN, Blaise J, Corcoran M, Hamilton MA (2018) Conserving and restoring the Caicos pine forests: the first decade. *Plant Diversity* (in press).

Danthu P, Ramaroson N, Rambeloarisoa G (2008) Seasonal dependence of rooting success in cuttings from natural forest trees in Madagascar. *Agroforestry Systems* 73, 47–53.

Darbyshire I, Anderson S, Asatryan A, *et al.* (2017) Important plant areas: revised selection criteria for a global approach to plant conservation. *Biodiversity and Conservation* 26, 1767–1800.

Dauber J, Biesmeijer JC, Gabriel D, *et al.* (2010) Effects of patch size and density on flower visitation and seed set of wild plants: a pan-European approach. *Journal of Ecology* 98, 188–196.

Davies AL, Watson F (2007) Understanding the changing value of natural resources: an integrated palaeoecological-historical investigation into grazing-woodland interactions by Loch Awe, Western Highlands of Scotland. *Journal of Biogeography* 34, 1777–1791.

Davies PM (2010) Climate change implications for river restoration in global bio-diversity hotspots. *Restoration Ecology* 18, 261–268.

Davis MB, Shaw RG (2001) Range shifts and adaptive responses to Quaternary climate change. *Science* 292, 673–679.

Daws MI, Koch JM (2015) Long-term restoration success of re-sprouter under-storey species is facilitated by protection from herbivory and a reduction in competition. *Plant Ecology* 216, 565–576.

De Frenne P, Rodriguez-Sanchez F, Coomes DA, *et al.* (2013) Microclimate moderates plant responses to macroclimate warming. *Proceedings of the National Academy of Sciences of the United States of America* 110, 18561–18565.

De la Montana E, Rey-Benayas JM, Carrascal LM (2006) Response of bird communities to silvicultural thinning of Mediterranean maquis. *Journal of Applied Ecology* 43, 651–659.

de la Pena-Domene M, Martinez-Garza C, Howe HF (2013) Early recruitment dynamics in tropical restoration. *Ecological Applications* 23, 1124–1134.

de la Pena-Domene M, Martinez-Garza C, Palmas-Perez S, Rivas-Alonso E, Howe HF (2014) Roles of birds and bats in early tropical-forest restoration. *PLoS ONE* 9, e104656.

De Motta MJ (2010) A history of Hawaiian plant propagation. *Sibbaldia* 8, 31–43.

De Poorter M (2007) *Invasive alien species and protected areas: a scoping report. Part 1. Scoping the scale and nature of invasive alien species threats to protected areas, impediments to invasive alien species management and means to address those impediments.* Auckland, New Zealand: Global Invasive Species Programme, Invasive Species Specialist Group.

de Rezende CL, Uezu A, Scarano FR, Dunn Araujo DS (2015) Atlantic Forest spontaneous regeneration at landscape scale. *Biodiversity and Conservation* 24, 2255–2272.

De Rouw A (1993) Regeneration by sprouting in slash and burn rice cultivation, Taï rain forest, Côte d'Ivoire. *Journal of Tropical Ecology* 9, 387–408.

De Steven D, Sharitz RR, Singer JH, Barton CD (2006) Testing a passive revegetation approach for restoring coastal plain depression wetlands. *Restoration Ecology* 14, 452–460.

Degen B, Blanc L, Caron H, *et al.* (2006) Impact of selective logging on genetic composition and demographic structure of four tropical tree species. *Biological Conservation* 131, 386–401.

DeMauro MM (1993) Relationship of breeding system to rarity in the Lakeside daisy (*Hymenoxys acaulis* var. *glabra*). *Conservation Biology* 7, 542–550.

DeMauro MM (1994) Development and implementation of a recovery program for the federal threatened Lakeside daisy (*Hymenoxys acaulis* var. *glabra*). In *Restoration of Endangered Species: Conceptual Issues, Planning, and Implementation*, Bowles ML, Whelan CJ (eds). Cambridge: Cambridge University Press, pp. 298–321.

Deng T, Abbott RJ, Li W, Sun H, Volis S (2019) Genetic diversity hotspots and refugia identified by mapping multi-plant species haplotype diversity in China (submitted).

Denham AJ (2008) Seed predation limits post-fire recruitment in the waratah (*Telopea speciosissima*). *Plant Ecology* 199, 9–19.

Dennis RLH, Shreeve TG, Van Dyck H (2003) Towards a functional resource-based concept for habitat: a butterfly biology viewpoint. *Oikos* 102, 417–426.

Denslow JS (1987) Tropical rainforest gaps and tree species diversity. *Annual Review of Ecology, Evolution, and Systematics* 18, 431–451.

Denslow JS, Ellison AM, Sanford RE (1998) Treefall gap size effects on above- and below-ground processes in a tropical wet forest. *Journal of Ecology* 86, 597–609.

Dent DH, DeWalt SJ, Denslow JS (2013) Secondary forests of Central Panama increase in similarity to old-growth forest over time in shade tolerance but not species composition. *Journal of Vegetation Science* 24, 530–542.

Deredec A, Courchamp F (2007) Importance of the Allee effect for reintroductions. *Ecoscience* 14, 440–451.

Devine WD, Harrington CA (2009) Western redcedar response to precommercial thinning and fertilization through 25 years posttreatment. *Canadian Journal of Forest Research* 39, 619–628.

Devine WD, Harrington CA, Leonard LP (2007) Post-planting treatments increase growth of Oregon white oak (*Quercus garryana* Dougl. ex Hook.) seedlings. *Restoration Ecology* 15, 212–222.

DeWalt SJ, Maliakal SK, Denslow JS (2003) Changes in vegetation structure and composition along a tropical forest chronosequence: implications for wildlife. *Forest Ecology and Management* 182, 139–151.

Dey DC, Gardiner ES, Schweitzer CJ, Kabrick JM, Jacobs DF (2012) Underplanting to sustain future stocking of oak (*Quercus*) in temperate deciduous forests. *New Forests* 43, 955–978.

Dhar S (1996) *Corypha taliera*: endangered palm extinct in the wild. *Palm Journal* 130, 10–11.

Diamond J (1975) The island dilemma: lessons of modern biogeographic studies for the design of natural reserves. *Biological Conservation* 7, 129–146.

Diamond JM (1972) Biogeographic kinetics: estimation of relaxation times for avifaunas of Southwest Pacific Islands. *Proceedings of the National Academy of Sciences of the United States of America* 69, 3199–3203.

Diaz S, Cabido M (1997) Plant functional types and ecosystem function in relation to global change. *Journal of Vegetation Science* 8, 463–474.

Dickinson MB, Whigham DF, Hermann SM (2000) Tree regeneration in felling and natural treefall disturbances in a semideciduous tropical forest in Mexico. *Forest Ecology and Management* 134, 137–151.

Dietl GP, Kidwell SM, Brenner M, *et al.* (2015) Conservation paleobiology: leveraging knowledge of the past to inform conservation and restoration. *Annual Review of Earth and Planetary Sciences* 43, 3.1–3.25.

Dillon R, Monks L, Coates D (2018) Establishment success and persistence of threatened plant translocations in south west Western Australia: an experimental approach. *Australian Journal of Botany* 66, 338–346.

Dinerstein E (2000) *A Workbook for Conducting Biological Assessments and Developing Biodiversity Visions for Ecoregion-based Conservation.* Washington, DC: World Wildlife Fund.

Diniz MF, Goncalves TV, Brito D (2017) Last of the green: identifying priority sites to prevent plant extinctions in Brazil. *Oryx* 51, 131–136.

Dixon KW (2009) Pollination and restoration. *Science* 325, 571–573.

Dizon AE, Lockyer C, Perrin WF (1992) Rethinking the stock concept: a phylogeographic approach. *Conservation Biology* 6, 24–36.

Dobrowski SZ (2011) A climatic basis for microrefugia: the influence of terrain on climate. *Global Change Biology* 17, 1022–1035.

Dobson AP, Bradshaw AD, Baker AJM (1997) Hopes for the future: restoration ecology and conservation biology. *Science* 277, 515–522.

Dodson EK, Ares A, Puettmann KJ (2012) Early responses to thinning treatments designed to accelerate late successional forest structure in young coniferous stands of western Oregon, USA. *Canadian Journal of Forest Research* 42, 345–355.

Doebeli M, Dieckmann U (2003) Speciation along environmental gradients. *Nature* 421, 259–264.

Dolezal J, St'astna P, Hara T, Srutek M (2004) Neighbourhood interactions and environmental factors influencing old-pasture succession in the Central Pyrenees. *Journal of Vegetation Science* 15, 101–108.

Donatti CI, Guimaraes PR, Galetti M, *et al.* (2011) Analysis of a hyper-diverse seed dispersal network: modularity and underlying mechanisms. *Ecology Letters* 14, 773–781.

Donoso DS, Grez AA, Simonetti JA (2004) Effects of forest fragmentation on the granivory of differently sized seeds. *Biological Conservation* 115, 63–70.

Doucet J-L, Kouadio YL, Monticelli D, Lejeune P (2009) Enrichment of logging gaps with moabi (*Baillonella toxisperma* Pierre) in a Central African rain forest. *Forest Ecology and Management* 258, 2407–2415.

Drayton B, Primack RB (2000) Rates of success in the reintroduction by four methods of several perennial plant species in eastern Massachusetts. *Rhodora* 102, 299–331.

Drayton B, Primack RB (2012) Success rates for reintroductions of eight perennial plant species after 15 years. *Restoration Ecology* 20, 299–303.

Drechsler M, Lamont BB, Burgman MA, *et al.* (1999) Modelling the persistence of an apparently immortal *Banksia* species after fire and land clearing. *Biological Conservation* 88, 249–259.

Drezner TD (2006) Plant facilitation in extreme environments: the non-random distribution of saguaro cacti (*Carnegiea gigantea*) under their nurse associates and the relationship to nurse architecture. *Journal of Arid Environments* 65, 46–61.

Duarte LDS, Dos-Santos MMG, Hartz SM, Pillar VD (2006) Role of nurse plants in Araucaria Forest expansion over grassland in south Brazil. *Austral Ecology* 31, 520–528.

Ducci F (2014) Species restoration through dynamic *ex situ* conservation: *Abies nebrodensis* as a model. In *Genetic Considerations in Ecosystem Restoration Using Native Tree Species. State of the World's Forest Genetic Resources: Thematic Study*, Bozzano M, Jalonen R, Thomas E, *et al.* (eds). Rome: FAO and Bioversity International, pp. 224–233.

Dudley N (2008) *Guidelines for Applying Protected Area Management Categories.* Gland, Switzerland: IUCN.

Dudley N, Boucher JL, Cuttelod A, Brooks TM, Langhammer PF (2014) *Applications of Key Biodiversity Areas: End-user Consultations.* Gland, Switzerland: IUCN.

Duffy DC, Meier AJ (1992) Do Appalachian herbaceous understories ever recover from clearcutting? *Conservation Biology* 6, 196–201.

Dugan P (2000) Assisted natural regeneration: methods, results and issues relevant to sustained participation by communities. In *Forest Restoration for Wildlife Conservation*, Elliott S, Kerby J, Blakesley D, *et al.* (eds). Chiang Mai, Thailand: International Tropical Timber Organisation and The Forest Restoration Research Unit, Chiang Mai University, pp. 195–199.

Dullinger S, Huelber K (2011) Experimental evaluation of seed limitation in alpine snowbed plants. *PLoS ONE* 6, e21537.

Dumroese RK, Williams MI, Stanturf JA, Clair JBS (2015) Considerations for restoring temperate forests of tomorrow: forest restoration, assisted migration, and bioengineering. *New Forests* 46, 947–964.

Dunn RR (2004) Recovery of faunal communities during tropical forest regeneration. *Conservation Biology* 18, 302–309.

Dupouey J-L, Dambrine E, Laffite J-D, Moares C (2002) Irreversible impact of past land use on forest soils and biodiversity. *Ecology* 83, 2978–2984.

Eckert CG, Samis KE, Lougheed SC (2008) Genetic variation across species' geographical ranges: the central-marginal hypothesis and beyond. *Molecular Ecology* 17, 1170–1188.

Edwards DP, Ansell FA, Ahmad AH, Nilus R, Hamer KC (2009) The value of rehabilitating logged rainforest for birds. *Conservation Biology* 23, 1628–1633.

Edwards DP, Laurance WF (2013) Biodiversity despite selective logging. *Science* 339, 646–647.

Edwards TC, Cutler DR, Zimmermann NE, Geiser L, Alegria J (2005) Model-based stratifications for enhancing the detection of rare ecological events. *Ecology* 86, 1081–1090.

Effiom EO, Nunez-Iturri G, Smith HG, Ottosson U, Olsson O (2013) Bushmeat hunting changes regeneration of African rainforests. *Proceedings of the Royal Society B: Biological Sciences* 280, 20130246.

Egan D, Howell EA (2001) *The Historical Ecology Handbook: A Restorationist's Guide to Reference Ecosystems.* Washington, DC: Island Press.

Ehrlen J, Eriksson O (2000) Dispersal limitation and patch occupancy in forest herbs. *Ecology* 81, 1667–1674.

Eken G, Bennun L, Brooks TM, *et al.* (2004) Key biodiversity areas as site conservation targets. *BioScience* 54, 1110–1118.

Eldridge DJ, Westoby M, Holbrook KG (1991) Soil-surface characteristics, microtopography and proximity to mature shrubs: effects on survival of several cohorts of *Atriplex vesicaria* seedlings. *Journal of Ecology* 79, 357–364.

Eldridge DJ, Zaady E, Shachak M (2002) Microphytic crusts, shrub patches and water harvesting in the Negev Desert: the Shikim system. *Landscape Ecology* 17, 587–597.

Elith J, Leathwick JR (2009) Species distribution models: ecological explanation and prediction across space and time. *Annual Review of Ecology, Evolution, and Systematics* 40, 677–697.

Elliott S, Blakesley D, Anusarnsunthorn V (1998) *Forests for the Future: Growing and Planting Native Trees for Restoring Forest Ecosystems.* Chiang Mai, Thailand: Biology Department, Science Faculty, Chiang Mai University.

Elliott S, Blakesley D, Hardwick K (2013) *Restoring Tropical Forests: A Practical Guide.* Richmond, UK: Royal Botanic Gardens Kew.

Elliott S, Navakitbumrung P, Kuarak C, *et al.* (2003) Selecting framework tree species for restoring seasonally dry tropical forests in northern Thailand based on field performance. *Forest Ecology and Management* 184, 177–191.

Emmett RS, Nye DE (2017) *The Environmental Humanities: A Critical Introduction.* Cambridge, MA: MIT Press.

Encino-Ruiz L, Lindig-Cisneros R, Gomez-Romero M, Blanco-Garcia A (2013) Performance of three tropical forest tree species in a test deciduous ecological restoration. *Botanical Sciences* 91, 107–114.

Endler JA (1977) *Geographic Variation, Speciation, and Clines.* Princeton, NJ: Princeton Univesity Press.

Endress BA, Chinea JD (2001) Landscape patterns of tropical forest recovery in the Republic of Palau. *Biotropica* 33, 555–565.

Engelbrecht BMJ, Comita LS, Condit R, *et al.* (2007) Drought sensitivity shapes species distribution patterns in tropical forests. *Nature* 447, 80–82.

Enright NJ, Miller BP, Perry GLW (2003) Demography of the long-lived conifer *Agathis ovata* in maquis and rainforest, New Caledonia. *Journal of Vegetation Science* 14, 625–636.

Enright NJ, Ogden J, Rigg LS (1999) Dynamics of forests with Araucariaceae in the western Pacific. *Journal of Vegetation Science* 10, 793–804.

Ensslin A, Sandner TM, Matthies D (2011) Consequences of ex situ cultivation of plants: genetic diversity, fitness and adaptation of the monocarpic *Cynoglossum officinale* L. in botanic gardens. *Biological Conservation* 144, 272–278.

Ensslin A, Tschope O, Burkart M, Joshi J (2015) Fitness decline and adaptation to novel environments in ex situ plant collections: current knowledge and future perspectives. *Biological Conservation* 192, 394–401.

Erefur C, Bergsten U, Lundmark T, de Chantal M (2011) Establishment of planted Norway spruce and Scots pine seedlings: effects of light environment, fertilisation, and orientation and distance with respect to shelter trees. *New Forests* 41, 263–276.

Eriksson O (1993) The species-pool hypothesis and plant community diversity. *Oikos* 68, 371–374.

Eriksson O, Ehrlen J (1992) Seed and microsite limitation of recruitment in plant populations. *Oecologia* 91, 360–364.

Etterson JR, Shaw RG (2001) Constraint to adaptive evolution in response to global warming. *Science* 294, 151–154.

Facelli JM (1994) Multiple indirect effects of plant litter affect the establishment of woody seedlings in old fields. *Ecology* 75, 1727–1735.

Facon B, Hufbauer RA, Tayeh A, *et al.* (2011) Inbreeding depression is purged in the invasive insect *Harmonia axyridis*. *Current Biology* 21, 424–427.

Fahselt D (2007) Is transplanting an effective means of preserving vegetation? *Canadian Journal of Botany* 85, 1007–1017.

Fajardo L, Caceres A, Arrindell P (2014) Arbuscular mycorrhizae, a tool to enhance the recovery and re-introduction of *Juglans venezuelensis* Manning, an endemic tree on the brink of extinction. *Symbiosis* 64, 63–71.

Falk DA (1987) Inegrated conservation strategies for endangered plants. *Natural Areas Journal* 7, 118–123.

Falk DA (1990a) Endangered forest resources in the U.S.: integrated strategies for conservation of rare species and genetic diversity. *Forest Ecology and Management* 35, 91–117.

Falk DA (1990b) Integrated strategies for conserving plant genetic diversity. *Annals of the Missouri Botanical Garden* 77, 38–47.

Falk DA (2017) Restoration ecology, resilience, and the axes of change. *Annals of the Missouri Botanical Garden* 102, 201–216.

Falk DA, Knapp EE, Guerrant EO (2001) *An Introduction to Restoration Genetics.* Washington, DC: Plant Conservation Alliance, Bureau of Land Management, US Department of Interior, US Environmental Protection Authority.

Falk DA, Olwell P (1992) Scientific and policy considerations in restoration and reintroduction of endangered species. *Rhodora* 94, 287–315.

Falk DA, Palmer MA, Zedler JB (2006) *Foundations of Restoration Ecology.* Washington, DC: Island Press.

Fant JB, Havens K, Kramer AT, *et al.* (2016) What to do when we can't bank on seeds: what botanic gardens can learn from the zoo community about conserving plants in living collections. *American Journal of Botany* 103, 1541–1543.

FAO (2010) *Global Forest Resources Assessment 2010.* Forestry paper 163. Rome: Food and Agriculture Organization.

FAO (2011) *The State of the World's Land and Water Resources for Food and Agriculture (SOLAW): Managing Systems at Risk.* Rome: Food and Agriculture Organization.

FAO/IPGRI (1994) *Genebank Standards.* Rome: Food and Agriculture Organization and International Plant Genetic Resources Institute.

Fashing PJ, Forrestel A, Scully C, Cords M (2004) Long-term tree population dynamics and their implications for the conservation of the Kakamega Forest, Kenya. *Biodiversity and Conservation* 13, 753–771.

Fastie CL (1995) Causes and ecosystem consequences of multiple pathways of primary succession at Glacier Bay, Alaska. *Ecology* 76, 1899–1916.

Fedriani JM, Zywiec M, Delibes M (2012) Thieves or mutualists? Pulp feeders enhance endozoochore local recruitment. *Ecology* 93, 575–587.

Feeley KJ, Davies SJ, Noor MNS, Kassim AR, Tan S (2007) Do current stem size distributions predict future population changes? An empirical test of intraspecific patterns in tropical trees at two spatial scales. *Journal of Tropical Ecology* 23, 191–198.

Feinsinger P, Tiebout HM, Young BE (1991) Do tropical bird-pollinated plants exhibit density-dependent interactions? Field experiments. *Ecology* 72, 1953–1963.

Fenster CB, Dudash MR (1994) Genetic considerations for plant population restoration and conservation. In *Restoration of Endangered Species: Conceptual Issues, Planning, and Implementation*, Bowles ML, Whelan CJ (eds). Cambridge: Cambridge University Press, pp. 34–63.

Fenu G, Cogoni D, Bacchetta G (2016) The role of fencing in the success of threatened plant species translocation. *Plant Ecology* 217, 207–217.

Fenu G, Mattana E, Congiu A, Bacchetta G (2010) The endemic vascular flora of Supramontes (Sardinia), a priority plant conservation area. *Candollea* 65, 347–358.

Fernández-Lugo S, de Nascimento L, Méndez J, *et al.* (2015) Seedling survival patterns in Macaronesian laurel forest: a long-term study in Tenerife (Canary Islands). *Forestry* 88, 121–130.

Ferrandiz AS, Solbes MJL, Esteve AE, Morales MAR, Navarrete EDS (2006) Effects of site preparation with micro-basins on *Pinus halepensis* Mill. afforestations in a semiarid ombroclimate. *Annals of Forest Science* 63, 15–22.

Ferrando-Pardo I, Ferrer-Gallego P, Laguna-Lumbreras E (2016) Assessing the conservation value of *ex situ* seed bank collections of endangered wild plants. *Israel Journal of Plant Sciences* 63, 333–346.

Ferrazzano S, Williamson PS (2013) Benefits of mycorrhizal inoculation in reintroduction of endangered plant species under drought conditions. *Journal of Arid Environments* 98, 123–125.

Ferson S, Burgman MA (1990) The dangers of being few: demographic risk analysis for rare species extinction. *New York State Museum Bulletin* 471, 129–132.

Feurdean A, Willis KJ (2008) The usefulness of a long-term perspective in assessing current forest conservation management in the Apuseni Natural Park, Romania. *Forest Ecology and Management* 256, 421–430.

Fiedler PL, Laven RD (1996) Selecting reintroduction sites. In *Restoring Diversity: Strategies for Reintroduction of Endangered Plants*, Falk DA, Millar CI, Olwell M (eds). Washington, DC: Island Press, pp. 157–170.

Filazzola A, Lortie CJ (2014) A systematic review and conceptual framework for the mechanistic pathways of nurse plants. *Global Ecology and Biogeography* 23, 1335–1345.

Fine PVA, Mesones I, Coley PD (2004) Herbivores promote habitat specialization by trees in Amazonian forests. *Science* 305, 663–665.

Finegan B (1996) Pattern and process in neotropical secondary rain forests: the first 100 years of succession. *Trends in Ecology and Evolution* 11, 119–124.

Fiorella K, Cameron A, Sechrest W, Winfree R, Kremen C (2010) Methodological considerations in reserve system selection: a case study of Malagasy lemurs. *Biological Conservation* 143, 963–973.

Fisher JB, Jayachandran K (2002) Arbuscular mycorrhizal fungi enhance seedling growth in two endangered plant species from South Florida. *International Journal of Plant Sciences* 163, 559–566.

Fjeldsa J, Lovett JC (1997) Geographical patterns of old and young species in African forest biota: the significance of specific montane areas as evolutionary centres. *Biodiversity and Conservation* 6, 325–346.

Fletcher R (2010) Neoliberal environmentality: towards a poststructuralist political ecology of the conservation debate. *Conservation and Society* 8, 171.

Flinn KM, Marks PL (2007) Agricultural legacies in forest environments: tree communities, soil properties, and light availability. *Ecological Applications* 17, 452–463.

Florens F (2013) Conservation in Mauritius and Rodrigues: challenges and achievements from two ecologically devastated oceanic islands. In *Conservation Biology: Voices From the Tropics*, Sodhi NS, Gibson L, Raven PH (eds). Chichester, UK: Wiley-Blackwell, pp. 40–50.

Florens FBV, Baider C (2013) Ecological restoration in a developing island nation: how useful is the science? *Restoration Ecology* 21, 1–5.

Florens FBV, Mauremootoo JR, Fowler SV, Winder L, Baider C (2010) Recovery of indigenous butterfly community following control of invasive alien plants in a tropical island's wet forests. *Biodiversity and Conservation* 19, 3835–3848.

Florens FV, Baider C, Martin GM, *et al.* (2016) Invasive alien plants progress to dominate protected and best-preserved wet forests of an oceanic island. *Journal for Nature Conservation* 34, 93–100.

Flores J, Jurado E (2003) Are nurse-protege interactions more common among plants from arid environments? *Journal of Vegetation Science* 14, 911–916.

Folke C, Carpenter S, Walker B, *et al.* (2004) Regime shifts, resilience, and biodiversity in ecosystem management. *Annual Review of Ecology, Evolution, and Systematics* 35, 557–581.

Forbes AS, Norton DA, Carswell FE (2015) Underplanting degraded exotic *Pinus* with indigenous conifers assists forest restoration. *Ecological Management and Restoration* 16, 41–49.

Forbes AS, Norton DA, Carswell FE (2016) Artificial canopy gaps accelerate restoration within an exotic *Pinus radiata* plantation. *Restoration Ecology* 24, 336–345.

Forest Restoration Research Unit (2006) *How to Plant a Forest: The Principles and Practice of Restoring Tropical Forests*. Chiang Mai, Thailand: Biology Department, Science Faculty, Chiang Mai University.

Forsyth SA (2003) Density-dependent seed set in the Haleakala silversword: evidence for an Allee effect. *Oecologia* 136, 551–557.

Forte Gil JA, Yabor L, Bellido Nadal A, *et al.* (2017) A methodological approach for testing the viability of seeds stored in short-term seed banks. *Notulae Scientia Biologicae* 9, 563–570.

Forup ML, Henson KSE, Craze PG, Memmott J (2008) The restoration of ecological interactions: plant-pollinator networks on ancient and restored heathlands. *Journal of Applied Ecology* 45, 742–752.

Forup ML, Memmott J (2005) The restoration of plant–pollinator interactions in hay meadows. *Restoration Ecology* 13, 265–274.

Foster BL, Tilman D (2003) Seed limitation and the regulation of community structure in oak savanna grassland. *Journal of Ecology* 91, 999–1007.

Foster JR, Reiners WA (1986) Size distribution and expansion of canopy gaps in a northern Appalachian spruce-fir forest. *Plant Ecology* 68, 109–114.

Fotinos TD, Namoff S, Lewis C, *et al.* (2015) Genetic evaluation of a reintroduction of Sargent's cherry palm, *Pseudophoenix sargentii*. *Journal of the Torrey Botanical Society* 142, 51–62.

Fowells HA (1965) *Silvics of Forest Trees of the United States. Agricultural Handbook No. 271.* Washington, DC: USDA Forest Service.

Fowler NL (1988) What is a safe site – neighbor, litter, germination date, and patch effects. *Ecology* 69, 947–961.

Foxcroft LC, Pyšek P, Richardson DM, Pergl J, Hulme PE (2014) The bottom line: impacts of alien plant invasions in protected areas. In *Plant Invasions in Protected Areas: Patterns, Problems and Challenges*, Foxcroft LC, Pyšek P, Richardson DM, Genovesi P (eds). Dordrecht, The Netherlands: Springer, pp. 19–41.

Fraga P, Viçens MM, Gradaille JLL (1997) Re-introduction of *Lysimachia minoricensis* (Primulaceae) in Minorca, Balearic Islands. *Re-introduction News* 13, 12–13.

Franceschinelli EV, Bawa KS (2000) The effect of ecological factors on the mating system of a South American shrub species (*Helicteres brevispira*). *Heredity* 84, 116–123

Franco AC, Nobel PS (1989) Effect of nurse plants on the microhabitat and growth of cacti. *Journal of Ecology* 77, 870–886.

Frankel OH, Soulé ME (1981) *Conservation and Evolution*. Cambridge: Cambridge University Press.

Frankham R (2010) Where are we in conservation genetics and where do we need to go? *Conservation Genetics* 11, 661–663.

Frankham R, Ballou JD, Eldridge MDB, *et al.* (2011) Predicting the probability of outbreeding depression. *Conservation Biology* 25, 465–475.

Frankham R, Bradshaw CJA, Brook BW (2014) Genetics in conservation management: revised recommendations for the 50/500 rules, Red List criteria and population viability analyses. *Biological Conservation* 170, 56–63.

Franklin IR (1980) Evolutionary change in small populations. In *Conservation Biology: An Evolutionary-Ecological Perspective*, Soulé ME, Wilcox BA (eds). Sunderland, MA: Sinauer Associates, pp. 135–149.

Franklin J (2009) *Mapping Species Distributions: Spatial Inference and Predictions.* Cambridge: Cambridge University Press.

Franklin JF, Spies TA, Van Pelt R, *et al.* (2002) Disturbances and structural development of natural forest ecosystems with silvicultural implications, using Douglas-fir forests as an example. *Forest Ecology and Management* 155, 399–423.

Franks AJ, Yates CJ, Hobbs RJ (2009) Defining plant functional groups to guide rare plant management. *Plant Ecology* 204, 207–216.

Fraser DJ, Bernatchez L (2001) Adaptive evolutionary conservation: towards a unified concept for defining conservation units. *Molecular Ecology* 10, 2741–2752.

Fraser LH, Harrower WL, Garris HW, *et al.* (2015) A call for applying trophic structure in ecological restoration. *Restoration Ecology* 23, 503–507.

Freitag S, Van Jaarsveld AS (1997) Relative occupancy, endemism, taxonomic distinctiveness and vulnerability: prioritizing regional conservation actions. *Biodiversity and Conservation* 6, 211–232.

Freitag S, Van Jaarsveld AS, Biggs HC (1997) Ranking priority biodiversity areas: an iterative conservation value-based approach. *Biological Conservation* 82, 263–272.

Fricke EC, Tewksbury JJ, Rogers HS (2014) Multiple natural enemies cause distance-dependent mortality at the seed-to-seedling transition. *Ecology Letters* 17, 593–598.

Fu Y, Chen S, Wu T, Hao J, Sima Y (2009) Comparison of seed modality and germination charateristics of compare of *Manglietia hookeri* and endangered *M. grandis. Journal of Northwest Forestry College* 24, 33–37.

Fukami T, Bezemer TM, Mortimer SR, van der Putten WH (2005) Species divergence and trait convergence in experimental plant community assembly. *Ecology Letters* 8, 1283–1290.

Fukue Y, Kado T, Lee SL, *et al.* (2007) Effects of flowering tree density on the mating system and gene flow in *Shorea leprosula* (Dipterocarpaceae) in Peninsular Malaysia. *Journal of Plant Research* 120, 413–420.

Fule PZ, Cocke AE, Heinlein TA, Covington WW (2004) Effects of an intense prescribed forest fire: is it ecological restoration? *Restoration Ecology* 12, 220–230.

Fule PZ, Covington WW, Moore MM (1997) Determining reference conditions for ecosystem management of southwestern ponderosa pine forests. *Ecological Applications* 7, 895–908.

Funk JL, Cleland EE, Suding KN, Zavaleta ES (2008) Restoration through reassembly: plant traits and invasion resistance. *Trends in Ecology and Evolution* 23, 695–703.

Fustec J, Lodé T, Le Jacques D, Cormier JP (2001) Colonization, riparian habitat selection and home range size in a reintroduced population of European beavers in the Loire. *Freshwater Biology* 46, 1361–1371.

Gallagher RV, Makinson RO, Hogbin PM, Hancock N (2015) Assisted colonization as a climate change adaptation tool. *Austral Ecology* 40, 12–20.

Game ET, Lipsett-Moore G, Saxon E, Peterson N, Sheppard S (2011) Incorporating climate change adaptation into national conservation assessments. *Global Change Biology* 17, 3150–3160.

Gamfeldt L, Snall T, Bagchi R, *et al.* (2013) Higher levels of multiple ecosystem services are found in forests with more tree species. *Nature Communications* 4, 1340.

Gao Z, Zhang C, Milne RI (2010) Size-class structure and variation in seed and seedling traits in relation to population size of an endangered species *Craigia yunnanensis* (Tiliaceae). *Australian Journal of Botany* 58, 214–223.

Garcia D, Obeso JR (2003) Facilitation by herbivore-mediated nurse plants in a threatened tree, *Taxus baccata*: local effects and landscape level consistency. *Ecography* 26, 739–750.

Garcia D, Zamora R, Hodar JA, Gomez JM, Castro J (2000) Yew (*Taxus baccata* L.) regeneration is facilitated by fleshy-fruited shrubs in Mediterranean environments. *Biological Conservation* 95, 31–38.

Garcia del Barrio JM, Aunon F, Sanchez de Ron D, Alia R (2013) Assessing regional species pools for restoration programs in Spain. *New Forests* 44, 559–576.

Garcia LC, Cianciaruso MV, Ribeiro DB, Maes dos Santos FA, Rodrigues RR (2015) Flower functional trait responses to restoration time. *Applied Vegetation Science* 18, 402–412.

Garcia LC, Hobbs RJ, Maees dos Santos FA, Rodrigues RR (2014) Flower and fruit availability along a forest restoration gradient. *Biotropica* 46, 114–123.

García-Hernández M, Toledo-Aceves T, López-Barrera F, Sosa VJ, Paz H (2019) Effects of environmental filters on early establishment of cloud forest trees along elevation gradients: implications for assisted migration. *Forest Ecology and Management* 432, 427–435.

Garcia-Moya E, McKell CM (1970) Contribution of shrubs to the nitrogen economy of a desert-wash plant community. *Ecology* 51, 81–88.

Garcia-Ramos G, Kirkpatrick M (1997) Genetic models of adaptation and gene flow in peripheral populations. *Evolution* 51, 21–28.

Gardner CJ (2011) IUCN management categories fail to represent new, multiple-use protected areas in Madagascar. *Oryx* 45, 336–346.

Gardner TA, Barlow J, Sodhi NS, Peres CA (2010) A multi-region assessment of tropical forest biodiversity in a human-modified world. *Biological Conservation* 143, 2293–2300.

Garner W, Steinberger Y (1989) A proposed mechanism for the formation of "Fertile Islands" in the desert ecosystem. *Journal of Arid Environments* 16, 257–262.

Garrido JL, Rey PJ, Herrera CM (2007) Regional and local variation in seedling emergence, mortality and recruitment of a perennial herb in Mediterranean mountain habitats. *Plant Ecology* 190, 109–121.

Garrity DP, Soekardi M, VanNoordwijk M, *et al.* (1996) The *Imperata* grasslands of tropical Asia: area, distribution, and typology. *Agroforestry Systems* 36, 3–29.

Garwood NC (1989) Tropocal soil seed banks: a review. In *Ecology of Soil Seed Banks*, Leck MA, Parker VT, Simpson RI (eds). San Diego, CA: Academic Press, pp. 149–209.

Gaskett AC (2011) Orchid pollination by sexual deception: pollinator perspectives. *Biological Reviews* 86, 33–75.

Gaston KJ (2003) *The Structure and Dynamics of Geographic Ranges.* Oxford, UK: Oxford University Press.

Gaston KJ, Jackson SE, Cantu-Salazar L, Cruz-Pinon G (2008) The ecological performance of protected areas. In *Annual Review of Ecology, Evolution, and Systematics* 39, 93–113.

Gathmann A, Tscharntke T (2002) Foraging ranges of solitary bees. *Journal of Animal Ecology* 71, 757–764.

Gavin DG, Fitzpatrick MC, Gugger PF, *et al.* (2014) Climate refugia: joint inference from fossil records, species distribution models and phylogeography. *New Phytologist* 204, 37–54.

Gemma JN, Koske RE, Habte M (2002) Mycorrhizal dependency of some endemic and endangered Hawaiian plant species. *American Journal of Botany* 89, 337–345.

Ghazoul J (2005) Pollen and seed dispersal among dispersed plants. *Biological Reviews* 80, 413–443.

Ghazoul J, Liston KA, Boyle TJB (1998) Disturbance-induced density-dependent seed set in *Shorea siamensis* (Dipterocarpaceae), a tropical forest tree. *Journal of Ecology* 86, 462–473.

Gibbs JP, Marquez C, Sterling EJ (2008) The role of endangered species reintroduction in ecosystem restoration: tortoise-cactus interactions on Espanola island, Galapagos. *Restoration Ecology* 16, 88–93.

Gibson AL, Espeland EK, Wagner V, Nelson CR (2016) Can local adaptation research in plants inform selection of native plant materials? An analysis of experimental methodologies. *Evolutionary Applications* 9, 1219–1228.

Gibson DJ, Baer SG, Klopf RP, *et al.* (2013) Limited effects of dominant species population source on community composition during community assembly. *Journal of Vegetation Science* 24, 429–440.

Gibson L, Lee TM, Koh LP, *et al.* (2011) Primary forests are irreplaceable for sustaining tropical biodiversity. *Nature* 478, 378–383.

Gilbert GS, Harms KE, Hamill DN, Hubbell SP (2001) Effects of seedling size, El Nino drought, seedling density, and distance to nearest conspecific adult on 6-year survival of *Ocotea whitei* seedlings in Panama. *Oecologia* 127, 509–516.

Gilbert GS, Hubbell SP, Foster RB (1994) Density and distance-to-adult effects of a canker disease of trees in a moist tropical forest. *Oecologia* 98, 100–108.

Gillespie TW, Lipkin B, Sullivan L, *et al.* (2012) The rarest and least protected forests in biodiversity hotspots. *Biodiversity and Conservation* 21, 3597–3611.

Gilman AC, Letcher SG, Fincher RM, *et al.* (2016) Recovery of floristic diversity and basal area in natural forest regeneration and planted plots in a Costa Rican wet forest. *Biotropica* 48, 798–808.

Gilman SE, Urban MC, Tewksbury J, Gilchrist GW, Holt RD (2010) A framework for community interactions under climate change. *Trends in Ecology and Evolution* 25, 325–331.

Gilpin M, Hanski I (1991) *Metapopulation Dynamics: Empirical and Theoretical Investigations.* London: Academic Press.

Gitay H, Noble IR (1997) What are functional types and how should we seek them? In *Plant Functional Types*, Smith TM, Shugart HH, Woodward FI (eds). Cambridge: Cambridge University Press, pp. 3–19.

Glowka L, Burhenne-Guilman F, Synge H, McNeely J, Gündling L (1994) *Guide to the Convention on Biological Diversity. Environment Policy and Law Paper No. 30.* Gland, Switzerland: International Union for the Conservation of Nature and Natural Resources.

Gockel H (1995) Die Trupp-Pflanzung, Ein neues Pflanzschema zur Begründung von Eichenbeständen. *Forst und Holz* 50, 570–575.

Godefroid S, Piazza C, Rossi G, *et al.* (2011) How successful are plant species reintroductions? *Biological Conservation* 144, 672–682.

Godefroid S, Van de Vyver A, Vanderborght T (2010) Germination capacity and viability of threatened species collections in seed banks. *Biodiversity and Conservation* 19, 1365–1383.

Gomes VM, Negreiros D, Fernandes GW, *et al.* (2018) Long-term monitoring of shrub species translocation in degraded Neotropical mountain grassland. *Restoration Ecology* 26, 91–96.

Gomez-Aparicio L (2009) The role of plant interactions in the restoration of degraded ecosystems: a meta-analysis across life-forms and ecosystems. *Journal of Ecology* 97, 1202–1214.

Gomez-Aparicio L, Gomez JM, Zamora R (2005a) Microhabitats shift rank in suitability for seedling establishment depending on habitat type and climate. *Journal of Ecology* 93, 1194–1202.

Gomez-Aparicio L, Zamora R, Gomez JM (2005b) The regeneration status of the endangered *Acer opalus* subsp. *granatense* throughout its geographical distribution in the Iberian Peninsula. *Biological Conservation* 121, 195–206.

Gomez-Aparicio L, Zamora R, Gomez JM, *et al.* (2004) Applying plant facilitation to forest restoration: a meta-analysis of the use of shrubs as nurse plants. *Ecological Applications* 14, 1128–1138.

Gondard H, Sandrine J, Aronson J, Lavorel S (2003) Plant functional types: a promising tool for management and restoration of degraded lands. *Applied Vegetation Science* 6, 223–234.

González-Tokman DM, Barradas VL, Boege K, *et al.* (2018) Performance of 11 tree species under different management treatments in restoration plantings in a tropical dry forest. *Restoration Ecology* 26, 642–649.

Gonzalez-Varo JP, Albaladejo RG, Aizen MA, Arroyo J, Aparicio A (2015) Extinction debt of a common shrub in a fragmented landscape. *Journal of Applied Ecology* 52, 580–589.

Gonzalez-Varo JP, Albaladejo RG, Aparicio A (2009) Mating patterns and spatial distribution of conspecific neighbours in the Mediterranean shrub *Myrtus communis* (Myrtaceae). *Plant Ecology* 203, 207–215.

Goosem M, Paz C, Fensham R, *et al.* (2016) Forest age and isolation affect the rate of recovery of plant species diversity and community composition in secondary rain forests in tropical Australia. *Journal of Vegetation Science* 27, 504–514.

Goosem SP, Tucker NIJ (1995) *Repairing the Rainforest: Theory and Practice of Rainforest Re-establishment in North Queensland's Wet Tropics.* Cairns, Australia: Wet Tropics Management Authority.

Gosling P (2007) *Raising Trees and Shrubs from Seed.* Edinburgh: Forestry Commission.

Gotelli NJ (1991) Metapopulation models: the rescue effect, the propagule rain, and the core-satelite hypothesis. *American Naturalist* 134, 768–776.

Götzenberger L, de Bello F, Bråthen KA, *et al.* (2012) Ecological assembly rules in plant communities: approaches, patterns and prospects. *Biological Reviews* 87, 111–127.

Govaerts R (2010) Safeguarding extinct plants in ex situ collections. *BGjournal* 7, 22–24.

Grant PBC, Samways MJ (2011) Micro-hotspot determination and buffer zone value for Odonata in a globally significant biosphere reserve. *Biological Conservation* 144, 772–781.

Gratzfeld J, Kozlowski G, Fazan L, *et al.* (2015) Whither rare relict trees in climate of rapid change? *BGjournal* 12, 21–25.

Grauel WT, Putz FE (2004) Effects of lianas on growth and regeneration of *Prioria copaifera* in Darien, Panama. *Forest Ecology and Management* 190, 99–108.

Gray A (2000) Adaptive ecosystem management in the Pacific Northwest: a case study from coastal Oregon. *Conservation Ecology* 4, 6.

Gray AN, Spies TA (1996) Gap size, within-gap position and canopy structure effects on conifer seedling establishment. *Journal of Ecology* 84, 635–645.

Grégoire Taillefer A, Wheeler TA (2013) Animal colonization of restored peatlands: inoculation of plant material as a source of insects. *Restoration Ecology* 21, 140–144.

Gregorio N, Herbohn J, Harrison S, Pasa A, Ferraren A (2017) Regulating the quality of seedlings for forest restoration: lessons from the National Greening Program in the Philippines. *Small-Scale Forestry* 16, 83–102.

Greuter W (1994) Extinctions in Mediterranean areas. *Philosophical Transactions of the Royal Society of London Series B: Biological Sciences* 344, 41–46.

Griffin-Noyes E (2012) Observations made while recreating a native Hawaiian forest. *Sibbaldia* 10, 45–56.

Griffith B, Scott JM, Carpenter JW, Reed C (1989) Translocation as a species conservation tool: status and strategy. *Science* 245, 477–480.

Griffith MP, Calonje M, Meerow AW, *et al.* (2015) Can a botanic garden cycad collection capture the genetic diversity in a wild population? *International Journal of Plant Sciences* 176, 1–10.

Griffith P, Lewis CE, Francisco-Ortega J (2011) Palm conservation at a botanic garden: a case study of the Keys thatch palm. *Palms* 55, 93–101.

Griffith TM, Watson MA (2006) Is evolution necessary for range expansion? Manipulating reproductive timing of a weedy annual transplanted beyond its range. *American Naturalist* 167, 153–164.

Griffiths CJ, Hansen DM, Jones CG, Zuel N, Harris S (2011) Resurrecting extinct interactions with extant substitutes. *Current Biology* 21, 762–765.

Grimm NB, Chapin FSI, Bierwagen B, *et al.* (2013) The impacts of climate change on ecosystem structure and function. *Frontiers in Ecology and the Environment* 11, 474–482.

Griscom BW, Ashton PMS (2003) Bamboo control of forest succession: *Guadua sarcocarpa* in Southeastern Peru. *Forest Ecology and Management* 175, 445–454.

Griscom HP, Ashton MS (2011) Restoration of dry tropical forests in Central America: a review of pattern and process. *Forest Ecology and Management* 261, 1564–1579.

Griscom HP, Griscom BW, Ashton MS (2009) Forest regeneration from pasture in the dry tropics of Panama: effects of cattle, exotic grass, and forested riparia. *Restoration Ecology* 17, 117–126.

Groffman P, Baron J, Blett T, *et al.* (2006) Ecological thresholds: the key to successful environmental management or an important concept with no practical application? *Ecosystems* 9, 1–13.

Grogan J, Jennings SB, Landis RM, *et al.* (2008) What loggers leave behind: impacts on big-leaf mahogany (*Swietenia macrophylla*) commercial populations and potential for post-logging recovery in the Brazilian Amazon. *Forest Ecology and Management* 255, 269–281.

Groom MJ (1998) Allee effects limit population viability of an annual plant. *American Naturalist* 151, 487–496.

Groves C, Valutis L, Vosick D, *et al.* (2000) *Designing a Geography of Hope: A Practitioner's Handbook for Ecoregional Conservation Planning.* Arlington, VA: The Nature Conservancy.

Groves CR, Game ET, Anderson MG, *et al.* (2012) Incorporating climate change into systematic conservation planning. *Biodiversity and Conservation* 21, 1651–1671.

Groves CR, Jensen DB, Valutis LL, *et al.* (2002) Planning for biodiversity conservation: putting conservation science into practice. *BioScience* 52, 499–512.

Grubb PJ (1977) Maintenance of species-richness in plant communities – importance of regeneration niche. *Biological Reviews* 52, 107–145.

GSPC (2010) *Plants 2020.* Global Strategy for Plant Conservation. www.plants2020.net/

Guariguata MR (1999) Early response of selected tree species to liberation thinning in a young secondary forest in Northeastern Costa Rica. *Forest Ecology and Management* 124, 255–261.

Guariguata MR, Ostertag R (2001) Neotropical secondary forest succession: changes in structural and functional characteristics. *Forest Ecology and Management* 148, 185–206.

Guariguata MR, Pinard MA (1998) Ecological knowledge of regeneration from seed in neotropical forest trees: implications for natural forest management. *Forest Ecology and Management* 112, 87–99.

Guarino ESG, Scariot AO (2012) Tree seedling survival and growth in logged and undisturbed seasonal deciduous forest fragments in central Brazil. *Journal of Forest Research* 17, 193–201.

Guerrant EOJ (1992) Genetic and demographic considerations in the sampling and reintroduction of rare plants. In *Conservation Biology: The Theory and Practice of Nature Conservation, Preservation and Management*, Fiedler PL, Jain SK (eds). New York: Chapman and Hall, pp. 321–344.

Guerrant EOJ (1996) Designing populations: demographic, genetic and horticultural dimensions. In *Restoring Diversity. Strategies for Reintroduction of Endangered Plants*, Falk DA, Millar CI, Olwell M (eds). Washington, DC: Island Press, pp. 171–207.

Guerrant EOJ, Fiedler PL (2004) Accounting for sample decline during ex situ storage and reintroduction. In *Ex Situ Plant Conservation: Supporting Species Survival in the Wild*, Guerrant EOJ, Havens K, Maunder M (eds). Washington, DC: Island Press, pp. 365–386.

Guerrant EOJ, Fiedler PL, Havens K, Maunder M (2004) Revised genetic sampling guidelines for conservation collections of rare and endangered plants. In *Ex Situ Plant Conservation: Supporting Species Survival in the Wild*, Guerrant EOJ, Havens K, Maunder M (eds). Washington, DC: Island Press, pp. 419–438.

Guerrant EOJ, Havens K, Vitt P (2014) Sampling for effective ex situ plant conservation. *International Journal of Plant Sciences* 175, 11–20.

Guerrant EOJ, Havens K, Vitt P, *et al.* (2015) Population structure integral to seed collection guidelines: a response to Hoban and Schlarbaum (2014). *Biological Conservation* 184, 465–466.

Guerrant EOJ, Kaye TN (2007) Reintroduction of rare and endangered plants: common factors, questions and approaches. *Australian Journal of Botany* 55, 362–370.

Guerrant EOJ, Pavlik BM (1997) Reintroduction of rare plants: genetics, demography and the role of *ex situ* conservation methods. In *Conservation Biology for the Coming Decade*, Fiedler PL, Kareiva PM (eds). New York: Chapman and Hall, pp. 80–108.

Guerrant EOJ, Raven A (2003) Supporting *in situ* conservation: the Berry Botanic Garden, an ex situ regional resource in an integrated conservation community. In *Seed Conservation: Turning Science into Practice*, Smith RD, Dickie JB, Linington SH, Pritchard HW, Probert RJ (eds). Richmond, UK: The Royal Botanic Gardens Kew, pp. 879–896.

Guevara S, Meave J, Morenocasasola P, Laborde J (1992) Floristic composition and structure of vegetation under isolated trees in neotropical pastures. *Journal of Vegetation Science* 3, 655–664.

Guevara S, Purata SE, Vandermaarel E (1986) The role of forest trees in tropical secondary succession. *Vegetatio* 66, 77–84.

Guidetti BY, Amico GC, Dardanelli S, Rodriguez-Cabal MA (2016) Artificial perches promote vegetation restoration. *Plant Ecology* 217, 935–942.

Guilherme FAG, Oliveira AT, Appolinario V, Bearzoti E (2004) Effects of flooding regime and woody bamboos on tree community dynamics in a section of tropical semideciduous forest in South-Eastern Brazil. *Plant Ecology* 174, 19–36.

Guimaraes IC, Proctor J (2007) Mechanisms of plant regeneration during succession after shifting cultivation in eastern Amazonia. *Plant Ecology* 192, 303–315.

Guisan A, Broennimann O, Engler R, *et al.* (2006) Using niche-based models to improve the sampling of rare species. *Conservation Biology* 20, 501–511.

Guisan A, Thuiller W (2005) Predicting species distribution: offering more than simple habitat models. *Ecology Letters* 8, 993–1009.

Guisan A, Tingley R, Baumgartner JB, *et al.* (2013) Predicting species distributions for conservation decisions. *Ecology Letters* 16, 1424–1435.

Gunaratne AMTA, Gunatilleke CVS, Gunatilleke IAUN, Madawala HMSP, Burslem DFRP (2014) Overcoming ecological barriers to tropical lower montane forest succession on anthropogenic grasslands: synthesis and future prospects. *Forest Ecology and Management* 329, 340–350.

Gunaratne AMTA, Gunatilleke CVS, Gunatilleke IAUN, Weerasinghe HMSP, Burslem DFRP (2010) Barriers to tree seedling emergence on human-induced grasslands in Sri Lanka. *Journal of Applied Ecology* 47, 157–165.

Gurd DB, Nudds TD, Rivard DH (2001) Conservation of mammals in eastern North American wildlife reserves: how small is too small? *Conservation Biology* 15, 1355–1363.

Gustafson DJ, Gibson DJ, Nickrent DL (2002) Genetic diversity and competitive abilities of *Dalea purpurea* (Fabaceae) from remnant and restored grasslands. *International Journal of Plant Sciences* 163, 979–990.

Hacke UG, Sperry JS, Pockman WT, Davis SD, McCulloh KA (2001) Trends in wood density and structure are linked to prevention of xylem implosion by negative pressure. *Oecologia* 126, 457–461.

Hackney EE, McGraw JB (2001) Experimental demonstration of an Allee effect in American ginseng. *Conservation Biology* 15, 129–136.

Haggar JP, Briscoe CB, Butterfield RP (1998) Native species: a resource for the diversification of forestry production in the lowland humid tropics. *Forest Ecology and Management* 106, 195–203.

Hall JS (2008) Seed and seedling survival of African mahogany (*Entandrophragma* spp.) in the Central African Republic: implications for forest management. *Forest Ecology and Management* 255, 292–299.

Hall M (2010) *Restoration and History: The Search for a Usable Environmental Past*. New York: Routledge.

Hall P, Orrell LC, Bawa KS (1994) Genetic diversity and mating system in a tropical tree *Carapa guianensis* (Meliaceae). *American Journal of Botany* 81, 1104–1111.

Halle S (2007) Science, art, or application: the "Karma" of restoration ecology. *Restoration Ecology* 15, 358–361.

Halme P, Allen KA, Auniņš A, *et al.* (2013) Challenges of ecological restoration: lessons from forests in northern Europe. *Biological Conservation* 167, 248–256.

Halme P, Monkkonen M, Kotiaho JS, Ylisirnio A-L, Markkanen A (2009) Quantifying the indicator power of an indicator species. *Conservation Biology* 23, 1008–1016.

Halpin PN (1997) Global climate change and natural-area protection: management responses and research directions. *Ecological Applications* 7, 828–843.

Halsey SJ, Bell TJ, McEachern K, Pavlovic NB (2015) Comparison of reintroduction and enhancement effects on metapopulation viability. *Restoration Ecology* 23, 375–384.

Hamilton MB (1994) *Ex situ* conservation of wild plant species: time to reassess the genetic assumptions and implications of seed banks. *Conservation Biology* 8, 39–49.

Hampe A, Arroyo J (2002) Recruitment and regeneration in populations of an endangered South Iberian Tertiary relict tree. *Biological Conservation* 107, 263–271.

Hampe A, Jump AS (2011) Climate relicts: past, present, future. *Annual Review of Ecology, Evolution, and Systematics* 42, 313–333.

Hampe A, Petit RJ (2005) Conserving biodiversity under climate change: the rear edge matters. *Ecology Letters* 8, 461–467.

Handel SN (1997) The role of plant-animal mutualisms in the design and restoration of natural communities. In *Restoration Ecology and Sustainable Development*, Urbanska KM, Webb NR, Edwards PJ (eds). Cambridge: Cambridge University Press, pp. 111–132.

Hanna C, Foote D, Kremen C (2013) Invasive species management restores a plant–pollinator mutualism in Hawaii. *Journal of Applied Ecology* 50, 147–155.

Hannah L, Midgley G, Andelman S, *et al.* (2007) Protected area needs in a changing climate. *Frontiers in Ecology and the Environment* 5, 131–138.

Hanski I, Ovaskainen O (2002) Extinction debt at extinction threshold. *Conservation Biology* 16, 666–673.

Hanski I, Simberloff D (1997) The metapopulation approach, its history, conceptual domain, and application to conservation. In *Metapopulation Biology: Ecology, Genetics, and Evolution*, Hanski I, Gilpin ME (eds). San Diego, CA: Academic Press, pp. 5–26.

Hardwick K, Healey J, Elliott S, Garwood N, Anusarnsunthorn V (1997) Understanding and assisting natural regeneration processes in degraded seasonal evergreen forests in northern Thailand. *Forest Ecology and Management* 99, 203–214.

Hardwick K, Healey JR, Elliott S, Blakesley D (2004) Research needs for restoring seasonal tropical forests in Thailand: accelerated natural regeneration. *New Forests* 27, 285–302.

Hardwick KA, Fiedler P, Lee LC, *et al.* (2011) The role of botanic gardens in the science and practice of ecological restoration. *Conservation Biology* 25, 265–275.

Harmon ME, Franklin JF (1989) Tree seedlings on logs in *Picea-Tsuga* forests of Oregon and Washington. *Ecology* 70, 48–59.

Harmon ME, Franklin JF, Swanson FJ, *et al.* (1986) Ecology of coarse woody debris in temperate ecosystems. *Advances in Ecological Research* 15, 133–302.

Harms KE, Condit R, Hubbell SP, Foster RB (2001) Habitat associations of trees and shrubs in a 50-ha neotropical forest plot. *Journal of Ecology* 89, 947–959.

Harms KE, Wright SJ, Calderon O, Hernandez A, Herre EA (2000) Pervasive density-dependent recruitment enhances seedling diversity in a tropical forest. *Nature* 404, 493–495.

Harpenslager SF, Lamers LPM, der HeideaJan T, Roelofs GM, Smoldersab AJP (2016) Harnessing facilitation: why successful re-introduction of *Stratiotes aloides* requires high densities under high nitrogen loading. *Biological Conservation* 195, 17–23.

Harper JL (1977) *Population Biology of Plants*. London: Academic Press.

Harper JL, McNaughton IH, Sagar GR, Clatworthy JN (1961) Evolution and ecology of closely related species living in same area. *Evolution* 15, 209–227.

Harrington RA, Ewel JJ (1997) Invasibility of tree plantations by native and non-indigenous plant species in Hawaii. *Forest Ecology and Management* 99, 153–162.

Harrington TB (2011) Overstory and understory relationships in longleaf pine plantations 14 years after thinning and woody control. *Canadian Journal of Forest Research* 41, 2301–2314.

Harris JA, Hobbs RJ, Higgs E, Aronson J (2006) Ecological restoration and global climate change. *Restoration Ecology* 14, 170–176.

Harrison RD, Tan S, Plotkin JB, *et al.* (2013) Consequences of defaunation for a tropical tree community. *Ecology Letters* 16, 687–694.

Harrison S, Noss R (2017) Endemism hotspots are linked to stable climatic refugia. *Annals of Botany* 119, 207–214.

Hartley MJ (2002) Rationale and methods for conserving biodiversity in plantation forests. *Forest Ecology and Management* 155, 81–95.

Haselwandter K (1997) Soil micro-organisms, mycorrhiza, and restoration ecology. In *Restoration Ecology and Sustainable Development*, Urbanska KM, Webb NR, Edwards PJ (eds). Cambridge: Cambridge University Press, pp. 65–80.

Havens K, Guerrant EOJ, Maunder M, Vitt P (2004) Guidelines for ex situ conservation collection management: minimizing risks. In *Ex Situ Plant*

*Conservation: Supporting Species Survival in the Wild*, Guerrant EOJ, Havens K, Maunder M (eds). Washington, DC: Island Press, pp. 454–473.

Havens K, Vitt P, Maunder M, Guerrant EOJ, Dixon K (2006) *Ex situ* plant conservation and beyond. *BioScience* 56, 525–531.

Havens K, Vitt P, Still S, *et al.* (2015) Seed sourcing for restoration in an era of climate change. *Natural Areas Journal* 35, 122–133.

Hawes JE, Peres CA (2014) Ecological correlates of trophic status and frugivory in neotropical primates. *Oikos* 123, 365–377.

Hawkes CV, Menges ES (1995) Density and seed production of a Florida endemic *Polygonella basiramia*, in relation to time since fire and open sand. *American Midland Naturalist* 133, 138–148.

Hayward MW (2009) Conservation management for the past, present and future. *Biodiversity and Conservation* 18, 765–775.

He L-Y, Tang CQ, Wu Z-L, *et al.* (2015) Forest structure and regeneration of the Tertiary relict *Taiwania cryptomerioide*s in the Gaoligong Mountains, Yunnan, southwestern China. *Phytocoenologia* 45, 135–155.

Hecht SB (1993) The logic of livestock and deforestation in Amazonia. *BioScience* 43, 687–695.

Hedrick PW (1995) Gene flow and genetic restoration: the Florida panther as a case study. *Conservation Biology* 9, 996–1007.

Heller NE, Zavaleta ES (2009) Biodiversity management in the face of climate change: a review of 22 years of recommendations. *Biological Conservation* 142, 14–32.

Helm A, Zobel M, Moles AT, Szava-Kovats R, Pärtel M (2015) Characteristic and derived diversity: implementing the species pool concept to quantify conservation condition of habitats. *Diversity and Distributions* 21, 711–721.

Hengeveld R, Haeck J (1982) The distribution of abundance. I. Measurements. *Journal of Biogeography* 9, 303–316.

Herault B, Honnay O, Thoen D (2005) Evaluation of the ecological restoration potential of plant communities in Norway spruce plantations using a life-trait based approach. *Journal of Applied Ecology* 42, 536–545.

Herault B, Thoen D, Honnay O (2004) Assessing the potential of natural woody species regeneration for the conversion of Norway spruce plantations on alluvial soils. *Annals of Forest Science* 61, 711–719.

Hereford J (2009) A quantitative survey of local adaptation and fitness trade-offs. *American Naturalist* 173, 579–588.

Herrera CM, Jordano P, Lopez-Soria L, Amat JA (1994) Recruitment of a mast-fruiting, bird-dispersed tree: bridging frugivore activity and seedling establishment. *Ecological Monographs* 64, 315–344.

Hewitt G (2000) The genetic legacy of the Quaternary ice ages. *Nature* 405, 907–913.

Heywood VH (2014) An overview of *in situ* conservation of plant species in the Mediterranean. *Flora Mediterranea* 24, 5–24.

Heywood VH (2016) *In situ* conservation of plant species: an unattainable goal? *Israel Journal of Plant Sciences* 63, 211–231.

Heywood VH (2017) Plant conservation in the Anthropocene – challenges and future prospects. *Plant Diversity* 39, 314–330.

Hickling R, Roy DB, Hill JK, Thomas CD (2005) A northward shift of range margins in British Odonata. *Global Change Biology* 11, 502–506.

Higgs E, Falk DA, Guerrini A, *et al.* (2014) The changing role of history in restoration ecology. *Frontiers in Ecology and the Environment* 12, 499–506.

Hijmans RJ, Graham CH (2006) The ability of climate envelope models to predict the effect of climate change on species distributions. *Global Change Biology* 12, 2272–2281.

Hirayama K, Ishida K, Tomaru N (2005) Effects of pollen shortage and self-pollination on seed production of an endangered tree, *Magnolia stellata*. *Annals of Botany* 95, 1009–1015.

Hirzel AH, Hausser J, Chessel D, Perrin N (2002) Ecological-niche factor analysis: how to compute habitat-suitability maps without absence data? *Ecology* 83, 2027–2036.

Hoban S, Schlarbaum S (2014) Optimal sampling of seeds from plant populations for ex-situ conservation of genetic biodiversity, considering realistic population structure. *Biological Conservation* 177, 90–99.

Hobbs RJ, Arico S, Aronson J, *et al.* (2006) Novel ecosystems: theoretical and management aspects of the new ecological world order. *Global Ecology and Biogeography* 15, 1–7.

Hobbs RJ, Hallett LM, Ehrlich PR, Mooney HA (2011) Intervention ecology: applying ecological science in the twenty-first century. *BioScience* 61, 442–450.

Hobbs RJ, Higgs E, Harris JA (2009) Novel ecosystems: implications for conservation and restoration. *Trends in Ecology and Evolution* 24, 599–605.

Hobbs RJ, Higgs ES, Hall CM (2013) *Novel Ecosystems: Intervening in the New Ecological World Order*. Oxford, UK: Wiley-Blackwell.

Hobbs RJ, Norton DA (1996) Towards a conceptual framework for restoration ecology. *Restoration Ecology* 4, 93–110.

Hodgson JA, Thomas CD, Wintle BA, Moilanen A (2009) Climate change, connectivity and conservation decision making: back to basics. *Journal of Applied Ecology* 46, 964–969.

Hoegh-Guldberg O, Hughes L, McIntyre S, *et al.* (2008) Assisted colonization and rapid climate change. *Science* 321, 345–346.

Hoekstra JM, Boucher TM, Ricketts TH, Roberts C (2005) Confronting a biome crisis: global disparities of habitat loss and protection. *Ecology Letters* 8, 23–29.

Hoekstra JM, Clark JA, Fagan WF, Boersma PD (2002) A comprehensive review of Endangered Species Act recovery plans. *Ecological Applications* 12, 630–640.

Hoffman N, Brown A (1992) *Orchids of South-west Australia.* Perth: University of Western Australia Press.

Hoffmann PM, Blum CT, Velazco SJE, Gill DJC, Borgo M (2015) Identifying target species and seed sources for the restoration of threatened trees in southern Brazil. *Oryx* 49, 425–430.

Hoffmann PM, Velazco SJE (2014) *How to Germinate Seed and Grow Tree Seedlings.* Cambridge: Global Trees Campaign, Fauna and Flora International.

Hofgaard A (1993) Structure and regeneration patterns in a virgin *Picea abies* forest in northern Sweden. *Journal of Vegetation Science* 4, 601–608.

Holl KD (1999) Factors limiting tropical rain forest regeneration in abandoned pasture: seed rain, seed germination, microclimate, and soil. *Biotropica* 31, 229–242.

Holl KD (2002) Effect of shrubs on tree seedling establishment in an abandoned tropical pasture. *Journal of Ecology* 90, 179–187.

Holl KD (2007) Old field vegetation succession in the neotropics. In *Old Fields. Dynamics and Restoration of Abandoned Farmland*, Cramer V, Hobbs R (eds). Washington, DC: Island Press, pp. 93–118.

Holl KD, Aide TM (2011) When and where to actively restore ecosystems? *Forest Ecology and Management* 261, 1558–1563.

Holl KD, Cairns J (2002) Monitoring and appraisal. In *Handbook of Ecological Restoration. Principles of Restoration*, Perrow MR, Davy AJ (eds). Cambridge: Cambridge University Press, pp. 409–432.

Holl KD, Loik ME, Lin EHV, Samuels IA (2000) Tropical montane forest restoration in Costa Rica: overcoming barriers to dispersal and establishment. *Restoration Ecology* 8, 339–349.

Holl KD, Reid JL, Miguel Chaves-Fallas J, Oviedo-Brenes F, Zahawi RA (2017) Local tropical forest restoration strategies affect tree recruitment more strongly than does landscape forest cover. *Journal of Applied Ecology* 54, 1091–1099.

Holl KD, Stout VM, Reid JL, Zahawi RA (2013) Testing heterogeneity-diversity relationships in tropical forest restoration. *Oecologia* 173, 569–578.

Holl KD, Zahawi RA, Cole RJ, Ostertag R, Cordell S (2011) Planting seedlings in tree islands versus plantations as a large-scale tropical forest restoration strategy. *Restoration Ecology* 19, 470–479.

Holsinger KE (1995) Conservation programs for endangered plant species. *Encyclopedia of Environmental Biology* 1, 385–400.

Holt RD, Gomulkiewicz R (1997) How does immigration influence local adaptation? A reexamination of a familiar paradigm. *American Naturalist* 149, 563–572.

Hooper DU, Chapin FS, Ewel JJ, *et al.* (2005a) Effects of biodiversity on ecosystem functioning: a consensus of current knowledge. *Ecological Monographs* 75, 3–35.

Hooper E, Condit R, Legendre P (2002) Responses of 20 native tree species to reforestation strategies for abandoned farmland in Panama. *Ecological Applications* 12, 1626–1641.

Hooper E, Legendre P, Condit R (2005b) Barriers to forest regeneration of deforested and abandoned land in Panama. *Journal of Applied Ecology* 42, 1165–1174.

Hopkins A, Wainwright J (1989) Changes in botanical composition and agricultural management of enclosed grassland in upland areas of England and Wales, 1970–86, and some conservation implications. *Biological Conservation* 47, 219–235.

Hornberg G, Ohlson M, Zackrisson O (1997) Influence of bryophytes and microrelief conditions on *Picea abies* seed regeneration patterns in boreal old-growth swamp forests. *Canadian Journal of Forest Research* 27, 1015–1023.

House SM (1992) Population density and fruit set in three dioecious tree species in Australian tropical rain forest. *Journal of Ecology* 80, 57–69.

Howe GT, Aitken SN, Neale DB, *et al.* (2003) From genotype to phenotype: unraveling the complexities of cold adaptation in forest trees. *Canadian Journal of Botany* 81, 1247–1266.

Howe HF (2016) Making dispersal syndromes and networks useful in tropical conservation and restoration. *Global Ecology and Conservation* 6, 152–178.

Howe HF, Martinez-Garza C (2014) Restoration as experiment. *Botanical Sciences* 92, 459–468.

Howe HF, Swallowed J (1982) Ecology of seed dispersal. *Annual Review of Ecology, Evolution, and Systematics* 13, 201–228.

Hubbell SP (1979) Tree dispersion, abundance and diversity in a tropical dry forest. *Science* 203, 1299–1309.

Hubbell SP (2001) *The Unified Neutral Theory of Biodiversity and Biogeography*. Princeton, NJ: Princeton Univesity Press.

Hubbell SP (2013) Tropical rain forest conservation and the twin challenges of diversity and rarity. *Ecology and Evolution* 3, 3263–3274.

Hubbell SP, Foster RB, O'Brien ST, *et al.* (1999) Light-gap disturbances, recruitment limitation, and tree diversity in a neotropical forest. *Science* 283, 554–557.

Huber PR, Greco SE, Thorne JH (2010) Spatial scale effects on conservation network design: trade-offs and omissions in regional versus local scale planning. *Landscape Ecology* 25, 683–695.

Hudson JR, Hanula JL, Horn S (2013) Removing Chinese privet from riparian forests still benefits pollinators five years later. *Biological Conservation* 167, 355–362.

Huenneke LF (1991) Ecological implications of genetic variation in plant populations. In *Genetics and Conservation of Rare Plants*, Falk DA, Holsinger KE (eds). Oxford, UK: Oxford University Press, pp. 31–44.

Huenneke LF, Vitousek PM (1990) Seedling and clonal recruitment of the invasive tree *Psidium cattleianum*: implications for management of native Hawaiian forests. *Biological Conservation* 53, 199–211.

Hufford KM, Mazer SJ (2003) Plant ecotypes: genetic differentiation in the age of ecological restoration. *Trends in Ecology and Evolution* 18, 147–155.

Hultine KR, Majure LC, Nixon VS, *et al.* (2016) The role of botanical gardens in the conservation of Cactaceae. *BioScience* 66, 1057–1065.

Hunt LP (2001) Heterogeneous grazing causes local extinction of edible perennial shrubs: a matrix analysis. *Journal of Applied Ecology* 38, 238–252.

Hunt VM, Jacobi SK, Knutson MG, *et al.* (2015) A data management system for long-term natural resource monitoring and management projects with multiple cooperators. *Wildlife Society Bulletin* 39, 464–471.

Hunter M (2007) Climate change and moving species: furthering the debate on assisted colonization. *Conservation Biology* 21, 1356–1358.

Hunter MLJ (1999) *Maintaining Biodiversity in Forest Ecosystems.* Cambridge: Cambridge University Press.

Hurka H (1994) Conservation genetics and the role of botanical gardens. In *Conservation Genetics*, Sandlund OT, Hindar K, Brown AHD (eds). Basel: Birkhäuser Verlag, pp. 371–380.

Husband BC, Campbell LG (2004) Population responses to novel environments: implications for *ex situ* plant conservation. In *Ex Situ Plant Conservation: Supporting Species Survival in the Wild*, Guerrant EOJ, Havens K, Maunder M (eds). Washington, DC: Island Press, pp. 231–266.

Huston M (1979) A general hypothesis of species diversity. *American Naturalist* 113, 81–101.

Huston M (1994) *Biological Diversity: The Coexistence of Species on Changing Landscapes.* Cambridge: Cambridge University Press.

Hylander K, Ehrlen J (2013) The mechanisms causing extinction debts. *Trends in Ecology and Evolution* 28, 341–346.

Ibanez I, Schupp EW (2002) Effects of litter, soil surface conditions, and microhabitat on *Cercocarpus ledifolius* Nutt. Seedling emergence and establishment. *Journal of Arid Environments* 52, 209–221.

Ingle NR (2003) Seed dispersal by wind, birds, and bats between Philippine montane rainforest and successional vegetation. *Oecologia* 134, 251–261.

Inman-Narahari F, Ostertag R, Asner GP, *et al.* (2014) Trade-offs in seedling growth and survival within and across tropical forest microhabitats. *Ecology and evolution* 4, 3755–3767.

Iriondo JM, Milla R, Volis S, Rubio de Casas R (2018) Reproductive traits and evolutionary divergence between Mediterranean crops and their wild relatives. *Plant Biology* 20, 78–88.

Isaac-Renton MG, Roberts DR, Hamann A, Spiecker H (2014) Douglas-fir plantations in Europe: a retrospective test of assisted migration to address climate change. *Global Change Biology* 20, 2607–2617.

Isagi Y, Tateno R, Matsuki Y, *et al.* (2007) Genetic and reproductive consequences of forest fragmentation for populations of *Magnolia obovata*. *Ecological Research* 22, 382–389.

ISSG (n.d.) Invasive Species Specialist Group. Gland, Switzerland: IUCN. www.issg.org/

Itoh A, Yamakura T, Kanzaki M, *et al.* (2002) Rooting ability of cuttings relates to phylogeny, habitat preference, and growth characteristics of tropical rainforest trees. *Forest Ecology and Management* 168, 275–287.

Iturriaga L, Jordan M, Roveraro C, Goreux A (1994) In vitro culture of *Sophora toromiro* (Papilionaceae), an endangered species. *Plant Cell Tissue and Organ Culture* 37, 201–204.

IUCN (2001) *UCN Red List Categories and Criteria: Version 3.1.* Gland, Switzerland and Cambridge, UK: IUCN Species Survival Commission.

IUCN (2002) *IUCN Technical Guidelines on the Management of* Ex-Situ *Populations for Conservation.* Gland, Switzerland and Cambridge, UK: IUCN Species Survival Commission.

IUCN (2016) *A Global Standard for the Identification of Key Biodiversity Areas, Version 1.0,* 1st edition. Gland, Switzerland and Cambridge, UK: IUCN Species Survival Commission.

IUCN/SSC (2013) *Guidelines for Reintroductions and Other Conservation Translocations. Version 1.0.* Gland, Switzerland and Cambridge, UK: IUCN Species Survival Commission.

IUCN/SSC (2014) *Guidelines on the Use of Ex Situ Management for Species Conservation. Version 2.0.* Gland, Switzerland and Cambridge, UK: IUCN Species Survival Commission.

Jabot F, Etienne RS, Chave J (2008) Reconciling neutral community models and environmental filtering: theory and an empirical test. *Oikos* 117, 1308–1320.

Jackson LL, Lopoukhine N, Hillyard D (1995) Ecological restoration: a definition and options. *Restoration Ecology* 3, 71–75.

Jacob AL, Lechowicz MJ, Chapman CA (2017) Non-native fruit trees facilitate colonization of native forest on abandoned farmland. *Restoration Ecology* 25, 211–219.

Jacobs DF, Dalgleish HJ, Nelson CD (2013) A conceptual framework for restoration of threatened plants: the effective model of American chestnut (*Castanea dentata*) reintroduction. *New Phytologist* 197, 378–393.

Jakovac CC, Pena-Claros M, Kuyper TW, Bongers F (2015) Loss of secondary-forest resilience by land-use intensification in the Amazon. *Journal of Ecology* 103, 67–77.

Jalonen R, Elliott S (2014) Framework species method. In *Genetic Considerations in Ecosystem Restoration Using Native Tree Species. State of the World's Forest Genetic Resources: Thematic Study.* Rome: FAO and Bioversity International, pp. 144–147.

Janzen DH (1990) An abandoned field is not a tree fall gap. *Vida Silvestre Neotropical* 2, 64–67.

Janzen DH (2001) Latent extinctions: the living dead. In *Encyclopedia of Biodiversity*, Levin SA (ed.). New York: Academic Press, pp. 689–699.

Janzen DH (2002) Tropical dry forest: area de conservacion Guanacaste, north-western Costa Rica. In *Handbook of Ecological Restoration*, Vol. II, Perrow M, Davy AJ (eds). Cambridge: Cambridge University Press, pp. 559–583.

Jennersten O (1988) Pollination in *Dianthus deltoides* (Caryophyllaceae): effect of habitat fragmentation on visitation and seed set. *Conservation Biology* 2, 359–366.

Jensen A, Meilby H (2012) Assessing the population status of a tree species using distance sampling: *Aquilaria crassna* (Thymelaeaceae) in Northern Laos. *International Journal of Forestry Research* 265831.

Johnson D, Beaulieu WT, Bever JD, Clay K (2012) Conspecific negative density dependence and forest diversity. *Science* 336, 904–907.

Johnson GR, Sorenson FC, Clair JBS, Cronnm RC (2004) Pacific Northwest forest tree seed zones – a template for native plants? *Native Plants Journal* 5, 131–140.

Johnson R, Stritch L, Olwell P, *et al.* (2010) What are the best seed sources for eco-system restoration on BLM and USFS lands? *Native Plants* 11, 117–131.

Jonášová M, van Hees A, Prach K (2006) Rehabilitation of monotonous exotic coniferous plantations: a case study of spontaneous establishment of different tree species. *Ecological Engineering* 28, 141–148.

Jones CC, del Moral R (2005) Effects of microsite conditions on seedling establishment on the foreland of Coleman Glacier, Washington. *Journal of Vegetation Science* 16, 293–300.

Jones CC, del Moral R (2009) Dispersal and establishment both limit colonization during primary succession on a glacier foreland. *Plant Ecology* 204, 217–230.

Jones ER, Wishnie MH, Deago J, Sautu A, Cerezo A (2004) Facilitating natural regeneration in *Saccharum spontaneum* (L.) grasslands within the Panama Canal Watershed: effects of tree species and tree structure on vegetation recruitment patterns. *Forest Ecology and Management* 191, 171–183.

Jones TA (2013) When local isn't best. *Evolutionary Applications* 6, 1109–1118.

Jordan PW, Nobel PS (1979) Infrequent establishment of seedlings of *Agave deserti* (Agavaceae) in the northwestern Sonoran Desert. *American Journal of Botany* 66, 1079–1084.

Jordano P, Garcia C, Godoy JA, Garcia-Castano JL (2007) Differential contribution of frugivores to complex seed dispersal patterns. *Proceedings of the National Academy of Sciences of the United States of America* 104, 3278–3282.

Joy DA, Young DR (2002) Promotion of mid-successional seedling recruitment and establishment by *Juniperus virginiana* in a coastal environment. *Plant Ecology* 160, 125–135.

Jusaitis M (2005) Translocation trials confirm specific factors affecting the establishment of three endangered plant species. *Ecological Management and Restoration* 6, 61–67.

Jusaitis M, Polomka L, Sorensen B (2004) Habitat specificity, seed germination and experimental translocation of the endangered herb *Brachycome muelleri* (Asteraceae). *Biological Conservation* 116, 251–266.

Kaiser-Bunbury CN, Bluethgen N (2015) Integrating network ecology with applied conservation: a synthesis and guide to implementation. *AoB Plants* 7, plv076.

Kaiser-Bunbury CN, Mougal J, Whittington AE, *et al.* (2017) Ecosystem restoration strengthens pollination network resilience and function. *Nature* 542, 223–227.

Kaiser-Bunbury CN, Traveset A, Hansen DM (2010) Conservation and restoration of plant-animal mutualisms on oceanic islands. *Perspectives in Plant Ecology, Evolution and Systematics* 12, 131–143.

Kalies EL, Chambers CL, Covington WW (2010) Wildlife responses to thinning and burning treatments in southwestern conifer forests: a meta-analysis. *Forest Ecology and Management* 259, 333–342.

Kalliovirta M, Ryttari T, Heikkinen RK (2006) Population structure of a threatened plant, *Pulsatilla patens*, in boreal forests: modelling relationships to overgrowth and site closure. *Biodiversity and Conservation* 15, 3095–3108.

Kamm U, Gugerli F, Rotach P, Edwards P, Holderegger R (2011) Genetic consequences of low local tree densities: implications for the management of naturally rare, insect pollinated species in temperate forests. *Forest Ecology and Management* 262, 1047–1053.

Kanowski J, Catterrall C, Proctor H, *et al.* (2005) Biodiversity values of timber plantations and restoration plantings for rainforest fauna in tropical and subtropical Australia. In *Reforestation in the Tropics and Subtropics of Australia Using Rainforest Tree Species*, Erskine PD, Lamb D, Bristow M (eds). Cairns, Australia: Rainforest CRC, pp. 183–205.

Kareiva P, Lalasz R, Marvier M (2012) Conservation in the Anthropocene beyond solitude and fragility. *Breakthrough Journal* Winter.

Kareiva P, Marvier M (2007) Conservation for the people. *Scientific American* 297, 50–57.

Kareiva P, Marvier M (2012) What is conservation science? *BioScience* 62, 962–969.

Kariuki M, Kooyman RM, Smith RGB, Wardell-Johnson G, Vanclay JK (2006) Regeneration changes in tree species abundance, diversity and structure in logged and unlogged subtropical rainforest over a 36-year period. *Forest Ecology and Management* 236, 162–176.

Kasenene JM, Murphy PG (1991) Post-logging tree mortality and major branch losses in Kibale Forest. *Forest Ecology and Management* 46, 295–307.

Kassen R (2002) The experimental evolution of specialists, generalists, and the maintenance of diversity. *Journal of Evolutionary Biology* 15, 173–190.

Kaufmann MR, Russell TG, Boyce DAJ, *et al.* (1994) *An ecological basis for ecosystem management.* US Forest Service General Technical Report RM-246.

Kawecki TJ, Ebert D (2004) Conceptual issues in local adaptation. *Ecology Letters* 7, 1225–1241.

Kay J, Strader AA, Murphy V, *et al.* (2011) *Palma corcho*: a case study in botanic garden conservation horticulture and economics. *Horttechnology* 21, 474–481.

Kaye T (2008) Vital steps toward success of endangered plant reintroductions. *Native Plants* 9, 313–322.

Kaye TN, Brandt A (2005) *Seeding and transplanting rare Willamette Valley prairie plants for population restoration.* Corvallis (OR): Institute for Applied Ecology. Report. Eugene, OR: USDI Bureau of Land Management.

Keane RE, Hessburg PF, Landres PB, Swanson FJ (2009) The use of historical range and variability (HRV) in landscape management. *Forest Ecology and Management* 258, 1025–1037.

Keddy PA (1992) Assembly and response rules: two goals for predictive community ecology. *Journal of Vegetation Science* 3, 157–164.

Keenan R, Lamb D, Woldring O, Irvine T, Jensen R (1997) Restoration of plant biodiversity beneath tropical tree plantations in Northern Australia. *Forest Ecology and Management* 99, 117–131.

Keenan RJ, Reams GA, Achard F, *et al.* (2015) Dynamics of global forest area: results from the FAO Global Forest Resources Assessment 2015. *Forest Ecology and Management* 352, 9–20.

Keim RF, Chambers JL, Hughes MS, *et al.* (2006) Long-term success of stump sprouts in high-graded baldcypress-water tupelo swamps in the Mississippi delta. *Forest Ecology and Management* 234, 24–33.

Keller M, Kollmann J, Edwards PJ (2000) Genetic introgression from distant provenances reduces fitness in local weed populations. *Journal of Applied Ecology* 37, 647–659.

Kelly AE, Goulden ML (2008) Rapid shifts in plant distribution with recent climate change. *Proceedings of the National Academy of Sciences of the United States of America* 105, 11823–11826.

Kelly CK, Smith HB, Buckley YM, *et al.* (2001) Investigations in commonness and rarity: a comparative analysis of co-occurring, congeneric Mexican trees. *Ecology Letters* 4, 618–627.

Kelso S, Heckmann K, Lawton J, Maentz G (1996) Endemic calciphiles of the Middle Arkansas Valley, Colorado. In *Southwestern Rare and Endangered Plants: Proceedings of the Second Conference. 1995 September 11–14; Flagstaff, Arizona. General Technical Report RM-GTR-283. US*, Maschinski J, Hammond HD, Holter L (eds). Fort Collins, CO: Department of Agriculture, Forest Service, Rocky Mountain Forest, and Range Experiment Station, pp. 270–280.

Kephart SR (2004) Inbreeding and reintroduction: progeny success in rare *Silene* populations of varied density. *Conservation Genetics* 5, 49–61.

Kephart SR, Paladino C (1997) Demographic change and microhabitat variability in a grassland endemic, *Silene douglasii* var *oraria* (Caryophyllaceae). *American Journal of Botany* 84, 179–189.

Keppel G, Mokany K, Wardell-Johnson GW, *et al.* (2015) The capacity of refugia for conservation planning under climate change. *Frontiers in Ecology and the Environment* 13, 106–112.

Keppel G, Morrison C, Watling D, Tuiwawa MV, Rounds IA (2012) Conservation in tropical Pacific Island countries: why most current approaches are failing. *Conservation Letters* 5, 256–265.

Kettle CJ (2010) Ecological considerations for using dipterocarps for restoration of lowland rainforest in Southeast Asia. *Biodiversity and Conservation* 19, 1137–1151.

Kettle CJ, Ennos RA, Jaffre T, Gardner M, Hollingsworth PM (2008) Cryptic genetic bottlenecks during restoration of an endangered tropical conifer. *Biological Conservation* 141, 1953–1961.

Kettle CJ, Ennos RA, Jaffre T, *et al.* (2012) Importance of demography and dispersal for the resilience and restoration of a critically endangered tropical conifer *Araucaria nemorosa*. *Diversity and Distributions* 18, 248–259.

Kettle CJ, Ghazoul J, Ashton P, *et al.* (2011) Seeing the fruit for the trees in Borneo. *Conservation Letters* 4, 184–191.

Keyser TL, Brown PM (2014) Long-term response of yellow-poplar to thinning in the southern Appalachian Mountains. *Forest Ecology and Management* 312, 148–153.

Kier G, Kreft H, Lee TM, *et al.* (2009) A global assessment of endemism and species richness across island and mainland regions. *Proceedings of the National Academy of Sciences of the United States of America* 106, 9322–9327.

Kirchner F, Robert A, Colas B (2006) Modelling the dynamics of introduced populations in the narrow-endemic *Centaurea corymbosa*: a demo-genetic integration. *Journal of Applied Ecology* 43, 1011–1021.

Kirkpatrick JB (1983) An iterative method for establishing priorities for the selection of nature reserves: an example from Tasmania. *Biological Conservation* 25, 127–134.

Kirkpatrick M, Barton NH (1997) Evolution of a species' range. *American Naturalist* 150, 1–23.

Kirksey E (2015) *Emergent Ecologies*. Durham, NC: Duke University Press.

Kitamura S, Yumoto T, Poonswad P, *et al.* (2002) Interactions between fleshy fruits and frugivores in a tropical seasonal forest in Thailand. *Oecologia* 133, 559–572.

Klein C, Wilson K, Watts M, *et al.* (2009) Incorporating ecological and evolutionary processes into continental-scale conservation planning. *Ecological Applications* 19, 206–217.

Knapp BO, Wang GG, Walker JL (2013) Effects of canopy structure and cultural treatments on the survival and growth of *Pinus palustris* Mill. seedlings underplanted in *Pinus taeda* L. stands. *Ecological Engineering* 57, 46–56.

Knapp BO, Wang GG, Walker JL, Cohen S (2006) Effects of site preparation treatments on early growth and survival of planted longleaf pine (*Pinus palustris* Mill.) seedlings in North Carolina. *Forest Ecology and Management* 226, 122–128.

Knight AT, Cowling RM, Campbell BM (2006) An operational model for implementing conservation action. *Conservation Biology* 20, 408–419.

Knight TM (2012) Using population viability analysis to plan reintroductions. In *Plant Reintroduction in a Changing Climate*, Maschinski J, Haskins KE (eds). Washington, DC: Island Press, pp. 155–170.

Kobe RK (1999) Light gradient partitioning among tropical tree species through differential seedling mortality and growth. *Ecology* 80, 187–201.

Koch C, Kollmann J (2012) Clonal re-introduction of endangered plant species: the case of German false tamarisk in pre-Alpine rivers. *Environmental Management* 50, 217–225.

Kodym A, Turner S, Delpratt J (2010) *In situ* seed development and *in vitro* regeneration of three difficult-to-propagate *Lepidosperma* species (Cyperaceae). *Australian Journal of Botany* 58, 107–114.

Koerner W, Dupouey JL, Dambrine E, Benoit M (1997) Influence of past land use on the vegetation and soils of present day forest in the Vosges mountains, France. *Journal of Ecology* 85, 351–358.

Kollmann J, Cordova JPP, Andersen RM (2008) Factors limiting regeneration of an endangered conifer in the highlands of Guatemala. *Journal for Nature Conservation* 16, 146–156.

Kolos A, Banaszuk P (2013) Mowing as a tool for wet meadows restoration: effect of long-term management on species richness and composition of sedge-dominated wetland. *Ecological Engineering* 55, 23–28.

Konuma A, Tsumura Y, Lee CT, Lee SL, Okuda T (2000) Estimation of gene flow in the tropical-rainforest tree *Neobalanocarpus heimii* (Dipterocarpaceae), inferred from paternity analysis. *Molecular Ecology* 9, 1843–1852.

Kooyman R (1996) *Growing Rainforest. Rainforest Restoration and Regeneration: Recommendations for the Humid Subtropical Region of Northern NSW and South–East Qld*. Brisbane, Australia: Greening Australia.

Kopecký M, Vojta J (2009) Land use legacies in post-agricultural forests in the Doupovské Mountains, Czech Republic. *Applied Vegetation Science* 12, 251–260.

Kormos CF, Zimmerman BL (2014) Response to: Putz *et al.* Sustaining conservation values in selectively logged tropical forests: the attained and the attainable. *Conservation Letters* 7, 143–144.

Kozlowski G, Frey D, Fazan L, Egli B, Pirintsos S (2012) *Zelkova abelicea*. In *IUCN Red List of Threatened Species version 2012.2*. Gland, Switzerland and Cambridge, UK: IUCN Species Survival Commission.

Kraft NJB, Baldwin BG, Ackerly DD (2010) Range size, taxon age, and hotspots of neoendemism in the California flora. *Diversity and Distributions* 16, 403–413.

Kramer A, Hird A, Shaw K, Dosmann M, Mims R (2011) *Conserving North America's Threatened Plants: Progress Report on Target 8 of the Global Strategy for Plant Conservation*. Glencoe, IL: Botanic Gardens Conservation International.

Kramer AT, Ison JL, Ashley MV, Howe HF (2008) The paradox of forest fragmentation genetics. *Conservation Biology* 22, 878–885.

Kroiss SJ, HilleRisLambers J (2015) Recruitment limitation of long-lived conifers: implications for climate change responses. *Ecology* 96, 1286–1297.

Krosby M, Tewksbury J, Haddad NM, Hoekstra J (2010) Ecological connectivity for a changing climate. *Conservation Biology* 24, 1686–1689.

Kueffer C, Beaver K, Mougal J (2013) Management of novel ecosystems in the Seychelles. In *Novel Ecosystems: Intervening in the New Ecological World Order*, Hobbs RJ, Higgs E, Hall C (eds). Oxford, UK: Wiley-Blackwell, pp. 228–238.

Kueffer C, Schumacher E, Dietz H, Fleischmann K, Edwards PJ (2010) Managing successional trajectories in alien-dominated, novel ecosystems by facilitating seedling regeneration: a case study. *Biological Conservation* 143, 1792–1802.

Kunin WE (1992) Density and reproductive success in wild populations of *Diplotaxis erucoides* (Brassicaceae). *Oecologia* 91, 129–133.

Kunin WE (1993) Sex and single mustard: population density and pollinator behaviour effects on seed-set. *Ecology* 74, 2145–2160.

Kunin WE (1997) Population size and density effects in pollination: pollinator foraging and plant reproductive success in experimental arrays of *Brassica kaber*. *Journal of Ecology* 85, 225–234.

Kunstler G, Curt T, Bouchaud M, Lepart J (2006) Indirect facilitation and competition in tree species colonization of sub-Mediterranean grasslands. *Journal of Vegetation Science* 17, 379–388.

Kuuluvainen T, Juntunen P (1998) Seedling establishment in relation to microhabitat variation in a windthrow gap in a boreal *Pinus sylvestris* forest. *Journal of Vegetation Science* 9, 551–562.

Kuusipalo J, Hadengganan S, Adjers G, Sagala APS (1997) Effect of gap liberation on the performance and growth of dipterocarp trees in a logged-over rainforest. *Forest Ecology and Management* 92, 209–219.

Kuusipalo J, Jafarsidik Y, Adjers G, Tuomela K (1996) Population dynamics of tree seedlings in a mixed dipterocarp rainforest before and after logging and crown liberation. *Forest Ecology and Management* 81, 85–94.

Kuussaari M, Bommarco R, Heikkinen RK, *et al.* (2009) Extinction debt: a challenge for biodiversity conservation. *Trends in Ecology and Evolution* 24, 564–571.

Laarmann D, Korjus H, Sims A, Kangur A, Stanturf JA (2013) Initial effects of restoring natural forest structures in Estonia. *Forest Ecology and Management* 304, 303–311.

Lack AJ (1991) Dead logs as a substrate for rain forest trees in Dominica. *Journal of Tropical Ecology* 7, 401–405.

Laffan SW, Crisp MD (2003) Assessing endemism at multiple spatial scales, with an example from the Australian vascular flora. *Journal of Biogeography* 30, 511–520.

Laguna E (2001) *The Micro-reserves as a Tool for Conservation of Threatened Plants in Europe.* Strasbourg: Council of Europe.

Laguna E, Deltoro VI, Perez-Botella J, *et al.* (2004) The role of small reserves in plant conservation in a region of high diversity in eastern Spain. *Biological Conservation* 119, 421–426.

Laguna E, Navarro A, Perez-Rovira P, Ferrando I, Ferrer-Gallego P (2016) Translocation of *Limonium perplexum* (Plumbaginaceae), a threatened coastal endemic. *Plant Ecology* 217, 1183–1194.

Laliberté E, Paquette A, Legendre P, Bouchard A (2009) Assessing the scale-specific importance of niches and other spatial processes on beta diversity: a case study from a temperate forest. *Oecologia* 159, 377–388.

Lamb D (1998) Large-scale ecological restoration of degraded tropical forest lands: the potential role of timber plantations. *Restoration Ecology* 6, 271–279.

Lamb D (2000) Some ecological principles for re-assembling forest ecosystems at degraded tropical sites. In *Forest Restoration for Wildlife Conservation*, Elliott S, Kerby J, Blakesley D, *et al.* (eds). Chiang Mai, Thailand: Biology Department, Science Faculty, Chiang Mai University, pp. 35–44.

Lamb D (2011) *Regreening the Bare Hills: Tropical Forest Conservation in the Asia-Pacific Region.* Dordrecht, The Netherlands: Springer.

Lamb D, Erskine PD, Parrotta JA (2005) Restoration of degraded tropical forest landscapes. *Science* 310, 1628–1632.

Lamb D, Siddique I, Erskine PD, Lex Engel V (2008) Tropical forest restoration experiences. In *Biodiversity: Structure and Function, from Encyclopedia of Life Support Systems (EOLSS), Developed under the Auspices of the UNESCO,* Barthlott W, Linsenmair KE, Porembski S (eds). Oxford,UK: EOLSS Publishers.

Lamb D, Stanturf J, Madsen P (2012) What is forest landscape restoration? In *Forest Landscape Restoration,* Stanturf J, Lamb D, Madsen P (eds). Dordrecht, The Netherlands: Springer, pp. 3–23.

Lambeck RJ (1997) Focal species: a multi-species umbrella for nature conservation. *Conservation Biology* 11, 849–856.

Lambeth CC, McCullough RB (1997) Genetic diversity in managed loblolly pine forests in the southeastern United States: perspective of the private industrial forest land owner. *Canadian Journal of Forest Research* 27, 409–414.

Lamont BB, Klinkhamer PGL, Witkowski ETF (1993) Population fragmentation may reduce fertility to zero in *Banksia goodii*: a demonstration of the Allee effect. *Oecologia* 94, 446–450.

Lamoreux JF, Morrison JC, Ricketts TH, *et al.* (2006) Global tests of biodiversity concordance and the importance of endemism. *Nature* 440, 212–214.

Lande R, Barrowclough GF (1987) Effective population size, genetic variation, and their use in population management. In *Viable Populations for Management,* Soulé ME (ed.). Cambridge: Cambridge University Press, pp. 87–124.

Landero JPC, Valiente-Banuet A (2010) Species-specificity of nurse plants for the establishment, survivorship, and growth of a columnar cactus. *American Journal of Botany* 97, 1289–1295.

Landres PB, Morgan P, Swanson FJ (1999) Overview of the use of natural variability concepts in managing ecological systems. *Ecological Applications* 9, 1179–1188.

Langhammer PF, Bakarr MI, Bennun LA, *et al.* (2007) *Identification and Gap Analysis of Key Biodiversity Areas: Targets for Comprehensive Protected Area Systems. IUCN Best Practice Protected Areas Guidelines Series No. 15.* Gland, Switzerland: IUCN.

Langlois A, Pellerin S (2016) Recovery of the endangered false hop sedge: a ten-year study. *Global Ecology and Conservation* 7, 214–224.

Laporte NT, Stabach JA, Grosch R, Lin TS, Goetz SJ (2007) Expansion of industrial logging in Central Africa. *Science* 316, 1451.

Larios L, Hallett LM, Suding KN (2017) Where and how to restore in a changing world: a demographic-based assessment of resilience. *Journal of Applied Ecology* 54, 1040–1050.

Larson AJ, Belote RT, Williamson MA, Aplet GH (2013) Making monitoring count: project design for active adaptive management. *Journal of Forestry* 111, 348–356.

Latham P, Tappeiner J (2002) Response of old-growth conifers to reduction in stand density in western Oregon forests. *Tree Physiology* 22, 137–146.

Laughlin DC (2014) Applying trait-based models to achieve functional targets for theory-driven ecological restoration. *Ecology Letters* 17, 771–784.

Laughlin DC, Strahan RT, Huffman DW, Sánchez Meador AJ (2017) Using trait-based ecology to restore resilient ecosystems: historical conditions and the future of montane forests in western North America. *Restoration Ecology* 25, S135–S146.

Laurance WF, Gascon C, Rankin-de Merona JM (1999) Predicting effects of habitat destruction on plant communities: a test of a model using Amazonian trees. *Ecological Applications* 9, 548–554.

Laurance WF, Goosem M, Laurance SGW (2009) Impacts of roads and linear clearings on tropical forests. *Trends in Ecology and Evolution* 24, 659–669.

Laurance WF, Nascimento HEM, Laurance SG, et al. (2006) Rapid decay of tree-community composition in Amazonian forest fragments. *Proceedings of the National Academy of Sciences of the United States of America* 103, 19010–19014.

Laurance WF, Perez-Salicrup D, Delamonica P, et al. (2001) Rain forest fragmentation and the structure of Amazonian liana communities. *Ecology* 82, 105–116.

Lauterbach D, Burkart M, Gemeinholzer B (2012) Rapid genetic differentiation between *ex situ* and their *in situ* source populations: an example of the endangered *Silene otites* (Caryophyllaceae). *Botanical Journal of the Linnean Society* 168, 64–75.

Lavergne S, Molofsky J (2007) Increased genetic variation and evolutionary potential drive the success of an invasive grass. *Proceedings of the National Academy of Sciences of the United States of America* 104, 3883–3888.

Lavorel S, Garnier E (2002) Predicting changes in community composition and ecosystem functioning from plant traits: revisiting the Holy Grail. *Functional Ecology* 16, 545–556.

Lawrence BA, Kaye TN (2011) Reintroduction of *Castilleja levisecta*: effects of ecological similarity, source population genetics, and habitat quality. *Restoration Ecology* 19, 166–176.

Lawton JH (1993) Range, population abundance, and conservation. *Trends in Ecology and Evolution* 8, 409–413.

Le Cadre S, Tully T, Mazer SJ, et al. (2008) Allee effects within small populations of *Aconitum napellus* ssp *lusitanicum*, a protected subspecies in northern France. *New Phytologist* 179, 1171–1182.

Le Lay G, Engler R, Franc E, Guisan A (2010) Prospective sampling based on model ensembles improves the detection of rare species. *Ecography* 33, 1015–1027.

Lebrija-Trejos E, Perez-Garcia EA, Meave JA, Bongers F, Poorter L (2010) Functional traits and environmental filtering drive community assembly in a species-rich tropical system. *Ecology* 91, 386–398.

Ledgard N (2001) The spread of lodgepole pine (*Pinus contorta*, Dougl.) in New Zealand. *Forest Ecology and Management* 141, 43–57.

Lee SL, Ng KKS, Saw LG, *et al.* (2006) Linking the gaps between conservation research and conservation management of rare dipterocarps: a case study of *Shorea lumutensis*. *Biological Conservation* 131, 72–92.

Leemans R (1991) Canopy gaps and establishment patterns of spruce (*Picea abies* (L.) Karst.) in two old-growth coniferous forests in central Sweden. *Vegetatio* 93, 157–165.

Lehmann A (1970) Tarvágás által okozott ökológiai változások az abaligeti karszton. *Pécsi Műszaki Szemle* 25, 15–21.

Leibold MA, McPeek MA (2006) Coexistence of the niche and neutral perspectives in community ecology. *Ecology* 87, 1399–1410.

Leimu R, Fisher M (2008) A meta-analysis of local adaptation in plants. *PLoS ONE* 3, e4010.

Leirana-Alcocer J, Parra-Tabla V (1999) Factors affecting the distribution, abundance and seedling survival of *Mammillaria gaumeri*, an endemic cactus of coastal Yucatan, Mexico. *Journal of Arid Environments* 41, 421–428.

Leisher C, Touval J, Hess SM, Boucher TM, Reymondin L (2013) Land and forest degradation inside protected areas in Latin America. *Diversity* 5, 779–795.

Leites LP, Robinson AP, Rehfeldt GE, Marshall JD, Crookston NL (2012) Height-growth response to climatic changes differs among populations of Douglas-fir: a novel analysis of historic data. *Ecological Applications* 22, 154–165.

Lenoir J, Hattab T, Pierre G (2017) Climatic microrefugia under anthropogenic climate change: implications for species redistribution. *Ecography* 40, 253–266.

Leroux SJ, Schmiegelow FKA, Lessard RB, Cumming SG (2007) Minimum dynamic reserves: a framework for determining reserve size in ecosystems structured by large disturbances. *Biological Conservation* 138, 464–473.

Leroy C, Caraglio Y (2003) Effect of tube shelters on the growth of young Turkish pines (*Pinus brutia* Ten. Pinaceae). *Annals of Forest Science* 60, 539–547.

Lesica P, Allendorf FW (1995) When are peripheral populations valuable for conservation. *Conservation Biology* 9, 753–760.

Levey DJ (1988) Tropical wet forest treefall gaps and distributions of understory birds and plants. *Ecology* 69, 1076–1089.

Ley-López JM, Avalos G (2017) Propagation of the palm flora in a lowland tropical rainforest in Costa Rica: fruit collection and germination patterns. *Tropical Conservation Science* 10, doi.org/10.1177/1940082917740703.

Li JQ, Romane FJ (1997) Effects of germination inhibition on the dynamics of *Quercus ilex* stands. *Journal of Vegetation Science* 8, 287–294.

Li X-Y, Zhao W-W, Song Y-X, Wang W, Zhang X-Y (2008) Rainfall harvesting on slopes using contour furrows with plastic-covered transverse ridges for growing *Caragana korshinskii* in the semiarid region of China. *Agricultural Water Management* 95, 539–544.

Liebsch D, Marques MCM, Goldenberg R (2008) How long does the Atlantic Rain Forest take to recover after a disturbance? Changes in species composition and ecological features during secondary succession. *Biological Conservation* 141, 1717–1725.

Lienert J (2004) Habitat fragmentation effects on finess of plant populations – a review. *Journal for Nature Conservation* 12, 53–72.

Lilja-Rothsten S, de Chantal M, Peterson C, *et al.* (2008) Microsites before and after restoration in managed *Picea abies* stands in southern Finland: effects of fire and partial cutting with dead wood creation. *Silva Fennica* 42, 165–176.

Lima RAF, Rother DC, Muler AE, Lepsch IF, Rodrigues RR (2012) Bamboo over-abundance alters forest structure and dynamics in the Atlantic Forest hotspot. *Biological Conservation* 147, 32–39.

Linder HP (2001) Plant diversity and endemism in sub-Saharan tropical Africa. *Journal of Biogeography* 28, 169–182.

Linhart YB, Grant MC (1996) Evolutionary significance of local genetic differentiation in plants. *Annual Review of Ecology, Evolution, and Systematics* 27, 237–277.

Liu JG, Linderman M, Ouyang ZY, *et al.* (2001) Ecological degradation in protected areas: the case of Wolong Nature Reserve for giant pandas. *Science* 292, 98–101.

Lloret F, Médail F, Brundu G, *et al.* (2005) Species attributes and invasion success by alien plants on Mediterranean islands. *Journal of Ecology* 93, 512–520.

Lloyd AH, Graumlich LJ (1997) Holocene dynamics of treeline forests in the Sierra Nevada. *Ecology* 78, 1199–1210.

Loarie SR, Carter BE, Hayhoe K, *et al.* (2008) Climate change and the future of California's endemic flora. *PLoS ONE* 3, e2502.

Loarie SR, Duffy PB, Hamilton H, *et al.* (2009) The velocity of climate change. *Nature* 462, 1052–1055.

Lof M, Dey DC, Navarro RM, Jacobs DF (2012) Mechanical site preparation for forest restoration. *New Forests* 43, 825–848.

Longman KA (2003) *Tropical Trees: Propagation and Planting Manuals. Volume 2: Raising Seedlings of Tropical Trees.* Canberra, Australia: Commonwealth Science Council.

Lonsdale D, Pautasso M, Holdenrieder O (2008) Wood-decaying fungi in the forest: conservation needs and management options. *European Journal of Forest Research* 127, 1–22.

Lopes JdCA, Jennings SB, Matni NM (2008) Planting mahogany in canopy gaps created by commercial harvesting. *Forest Ecology and Management* 255, 300–307.

Lopez-A DM, Bock BC, Bedoya G (2008) Genetic structure in remnant populations of an endangered Andean *Magnolia*. *Biotropica* 40, 375–379.

López-Pujol J, Zhang F-M, Ge S (2006) Plant biodiversity in China: richly varied, endangered, and in need of conservation. *Biodiversity and Conservation* 15, 3983–4026.

Lott R, Sexton G, Novak M (2005) Seed and seedling supply for farm forestry projects in the tropics and subtropics of eastern Australia. In *Reforestation in the Tropics and Subtropics of Australia Using Rainforest Tree Species*, Erskine PD, Lamb D, Bristow M (eds). Cairns, Australia: Rainforest CRC, pp. 24–48.

Lowry PP, Schatz GE, Phillipson PB (1997) The classification of natural and anthropogenic vegetation in Madagascar. In *Natural Change and Human Impact in Madagascar*, Goodman SM, Patterson BD (eds). Washington, DC: Smithsonian Institution Press, pp. 93–123.

Lu Y, Ranjitkar S, Xu J-C, *et al.* (2016) Propagation of native tree species to restore subtropical evergreen broad-leaved forests in SW China. *Forests* 7, 12.

Ludwig JA, Tongway DJ (1995) Spatial organisation of landscapes and its function in semi-arid woodlands, Australia. *Landscape Ecology* 10, 51–63.

Ludwig N, Lavergne C, Sevathian J-C (2010) Notes on the conservation status of Mauritian palms. *Palms* 54, 77–93.

Lugo AE (1997) The apparent paradox of reestablishing species richness on degraded lands with tree monocultures. *Forest Ecology and Management* 99, 9–19.

Lugo AE, Helmer E (2004) Emerging forests on abandoned land: Puerto Rico's new forests. *Forest Ecology and Management* 190, 145–161.

Lunt ID, Byrne M, Hellmann JJ, *et al.* (2013) Using assisted colonisation to conserve biodiversity and restore ecosystem function under climate change. *Biological Conservation* 157, 172–177.

Lussetti D, Axelsson E, Ilstedt U, Falck J, Karlsson A (2016) Supervised logging and climber cutting improves stand development: 18 years of post-logging data in a tropical rain forest in Borneo. *Forest Ecology and Management* 381, 335–346.

Maas-Hebner KG, Emmingham WH, Larson DJ, Chan SS (2005) Establishment and growth of native hardwood and conifer seedlings underplanted in thinned Douglas-fir stands. *Forest Ecology and Management* 208, 331–345.

MacArthur RH, Wilson ED (1967) *The Theory of Island Biogeography*. Princeton, NJ: Princeton Univesity Press.

Maestre FT, Bautista S, Cortina J, Bellot J (2001) Potential for using facilitation by grasses to establish shrubs on a semiarid degraded steppe. *Ecological Applications* 11, 1641–1655.

Maggini R, Lehmann A, Zbinden N, *et al.* (2014) Assessing species vulnerability to climate and land use change: the case of the Swiss breeding birds. *Diversity and Distributions* 20, 708–719.

Maginnis S, Jackson W (2007) What is FLR and how does it differ from current approaches? In *The Forest Landscape Restoration Handbook*, Rietbergen-McCracken JSMAS (ed.). London, Earthscan, pp. 5–20.

Maina GG, Howe HF (2000) Inherent rarity in community restoration. *Conservation Biology* 14, 1335–1340.

Manchester SR, Chen ZD, Lu AM, Uemura K (2009) Eastern Asian endemic seed plant genera and their paleogeographic history throughout the Northern Hemisphere. *Journal of Systematics and Evolution* 47, 1–42.

Mangan SA, Schnitzer SA, Herre EA, *et al.* (2010) Negative plant-soil feedback predicts tree-species relative abundance in a tropical forest. *Nature* 466, 752–755.

Manokaran N (1998) Effects, 34 years later, of selective logging in the lowland dipterocarp forest at Pasoh, Peninsular Malaysia, and implications on present day logging in the hill forests. In *Conservation, Management and Development of Forest Resources*, Lee SS, May DY, Gauld ID, Bishoff J (eds). Kepong, Malaysia: Forest Research Institute of Malaysia, pp. 41–60.

Mansourian S, Lamb D, Gilmour D (2005) Overview of technical approaches to restoring tree cover at the site level. In *Forest Restoration in Landscapes: Beyond Planting Trees*, Mansourian S, Vallauri D, Dudley D (eds). New York: Springer, pp. 241–249.

Margules CR, Nicholls AO, Pressey RL (1988) Selecting networks of reserves to maximize biological diversity. *Biological Conservation* 43, 63–76.

Margules CR, Pressey RL (2000) Systematic conservation planning. *Nature* 405, 243–253.

Marris E (2011) *Rambunctious Garden: Saving Nature in a Postwild World.* New York: Bloomsbury.

Marris E, Kareiva P, Mascaro J, Ellis E (2011) Hope in the Age of Man. *The New York Times*, 7 December.

Marsico TD, Hellmann JJ (2009) Dispersal limitation inferred from an experimental translocation of *Lomatium* (Apiaceae) species outside their geographic ranges. *Oikos* 118, 1783–1792.

Martin PA, Newton AC, Bullock JM (2013) Carbon pools recover more quickly than plant biodiversity in tropical secondary forests. *Proceedings of the Royal Society B: Biological Sciences* 280, 20132236.

Martinez-Garza C, Howe HF (2003) Restoring tropical diversity: beating the time tax on species loss. *Journal of Applied Ecology* 40, 423–429.

Martinez-Garza C, Tobon W, Campo J, Howe HF (2013) Drought mortality of tree seedlings in an eroded tropical pasture. *Land Degradation and Development* 24, 287–295.

Martinez-Izquierdo L, Garcia MM, Powers JS, Schnitzer SA (2016) Lianas suppress seedling growth and survival of 14 tree species in a Panamanian tropical forest. *Ecology* 97, 215–224.

Martinez-Ramos M, Pingarroni A, Rodriguez-Velazquez J, *et al.* (2016) Natural forest regeneration and ecological restoration in human-modified tropical landscapes. *Biotropica* 48, 745–757.

Martins J, Moreira O, Silva L, Moura M (2011) Vegetative propagation of the endangered Azorean tree, *Picconia azorica*. *Arquipelago Life and Marine Sciences* 28, 39–46.

Marvier M, Kareiva P (2014) The evidence and values underlying "new conservation". *Trends in Ecology and Evolution* 29, 131–132.

Marvier M, Wong H (2012) Resurrecting the conservation movement. *Journal of Environmental Studies and Sciences* 2, 291–295.

Mascaro J (2011) Eighty years of succession in a noncommercial plantation on Hawai'i Island: are native species returning? *Pacific Science* 65, 1–15.

Mascaro J, Becklund KK, Hughes RF, Schnitzer SA (2008) Limited native plant regeneration in novel, exotic-dominated forests on Hawai'i. *Forest Ecology and Management* 256, 593–606.

Maschinski J, Albrecht MA (2017) Center for Plant Conservation's Best Practice Guidelines for the reintroduction of rare plants. *Plant Diversity* 39, 390–395.

Maschinski J, Baggs JE, Quintana-Ascencio PE, Menges ES (2006) Using population viability analysis to predict the effects of climate change on the extinction risk of an endangered limestone endemic shrub, Arizona cliffrose. *Conservation Biology* 20, 218–228.

Maschinski J, Baggs JE, Sacchi CF (2004) Seedling recruitment and survival of an endangered limestone endemic in its natural habitat and experimental reintroduction sites. *American Journal of Botany* 91, 689–698.

Maschinski J, Duquesnel J (2006) Successful reintroductions of the endangered long-lived Sargent's cherry palm, *Pseudophoenix sargentii*, in the Florida Keys. *Biological Conservation* 134, 122–129.

Maschinski J, Falk DA, Wright SJ, *et al.* (2012) Optimal locations for plant reintroductions in a changing world. In *Plant Reintroduction in a Changing Climate: Promises and Perils*, Maschinski J, Haskins KE (eds). Washington, DC: Island Press, pp. 109–130.

Maschinski J, Quintana-Asciencio PF (2016) Implications of population and metapopulation theory for restoration science and practice. In *Foundations of*

*Restoration Ecology*, Palmer M, Zedler J, Falk D (eds). Washington, DC: Island Press, pp. 182–215.

Maschinski J, Wright SJ, Koptur S, Pinto-Torres EC (2013) When is local the best paradigm? Breeding history influences conservation reintroduction survival and population trajectories in times of extreme climate events. *Biological Conservation* 159, 277–284.

Matlack GR (1994) Plant species migration in a mixed-history forest landscape in eastern North America. *Ecology* 75, 1491–1502.

Maunder M (1992) Plant reintroduction: an overview. *Biodiversity and Conservation* 1, 51–61.

Maunder M, Cowan RS, Stranc P, Fay MF (2001a) The genetic status and conservation management of two cultivated bulb species extinct in the wild: *Tecophilaea cyanocrocus* (Chile) and *Tulipa sprengeri* (Turkey). *Conservation Genetics* 2, 193–201.

Maunder M, Culham A, Alden B, *et al.* (2000) Conservation of the toromiro tree: case study in the management of a plant extinct in the wild. *Conservation Biology* 14, 1341–1350.

Maunder M, Guerrant EO, Havens K, Dixon KW (2004a) Realizing the full potential of ex situ contributions to global plant conservation. In *Ex Situ Plant Conservation: Supporting Species Survival in the Wild*, Guerrant EOJ, Havens K, Maunder M (eds). Washington, DC: Island Press, pp. 389–418.

Maunder M, Higgens S, Culham A (2001b) The effectiveness of botanic garden collections in supporting plant conservation: a European case study. *Biodiversity and Conservation* 10, 383–401.

Maunder M, Hughes C, Hawkins JA, Culham A (2004b) Hybridization in ex situ plant collections: conservation concerns liabilities and opportunities. In *Ex Situ Plant Conservation: Supporting Species Survival in the Wild*, Guerrant EOJ, Havens K, Maunder M (eds). Washington, DC: Island Press, pp. 325–364.

Maunder M, Page W, Mauremootoo J, *et al.* (2002) The decline and conservation management of the threatened endemic palms of the Mascarene Islands. *Oryx* 36, 56–65.

Mawdsley JR, O'Malley R, Ojima DS (2009) A review of climate-change adaptation strategies for wildlife management and biodiversity conservation. *Conservation Biology* 23, 1080–1089.

Mayor X, Roda F (1993) Growth response of holm oak (*Quercus ilex* L) to commercial thinning in the Montseny mountains (NE Spain). *Annales Des Sciences Forestieres* 50, 247–256.

McAuliffe JR (1986) Herbivore-limited establishment of a Sonoran Desert tree, *Cerciduium microphyllum*. *Ecology* 67, 276–280.

McCarroll J, Chambers FM, Webb JC, Thom T (2016) Using palaeoecology to advise peatland conservation: an example from West Arkengarthdale, Yorkshire, UK. *Journal for Nature Conservation* 30, 90–102.

McCarthy J (2001) Gap dynamics of forest trees: a review with particular attention to boreal forests. *Environmental Reviews* 9, 1–59.

McColley SD, Tyers DB, Sowell BF (2012) Aspen and willow restoration using beaver on the Northern Yellowstone winter range. *Restoration Ecology* 20, 450–455.

McConkey KR, Prasad S, Corlett RT, *et al.* (2012) Seed dispersal in changing landscapes. *Biological Conservation* 146, 1–13.

McDonald T, Gann GD, Jonson K, Dixon KW (2016) *International Standards for the Practice of Ecological Restoration: Including Principles and Key Concepts.* Washington, DC: Society for Ecological Restoration.

McDougall KL, Morgan JW (2005) Establishment of native grassland vegetation at Organ Pipes National Park near Melbourne, Victoria: vegetation changes from 1989 to 2003. *Ecological Management and Restoration* 6, 34–42.

McDowell N, Brooks JR, Fitzgerald SA, Bond BJ (2003) Carbon isotope discrimination and growth response of old *Pinus ponderosa* trees to stand density reductions. *Plant, Cell and Environment* 26, 631–644.

McEuen AB, Curran LM (2004) Seed dispersal and recruitment limitation across spatial scales in temperate forest fragments. *Ecology* 85, 507–518.

McGill BJ, Etienne RS, Gray JS, *et al.* (2007) Species abundance distributions: moving beyond single prediction theories to integration within an ecological framework. *Ecology Letters* 10, 995–1015.

McIntyre S, Hobbs R (1999) A framework for conceptualizing human effects on landscapes and its relevance to management and research models. *Conservation Biology* 13, 1282–1292.

McKay JK, Christian CE, Harrison S, Rice KJ (2005) "How local is local?" A review of practical and conceptual issues in the genetics of restoration. *Restoration Ecology* 13, 432–440.

McKinney LV, Nielsen LR, Collinge DB, *et al.* (2014) The ash dieback crisis: genetic variation in resistance can prove a long-term solution. *Plant Pathology* 63, 485–499.

McLachlan JS, Hellmann JJ, Schwartz MW (2007) A framework for debate of assisted migration in an era of climate change. *Conservation Biology* 21, 297–302.

McNamara S, Tinh DV, Erskine PD, *et al.* (2006) Rehabilitating degraded forest land in central Vietnam with mixed native species plantings. *Forest Ecology and Management* 233, 358–365.

Médail F, Diadema K (2009) Glacial refugia influence plant diversity patterns in the Mediterranean Basin. *Journal of Biogeography* 36, 1333–1345.

Médail F, Quézel P (1997) Hot-spots analysis for conservation of plant biodiversity in the Mediterranean Basin. *Annals of Missouri Botanical Garden* 84, 112–127.

Médail F, Quézel P (1999) Biodiversity hotspots in the Mediterranean basin: setting global conservation priorities. *Conservation Biology* 13, 1510–1513.

Medeiros AC, von Allmen EI, Chimera CG (2014) Dry forest restoration and unassisted native tree seedling recruitment at Auwahi, Maui. *Pacific Science* 68, 33–45.

Mee JA, Moore JS (2014) The ecological and evolutionary implications of microrefugia. *Journal of Biogeography* 41, 837–841.

Menges ES (2008) Restoration demography and genetics of plants: when is a translocation successful? *Australian Journal of Botany* 56, 187–196.

Menges ES, Guerrant E, Hamze S (2004) Effects of seed collection on the extinction risk of perennial plants. In *Ex Situ Plant Conservation: Supporting Species Survival in the Wild*, Guerrant EOJ, Havens K, Maunder M (eds). Washington, DC: Island Press, pp. 305–324.

Menges ES, Kimmich J (1996) Microhabitat and time-since-fire: effects on demography of *Eryngium cuneifolium* (Apiaceae), a Florida scrub endemic plant. *American Journal of Botany* 83, 185–191.

Menges ES, Smith SA, Weekley CW (2016) Adaptive introductions: how multiple experiments and comparisons to wild populations provide insights into requirements for long-term introduction success of an endangered shrub. *Plant Diversity* 38, 238–246.

Menz MHM, Phillips RD, Winfree R, *et al.* (2011) Reconnecting plants and pollinators: challenges in the restoration of pollination mutualisms. *Trends in Plant Science* 16, 4–12.

Merritt DJ, Dixon KW (2011) Restoration seed banks: a matter of scale. *Science* 332, 424–425.

Mesquita RCG, Ickes K, Ganade G, Williamson GB (2001) Alternative successional pathways in the Amazon Basin. *Journal of Ecology* 89, 528–537.

Mesquita RDCG, Massoca PEDS, Jakovac CC, Bentos TV, Williamson GB (2015) Amazon rain forest succession: stochasticity or land-use legacy? *BioScience* 65, 849–861.

Metzger F, Schultz J (1984) Understory response to 50 years of management of northern hardwood forest in Upper Michigan. *American Midland Naturalist* 112, 209–223.

Meyer E, Jensen N, Fraga N (2014) Seed banking California's rare plants. *California Fish and Game* 100, 79–85.

Michalski SG, Durka W (2012) Assessment of provenance delineation by genetic differentiation patterns and estimates of gene flow in the common grassland plant *Geranium pratense*. *Conservation Genetics* 13, 581–592.

Milchunas DG, Noy-Meir I (2002) Grazing refuges, external avoidance of herbivory, and plant diversity. *Oikos* 99, 113–130.

Millar CI (1998) Reconsidering the conservation of Monterey pine. *Fremontia* 26, 12–16.

Millar CI, Brubaker LB (2006) Climate change and paleoecology: new contexts for restoration ecology. In *Foundations of Restoration Ecology*, Falk DA, Palmer MA, Zedler JB (eds). Washington, DC: Society for Ecological, Restoration, International: Island Press, pp. 315–340.

Millar CI, Stephenson NL, Stephens SL (2007) Climate change and forests of the future: managing in the face of uncertainty. *Ecological Applications* 17, 2145–2151.

Millar CI, Woolfenden WB (1999) The role of climate change in interpreting historical variability. *Ecological Applications* 9, 1207–1216.

Miller C, Urban DL (1999) A model of surface fire, climate, and forest pattern in the Sierra Nevada, California. *Ecological Modelling* 114, 113–135.

Millet J, Tran N, Ngoc NV, Thi TT, Prat D (2013) Enrichment planting of native species for biodiversity conservation in a logged tree plantation in Vietnam. *New Forests* 44, 369–383.

Milton SJ, Bond WJ, Du Plessis MA, *et al.* (1999) A protocol for plant conservation by translocation in threatened lowland Fynbos. *Conservation Biology* 13, 735–743.

Mimura M, Aitken SN (2007) Adaptive gradients and isolation-by-distance with postglacial migration in *Picea sitchensis*. *Heredity* 99, 224–232.

Mimura M, Aitken SN (2010) Local adaptation at the range peripheries of Sitka spruce. *Journal of Evolutionary Biology* 23, 249–258.

Mittell EA, Nakagawa S, Hadfield JD (2015) Are molecular markers useful predictors of adaptive potential? *Ecology Letters* 18, 772–778.

Mittermeier RA, Myers N, Thomsen JB, da Fonseca GAB, Olivieri S (1998) Biodiversity hotspots and major tropical wilderness areas: approaches to setting conservation priorities. *Conservation Biology* 12, 516–520.

Miyawaki A (1993) Restoration of native forests from Japan to Malaysia. In *Restoration of Tropical Forest Ecosystem*, Lieth H, Lohmann M (eds). Dordrecht, The Netherlands: Kluwer Academic Publishers, pp. 5–24.

Miyawaki A (2014) Miyawaki method. In *Genetic Considerations in Ecosystem Restoration Using Native Tree Species. State of the World's Forest Genetic*

*Resources: Thematic Study*, Bozzano M, Jalonen R, Thomas E, *et al.* (eds). Rome: Food and Agriculture Organization and Bioversity International, pp. 133–137.

Moerke AH, Lamberti GA (2004) Restoring stream ecosystems: lessons from a midwestern state. *Restoration Ecology* 12, 327–334.

Moilanen A, Kujala H, Leathwick JR (2009a) The zonation framework and software for conservation prioritization. In *Spatial Conservation Prioritization: Quantitative Methods and Computational Tools*, Moilanen A, Wilson K, Possingham HP (eds). Oxford, UK: Oxford University Press, pp. 196–210.

Moilanen A, Wilson KA, Possingham HP (2009b) *Spatial Conservation Prioritization: Quantitative Methods and Computational Tools*. Oxford, UK: Oxford University Press.

Moir ML, Vesk PA, Brennan KE, *et al.* (2010) Current constraints and future directions in estimating coextinction. *Conservation Biology* 24, 682–690.

Molofsky J, Augspurger CK (1992) The effect of leaf litter on early seedling establishment in a tropical forest. *Ecology* 73, 68–77.

Monks L, Coates D, Bell T, Bowles ML (2012) Determining success criteria for reintroductions of threatened long-lived plants. In *Plant Reintroduction in a changing Climate: Promises and Perils*, Maschinski J, Haskins KE (eds). Washington, DC: Island Press/Center for Resource Economics, pp. 189–208.

Monks LT, Coates D (2002) The translocation of two critically endangered *Acacia* species. *Conservation Science Western Australia* 4, 54–61.

Montagnini F, Eibl B, Grance L, Maiocco D, Nozzi D (1997) Enrichment planting in overexploited subtropical forests of the Paranaense region of Misiones, Argentina. *Forest Ecology and Management* 99, 237–246.

Montalvo AM, Williams SL, Rice KJ, *et al.* (1997) Restoration biology: a population biology perspective. *Restoration Ecology* 5, 277–290.

Montero-Castano A, Vila M (2012) Impact of landscape alteration and invasions on pollinators: a meta-analysis. *Journal of Ecology* 100, 884–893.

Montes-Hernandez B, López-Barrera F (2013) Seedling establishment of *Quercus insignis*: a critically endangered oak tree species in southern Mexico. *Forest Ecology and Management* 310, 927–934.

Montoya D, Zavala MA, Rodriguez MA, Purves DW (2008) Animal versus wind dispersal and the robustness of tree species to deforestation. *Science* 320, 1502–1504.

Moore CT, Lonsdorf EV, Knutson MG, Laskowski HP, Lor SK (2011) Adaptive management in the US National Wildlife Refuge System: science-management partnerships for conservation delivery. *Journal of Environmental Management* 92, 1395–1402.

Moore MM, Covington WW, Fule PZ (1999) Reference conditions and ecological restoration: a southwestern ponderosa pine perspective. *Ecological Applications* 9, 1266–1277.

Moreira O, Martins J, Silva L, Moura M (2009) Propagation of the endangered Azorean cherry *Prunus azorica* using stem cuttings and air layering. *Arquipelago Life and Marine Sciences* 26, 9–14.

Morelli TL, Daly C, Dobrowski SZ, *et al.* (2016) Managing climate change refugia for climate adaptation. *PLoS ONE* 11, e0159909.

Morgan JW (1999) Have tubestock plantings successfully established populations of rare grassland species into reintroduction sites in western Victoria? *Biological Conservation* 89, 235–243.

Morgan P, Aplet GH, Haufler JB, *et al.* (1994) Historical range of variability: a useful tool for evaluating ecosystem change. *Journal of Sustainable Forestry* 2, 87–111.

Mori AS, Kitagawa R (2014) Retention forestry as a major paradigm for safeguarding forest biodiversity in productive landscapes: a global meta-analysis. *Biological Conservation* 175, 65–73.

Moritz C (1999) A molecular perspective of biodiversity. In *The Biology of Biodiversity*, Kato S (ed.). Tokyo: Springer Verlag, pp. 21–34.

Moritz C (2002) Strategies to protect biological diversity and the evolutionary processes that sustain it. *Systematic Biology* 51, 238–254.

Moro MJ, Pugnaire FI, Haase P, Puigdefabregas J (1997) Effect of the canopy of *Retama sphaerocarpa* on its understorey in a semiarid environment. *Functional Ecology* 11, 425–431.

Mosblech NAS, Bush MB, van Woesik R (2011) On metapopulations and microrefugia: palaeoecological insights. *Journal of Biogeography* 38, 419–429.

Mostacedo CB, Fredericksen TS (1999) Regeneration status of important tropical forest tree species in Bolivia: assessment and recommendations. *Forest Ecology and Management* 124, 263–273.

Motzkin G, Foster D, Allen A, Harrod J, Boone R (1996) Controlling site to evaluate history: vegetation patterns of a New England sand plain. *Ecological Monographs* 66, 345–365.

Motzkin G, Foster DR (2002) Grasslands, heathlands and shrublands in coastal New England: historical interpretations and approaches to conservation. *Journal of Biogeography* 29, 1569–1590.

Muller SD, Miramont C, Bruneton H, *et al.* (2012) A palaeoecological perspective for the conservation and restoration of wetland plant communities in the central French Alps, with particular emphasis on alder carr vegetation. *Review of Palaeobotany and Palynology* 171, 124–139.

Munzbergova Z, Milden M, Ehrlen J, Herben T (2005) Population viability and reintroduction strategies: a spatially explicit landscape-level approach. *Ecological Applications* 15, 1377–1386.

Murawski DA, Hamrick JL (1991) The effect of the density of flowering individuals on the mating systems of nine tropical tree species. *Heredity* 67, 167–174.

Murawski DA, Hamrick JL (1992) The mating system of *Cavanillesia platanifolia* under extremes of flowering-tree density: a test of predictions. *Biotropica* 24, 99–101.

Murawski DA, Hamrick JL, Hubbell SP, Foster RB (1990) Mating systems of two Bombacaceous trees of a neotropical moist forest. *Oecologia* 82, 501–506.

Murcia C (1997) Evaluation of Andean alder as a catalyst for the recovery of tropical cloud forests in Colombia. *Forest Ecology and Management* 99, 163–170.

Murray C, Marmorek D (2003) Adaptive management and ecological restoration. In *Ecological Restoration of Southwestern Ponderosa Pine Forests*, Friederici P (ed.). Washington, DC: Island Press, pp. 417–428.

Murray-Smith C, Brummitt NA, Oliveira-Filho AT, *et al.* (2009) Plant diversity hotspots in the Atlantic coastal forests of Brazil. *Conservation Biology* 23, 151–163.

Myers GP, Newton AC, Melgarejo O (2000a) The influence of canopy gap size on natural regeneration of Brazil nut (*Bertholletia excelsa*) in Bolivia. *Forest Ecology and Management* 127, 119–128.

Myers JA, Chase JM, Jimenez I, *et al.* (2013) Beta-diversity in temperate and tropical forests reflects dissimilar mechanisms of community assembly. *Ecology Letters* 16, 151–157.

Myers JA, Harms KE (2009) Seed arrival, ecological filters, and plant species richness: a meta-analysis. *Ecology Letters* 12, 1250–1260.

Myers N (1989) Threatened biotas: "hotspots" in tropical forests. *Environmentalist* 8, 1–20.

Myers N (1990) The biodiversity challenge: expanded hotspots analysis. *Environmentalist* 10, 243–256.

Myers N, Mittermeier RA, Mittermeier CG, da Fonseca GAB, Kent J (2000b) Biodiversity hotspots for conservation priorities. *Nature* 403, 853–858.

Myster RW (2004) Post-agricultural invasion, establishment, and growth of neotropical trees. *Botanical Review* 70, 381–402.

Myster RW (2007) Early successional pattern and process after sugarcane, banana, and pasture cultivation in Ecuador. *New Zealand Journal of Botany* 45, 101–110.

Myster RW, Pickett STA (1994) A comparison of rate of succession over 18 yr in 10 contrasting old fields. *Ecology* 75, 387–392.

Naito Y, Kanzaki M, Iwata H, *et al.* (2008) Density-dependent selfing and its effects on seed performance in a tropical canopy tree species, *Shorea acuminata* (Dipterocarpaceae). *Forest Ecology and Management* 256, 375–383.

Nakashizuka T (1989) Role of uprooting in composition and dynamics of an old-growth forest in Japan. *Ecology* 70, 1273–1278.

Narukawa Y, Iida S, Tanouchi H, Abe S, Yamamoto S (2003) State of fallen logs and the occurrence of conifer seedlings and saplings in boreal and subalpine old-growth forests in Japan. *Ecological Research* 18, 267–277.

Narukawa Y, Yamamoto S (2003) Development of conifer seedlings roots on soil and fallen logs in boreal and subalpine coniferous forests of Japan. *Forest Ecology and Management* 175, 131–139.

Nascimento HEM, Andrade ACS, Camargo JLC, *et al.* (2006) Effects of the surrounding matrix on tree recruitment in Amazonian forest fragments. *Conservation Biology* 20, 853–860.

Natlandsmyr B, Hjelle KL (2016) Long-term vegetation dynamics and land-use history: providing a baseline for conservation strategies in protected *Alnus glutinosa* swamp woodlands. *Forest Ecology and Management* 372, 78–92.

Nattero J, Malerba R, Medel R, Cocucci A (2011) Factors affecting pollinator movement and plant fitness in a specialized pollination system. *Plant Systematics and Evolution* 296, 77–85.

Navarro-Cerrillo RM, Griffith DM, Jose Ramirez-Soria M, *et al.* (2011) Enrichment of big-leaf mahogany (*Swietenia macrophylla* King) in logging gaps in Bolivia: the effects of planting method and silvicultural treatments on long-term seedling survival and growth. *Forest Ecology and Management* 262, 2271–2280.

Nave AG, Rodrigues RR (2007) Combination of species into filling and diversity groups as forest restoration methodology. In *High Diversity Forest Restoration in Degraded Areas*, Rodrigues RR, Martins SV, Gandolfi S (eds). New York: Nova Science Publishers, pp. 103–126.

Neale JR (2012) Genetic considerations in rare plant reintroduction: practical applications (or How are we doing?). In *Plant Reintroduction in a Changing Climate: Promises and Perils*, Maschinski J, Haskins KE (eds). Washington, DC: Island Press/Center for Resource Economics, pp. 71–88.

Neghme C, Santamaría L, Calviño-Cancela M (2017) Strong dependence of a pioneer shrub on seed dispersal services provided by an endemic endangered lizard in a Mediterranean island ecosystem. *PLoS ONE* 12, e0183072.

Nepstad D, Uhl C, Serrao EA (1990) Surmounting barriers to forest regeneration in abandoned, highly degraded pastures: a case study from Paragominas, Pará, Brazil. In *Alternatives to Deforestation: Steps Towards Sustainable Use of the Amazon Rain Forest*, Anderson AB (ed.). New York: Columbia University Press, pp. 215–229.

Nepstad DC, Uhl C, Pereira CA, daSilva JMC (1996) A comparative study of tree establishment in abandoned pasture and mature forest of eastern Amazonia. *Oikos* 76, 25–39.

Newmark WD, Jenkins CN, Pimm SL, McNeally PB, Halley JM (2017) Targeted habitat restoration can reduce extinction rates in fragmented forests. *Proceedings of the National Academy of Sciences of the United States of America* 114, 9635–9640.

Newsome TA, Brown KR, Nemec AFL (2016) Effects of opening size and microsite on performance of planted tree seedlings in high-elevation Engelmann spruce-subalpine fir forests managed as mountain caribou habitat in British Columbia. *Forest Ecology and Management* 370, 31–44.

Newton M, Cole EC (1987) A sustained-yield scheme for oldgrowth Douglas-fir. *Western Journal of Applied Forestry* 2, 22–25.

Ngo TL, Hölscher D (2014) The fate of five rare tree species after logging in a tropical limestone forest (Xuan Son National Park, northern Vietnam). *Tropical Conservation Science* 7, 326–341.

Nicholls AO, Margules CR (1993) An upgraded reserve selection algorithm. *Biological Conservation* 64, 165–169.

Niering WA, Lowe CH, Whittaker R, Whittaker RH (1963) The saguaro: a population in relation to environment. *Science* 142, 15–23.

Noack M (2011) Growth and nutrition of *Quercus petraea* underplanted in artificial pine stands under conversion in the northeastern German Lowlands. *Forest Systems* 20, 423–436.

Nobel PS (1980) Morphology, nurse plants, and minimum apical temperatures for young *Carnegiea gigantea*. *Botanical Gazette* 141, 188–191.

Noël F, Prati D, van Kleunen M, *et al.* (2011) Establishment success of 25 rare wetland species introduced into restored habitats is best predicted by ecological distance to source habitats. *Biological Conservation* 144, 602–609.

Norden N, Angarita HA, Bongers F, *et al.* (2015) Successional dynamics in Neotropical forests are as uncertain as they are predictable. *Proceedings of the National Academy of Sciences of the United States of America* 112, 8013–8018.

Noss RF (1990) Indicators for monitoring biodiversity: a hierarchical approach. *Conservation Biology* 4, 355–364.

Noss RF (2001) Beyond Kyoto: forest management in a time of rapid climate change. *Conservation Biology* 15, 578–590.

Noss RF, Dinerstein E, Gilbert B, *et al.* (1999a) Core areas: where nature reigns. In *Continental Conservation: Scientific Foundations of Regional Reserve Networks*, Soulé ME, Terborgh J (eds). Washington, DC: Island Press, pp. 99–128.

Noss RF, Strittholt JR, Vance-Borland K, Carroll C, Frost P (1999b) A conservation plan for the Klamath-Siskiyou ecoregion. *Natural Areas Journal* 19, 392–411.

Nunez CI, Aizen MA, Ezcurra C (1999) Species associations and nurse plant effects in patches of high-Andean vegetation. *Journal of Vegetation Science* 10, 357–364.

Nunez-Iturri G, Olsson O, Howe HF (2008) Hunting reduces recruitment of primate-dispersed trees in Amazonian Peru. *Biological Conservation* 141, 1536–1546.

Nyberg B (1999) *An Introductory Guide to Adaptive Management.* Vancouver, Canada: BC Forest Service.

Nyland RD (2002) *Silviculture: Concepts and Applications.* New York: McGraw-Hill.

O'Brien EK, Mazanec RA, Krauss SL (2007) Provenance variation of ecologically important traits of forest trees: implications for restoration. *Journal of Applied Ecology* 44, 583–593.

O'Hara KL, Nesmith JCB, Leonard L, Porter DJ (2010) Restoration of old forest features in coast redwood forests using early-stage variable-density thinning. *Restoration Ecology* 18, 125–135.

Oedekoven K (1980) The vanishing forest. *Environmental Policy and Law* 6, 184–185.

Ogden J, Stewart GH (1995) Community dynamics of the New Zealand conifers. In *Ecology of the Southern Conifers*, Enright N, Hill RS (eds). Washington, DC: Smithsonian Institution Press, pp. 81–119.

Ohlemueller R, Anderson BJ, Araujo MB, *et al.* (2008) The coincidence of climatic and species rarity: high risk to small-range species from climate change. *Biology Letters* 4, 568–572.

Okuda T, Suzuki M, Adachi N, *et al.* (2003) Effect of selective logging on canopy and stand structure and tree species composition in a lowland dipterocarp forest in peninsular Malaysia. *Forest Ecology and Management* 175, 297–320.

Oliveira-Filho AT, Vilela EA, Gavilanes ML, Carvalho DA (1994) Effect of flooding regime and understorey bamboos on the physiognomy and tree species composition of a tropical semideciduous forest in southeastern Brazil. *Vegetatio* 113, 99–124.

Olivier PI, van Aarde RJ, Lombard AT (2013) The use of habitat suitability models and species-area relationships to predict extinction debts in coastal forests, South Africa. *Diversity and Distributions* 19, 1353–1365.

Olson D, DellaSala DA, Noss RF, *et al.* (2012) Climate change refugia for biodiversity in the Klamath-Siskiyou ecoregion. *Natural Areas Journal* 32, 65–74.

Olson DM, Dinerstein E (1998) The Global 200: a representation approach to conserving the Earth's most biologically valuable ecoregions. *Conservation Biology* 12, 502–515.

Olson DM, Dinerstein E, Wikramanayake ED, *et al.* (2001) Terrestrial ecoregions of the worlds: a new map of life on Earth. *BioScience* 51, 933–938.

Omernik JM (1987) Ecoregions of the conterminous United States. *Annals of the Association of American Geographers* 77, 118–125.

Omernik JM (1995) Ecoregions: a spatial framework for environmental management. In *Biological Assessment and Criteria: Tools for Water Resource Planning and Decision Making*, Davis WS, Simon TP (eds). Boca Raton, FL: Lewis Publishing, pp. 49–62.

Omernik JM (2004) Perspectives on the nature and definition of ecological regions. *Environmental Management* 34, S27–S38.

Omernik JM, Bailey RG (1997) Distinguishing between watersheds and ecoregions. *Journal of the American Water Resources Association* 33, 935–949.

Onaindia M, Ametzaga-Arregi I, San Sebastián M, *et al.* (2013) Can understorey native woodland plant species regenerate under exotic pine plantations using natural succession? *Forest Ecology and Management* 308, 136–144.

Onal H (2004) First-best, second-best, and heuristic solutions in conservation reserve site selection. *Biological Conservation* 115, 55–62.

Oostermeijer JGB (2000) Population viability analysis of the rare *Gentiana pneumonanthe*: the importance of genetics, demography, and reproductive biology. In *Genetics, Demography and Viability of Fragmented Populations*, Young AG, Clarke GM (eds). Cambridge: Cambridge University Press, pp. 313–334.

Oostermeijer JGB, Luijten SH, den Nijs JCM (2003) Integrating demographic and genetic approaches in plant conservation. *Biological Conservation* 113, 389–398.

Oren R, Waring RH, Stafford SG, Barrett JW (1987) Twenty-four years of ponderosa pine growth in relation to canopy leaf area and understory competition. *Forest Science* 33, 538–547.

Orme CDL, Davies RG, Burgess M, *et al.* (2005) Global hotspots of species richness are not congruent with endemism or threat. *Nature* 436, 1016–1019.

Orsenigo S, Gentili R, Smolders AJP, *et al.* (2017) Reintroduction of a dioecious aquatic macrophyte (*Stratiotes aloides* L.) regionally extinct in the wild. Interesting answers from genetics. *Aquatic Conservation: Marine and Freshwater Ecosystems* 27, 10–23.

Ortega-Pieck A, López-Barrera F, Ramírez-Marcial N, García-Franco JG (2011) Early seedling establishment of two tropical montane cloud forest tree species: the role of native and exotic grasses. *Forest Ecology and Management* 261, 1336–1343.

Ostertag R, Giardina CP, Cordell S (2008) Understory colonization of *Eucalyptus* plantations in Hawaii in relation to light and nutrient levels. *Restoration Ecology* 16, 475–485.

Oswald BP, Neuenschwander LF (1993) Microsite variability and safe site description for western larch germination and establishment. *Bulletin of the Torrey Botanical Club* 120, 148–156.

Otsu C, Iijima H, Nagaike T, Hoshino Y (2017) Evidence of extinction debt through the survival and colonization of each species in semi-natural grasslands. *Journal of Vegetation Science* 28, 464–474.

Otto R, Garcia-del-Rey E, Mendez J, Maria Fernandez-Palacios J (2012) Effects of thinning on seed rain, regeneration and understory vegetation in a *Pinus canariensis* plantation (Tenerife, Canary Islands). *Forest Ecology and Management* 280, 71–81.

Otto R, Garzón-Machado V, Arco M, *et al.* (2017) Unpaid extinction debts for endemic plants and invertebrates as a legacy of habitat loss on oceanic islands. *Diversity and Distributions* 23, 1031–1041.

Packer A, Clay K (2000) Soil pathogens and spatial patterns of seedling mortality in a temperate tree. *Nature* 404, 278–281.

Padilla FM, Miranda J, Ortega R, *et al.* (2011) Does shelter enhance early seedling survival in dry environments? A test with eight Mediterranean species. *Applied Vegetation Science* 14, 31–39.

Padilla FM, Pugnaire FI (2006) The role of nurse plants in the restoration of degraded environments. *Frontiers in Ecology and the Environment* 4, 196–202.

Palacios G, Navarro Cerrillo RM, del Campo A, Toral M (2009) Site preparation, stock quality and planting date effect on early establishment of Holm oak (*Quercus ilex* L.) seedlings. *Ecological Engineering* 35, 38–46.

Palmer M, Smart J (2001) *Guidelines to the Selection of Important Plant Areas in Europe.* London: Planta Europa.

Palmiotto PA (1993) Initial response of *Shroea* wildings transplanted in gaps and understory microsites in a lowland rainforest. *Journal of Tropical Forest Science* 5, 403–415.

Paquette A, Bouchard A, Cogliastro A (2006) Survival and growth of underplanted trees: a meta-analysis across four biomes. *Ecological Applications* 16, 1575–1589.

Paquette A, Messier C (2011) The effect of biodiversity on tree productivity: from temperate to boreal forests. *Global Ecology and Biogeography* 20, 170–180.

Park A, Justiniano MJ, Fredericksen TS (2005) Natural regeneration and environmental relationships of tree species in logging gaps in a Bolivian tropical forest. *Forest Ecology and Management* 217, 147–157.

Parker WC, Dey DC, Newmaster SG, Elliott KA, Boysen E (2001) Managing succession in conifer plantations: converting young red pine (*Pinus resinosa* Ait.) plantations to native forest types by thinning and underplanting. *Forestry Chronicle* 77, 721–734.

Parker WC, Elliott KA, Dey DC, Boysen E (2008) Restoring southern Ontario forests by managing succession in conifer plantations. *Forestry Chronicle* 84, 83–94.

Parmesan C (2006) Ecological and evolutionary responses to recent climate change. *Annual Review of Ecology, Evolution, and Systematics* 37, 637–669.

Parmesan C, Ryrholm N, Stefanescu C, *et al.* (1999) Poleward shifts in geographical ranges of butterfly species associated with regional warming. *Nature* 399, 579–583.

Parrotta JA (1993) Secondary forest regeneration on degraded tropical lands. In *Restoration of Tropical Forest Ecosystems*, Lieth H, Lohmann M (eds). Dordrecht, The Netherlands: Kluwer Academic Publishers, pp. 63–73.

Parrotta JA (2000) Catalyzing natural forest restoration on degraded tropical landscapes. In *Forest Restoration for Wildlife Conservation*, Elliott S, Kerby J, Blakesley D, *et al.* (eds). Chiang Mai, Thailand: International Tropical Timber Organization and The Forest Restoration Research Unit, Chiang Mai University, pp. 45–56.

Parrotta JA, Knowles OH (2001) Restoring tropical forests on lands mined for bauxite: examples from the Brazilian Amazon. *Ecological Engineering* 17, 219–239.

Parrotta JA, Turnbull JW, Jones N (1997) Catalyzing native forest regeneration on degraded tropical lands. *Forest Ecology and Management* 99, 1–7.

Partel M, Szava-Kovats R, Zobel M (2011) Dark diversity: shedding light on absent species. *Trends in Ecology and Evolution* 26, 124–128.

Partel M, Zobel M, Zobel K, van der Maarel E (1996) The species pool and its relation to species richness: evidence from Estonian plant communities. *Oikos* 75, 111–117.

Paul JR, Randle AM, Chapman CA, Chapman LJ (2004) Arrested succession in logging gaps: is tree seedling growth and survival limiting? *African Journal of Ecology* 42, 245–251.

Pausas JG, Bladé C, Valdecantos A, *et al.* (2004) Pines and oaks in the restoration of Mediterranean landscapes of Spain: new perspectives for an old practice – a review. *Plant Ecology* 171, 209–220.

Pautasso M, Aas G, Queloz V, Holdenrieder O (2013) European ash (*Fraxinus excelsior*) dieback: a conservation biology challenge. *Biological Conservation* 158, 37–49.

Pavlik BM (1996) Defining and measuring success. In *Restoring Diversity. Strategies for Reintroduction of Endangered Plants*, Falk DA, Millar CI, Olwell M (eds). Washington, DC: Island Press, pp. 127–155.

Pavlik BM (1997) Perspectives, tools, and institutions for conserving rare plants. *The Southwestern Naturalist* 42, 375–383.

Pavlik BM, Nickrent DL, Howald AM (1993) The recovery of an endangered plant. I. Creating a new population of *Amsinckia grandiflora*. *Conservation Biology* 7, 510–526.

Payn T, Carnus J-M, Freer-Smith P, *et al.* (2015) Changes in planted forests and future global implications. *Forest Ecology and Management* 352, 57–67.

Pearce F (2015) *The New Wild: Why Invasive Species Will Be Nature's Salvation.* Boston, MA: Beacon Press.

Pechackova S, Hadincova V, Muenzbergova Z, Herben T, Krahulec F (2010) Restoration of species-rich, nutrient-limited mountain grassland by mowing and fertilization. *Restoration Ecology* 18, 166–174.

Pedrono M, Griffiths OL, Clausen A, *et al.* (2013) Using a surviving lineage of Madagascar's vanished megafauna for ecological restoration. *Biological Conservation* 159, 501–506.

Peet RK (1992) Community structure and ecosystem function. In *Plant Succession: Theory and Prediction*, Glenn-Lewin DC, Peet RK, Veblens TT (eds). London: Chapman and Hall, pp. 103–151.

Pena-Claros M, Fredericksen TS, Alarcon A, *et al.* (2008) Beyond reduced-impact logging: silvicultural treatments to increase growth rates of tropical trees. *Forest Ecology and Management* 256, 1458–1467.

Percy DM, Cronk QCB (1997) Conservation in relation to mating system in *Nesohedyotis arborea* (Rubiaceae), a rare endemic tree from St Helena. *Biological Conservation* 80, 135–145.

Peres CA, Palacios E (2007) Basin-wide effects of game harvest on vertebrate population densities in Amazonian forests: implications for animal-mediated seed dispersal. *Biotropica* 39, 304–315.

Pérez-Méndez N, Jordano P, Valido A (2015) Downsized mutualisms: consequences of seed dispersers' body-size reduction for early plant recruitment. *Perspectives in Plant Ecology, Evolution and Systematics* 17, 151–159.

Perez-Salicrup DR (2001) Effect of liana cutting on tree regeneration in a liana forest in Amazonian Bolivia. *Ecology* 82, 389–396.

Perry DJ, Knowles P (1990) Evidence of high self-fertilization in natural populations of eastern white cedar (*Thuja occidentalis*). *Canadian Journal of Botany* 68, 663–668.

Peterken GF (1996) *Natural Woodland: Ecology and Conservation in Northern Temperate Regions*. Cambridge: Cambridge University Press.

Peters RL, Darling JDS (1985) The greenhouse effect and nature reserves. *BioScience* 35, 707–717.

Peterson CJ, Carson WP, McCarthy BC, Pickett STA (1990) Microsite variation and soil dynamics within newly created treefall pits and mounds. *Oikos* 58, 39–46.

Peterson CJ, Dosch JJ, Carson WP (2014) Pasture succession in the Neotropics: extending the nucleation hypothesis into a matrix discontinuity hypothesis. *Oecologia* 175, 1325–1335.

Peterson CL, Kaufmann GS, Vandello C, Richardson ML (2013) Parent genotype and environmental factors influence introduction success of the critically endangered savannas mint (*Dicerandra immaculata* var. *savannarum*). *PLoS ONE* 8, e61429.

Petit RJ, Aguinagalde I, de Beaulieu JL, *et al.* (2003) Glacial refugia: hotspots but not melting pots of genetic diversity. *Science* 300, 1563–1565.

Philippi T (2005) Adaptive cluster sampling for estimation of abundances within local populations of low-abundance plants. *Ecology* 86, 1091–1100.

Philipson CD, Saner P, Marthews TR, *et al.* (2012) Light-based regeneration niches: evidence from 21 dipterocarp species using size-specific RGRs. *Biotropica* 44, 627–636.

Phillips RD, Faast R, Bower CC, Brown GR, Peakall R (2009) Implications of pollination by food and sexual deception for pollinator specificity, fruit set, population genetics and conservation of *Caladenia* (Orchidaceae). *Australian Journal of Botany* 57, 287–306.

Phillips SJ, Anderson RP, Schapire RE (2006) Maximum entropy modeling of species geographic distributions. *Ecological Modelling* 190, 231–259.

Pickett STA, Thompson JN (1978) Patch dynamics and the design of nature reserves. *Biological Conservation* 13, 27–37.

Piiroinen T, Nyeko P, Roininen H (2015) Natural establishment of indigenous trees under planted nuclei: a study from a clear-felled pine plantation in an afrotropical rain forest. *Forest Ecology and Management* 345, 21–28.

Piiroinen T, Valtonen A, Roininen H (2017) The seed-to-seedling transition is limited by ground vegetation and vertebrate herbivores in a selectively logged rainforest. *Forest Ecology and Management* 384, 137–146.

Pimm SL, Raven P (2000) Biodiversity: extinction by numbers. *Nature* 403, 843–845.

Pineiro J, Maestre FT, Bartolome L, Valdecantos A (2013) Ecotechnology as a tool for restoring degraded drylands: a meta-analysis of field experiments. *Ecological Engineering* 61, 133–144.

Piotto D, Montagnini F, Thomas W, Ashton M, Oliver C (2009) Forest recovery after swidden cultivation across a 40-year chronosequence in the Atlantic Forest of southern Bahia, Brazil. *Plant Ecology* 205, 261–272.

Plantlife International (2004) *Identifying and Protecting the World's Most Important Plant Areas.* Salisbury, UK: Plantlife International.

Plassmann K, Jones MLM, Edwards-Jones G (2010) Effects of long-term grazing management on sand dune vegetation of high conservation interest. *Applied Vegetation Science* 13, 100–112.

Plumptre AJ (1996) Changes following 60 years of selective timber harvesting in the Budongo Forest Reserve, Uganda. *Forest Ecology and Management* 89, 101–113.

Poiani KA, Richter BD, Anderson MG, Richter HE (2000) Biodiversity conservation at multiple scales: functional sites, landscapes, and networks. *BioScience* 50, 133–146.

Poorter L, Bongers F, van Rompaey R, de Klerk M (1996) Regeneration of canopy tree species at five sites in West African moist forest. *Forest Ecology and Management* 84, 61–69.

Poorter L, Ongers FB, Aide TM, *et al.* (2016) Biomass resilience of Neotropical secondary forests. *Nature* 530, 211–214.

Power RL, Haney A (1998) Adaptive management: a solution to restoration uncertainties. *Transactions of the Wisconsin Academy of Sciences, Arts, and Letters* 86, 177–188.

Powers JS, Becknell JM, Irving J, Perez-Aviles D (2009) Diversity and structure of regenerating tropical dry forests in Costa Rica: geographic patterns and environmental drivers. *Forest Ecology and Management* 258, 959–970.

Pressey RL, Bottrill MC (2009) Approaches to landscape- and seascape-scale conservation planning: convergence, contrasts, and challenges. *Oryx* 43, 464–475.

Pressey RL, Ferrier S, Hager TC, *et al.* (1996) How well protected are the forests of north-eastern New South Wales? Analyses of forest environments in relation to formal protection measures, land tenure, and vulnerability to clearing. *Forest Ecology and Management* 85, 311–333.

Pressey RL, Humphries CJ, Margules CR, Vane-wright RI, Williams PH (1993) Beyond opportunism: key principles for systematic reserve selection. *Trends in Ecology and Evolution* 8, 124–128.

Pressey RL, Johnson IR, Wilson PD (1994) Shades of irreplaceability: towards a measure of the contribution of sites to a reservation goal. *Biodiversity and Conservation* 3, 242–262.

Price TD, Kirkpatrick M (2009) Evolutionarily stable range limits set by interspecific competition. *Proceedings of the Royal Society B: Biological Sciences* 276, 1429–1434.

Primack RB, Miao SL (1992) Dispersal can limit local plant distribution. *Conservation Biology* 6, 513–519.

Primack RB, Miller-Rushing AJ (2009) The role of botanical gardens in climate change research. *New Phytologist* 182, 303–313.

Pritchard HW, Moat JF, Ferraz JBS, *et al.* (2014) Innovative approaches to the preservation of forest trees. *Forest Ecology and Management* 333, 88–98.

Prober S, Byrne M, McLean E, *et al.* (2015) Climate-adjusted provenancing: a strategy for climate-resilient ecological restoration. *Frontiers in Ecology and Evolution* 3, article 65.

Puettker T, Bueno A, Prado PI, Pardini R (2015) Ecological filtering or random extinction? Beta-diversity patterns and the importance of niche-based and neutral processes following habitat loss. *Oikos* 124, 206–215.

Pugnaire FI, Haase P, Puigdefabregas J (1996) Facilitation between higher plant species in a semiarid environment. *Ecology* 77, 1420–1426.

Pujol B, Pannell JR (2008) Reduced responses to selection after species range expansion. *Science* 321, 96.

Pujol B, Zhou SR, Sanchez Vilas J, Pannell JR (2009) Reduced inbreeding depression after species range expansion. *Proceedings of the National Academy of Sciences of the United States of America* 106, 15379–15383.

Pulido FJ, Diaz M (2005) Regeneration of a Mediterranean oak: a whole-cycle approach. *Ecoscience* 12, 92–102.

Pulliam HR, Babbitt B (1997) Ecology: science and the protection of endangered species. *Science* 275, 499–500.

Putz FE (1983) Treefall pits and mounds, buried seeds, and the importance of soil disturbance to pioneer trees on Barro Colorado Island, Panama. *Ecology* 64, 1069–1074.

Putz FE (1984) How trees avoid and shed lianas. *Biotropica* 16, 19–23.

Pyšek P, Genovesi P, Pergl J, Monaco A, Wild J (2014) Invasion of protected areas in Europe: an old continent facing new problems. In *Plant Invasions in Protected Areas: Patterns, Problems and Challenges*, Foxcroft LC, Pyšek P, Richardson DM (eds). Dordrecht, The Netherlands: Springer, pp. 209–240.

Pyšek P, Liska J (1991) Colonization of *Sibbaldia tetrandra* cushions on alpine scree in the Palmiro-Alai mountains, Central Asia. *Arctic and Alpine Research* 23, 263–272.

Qian S, Tang CQ, Yi S, *et al.* (2018) Conservation and development in conflict: regeneration of wild *Davidia involucrata* (Nyssaceae) communities weakened by bamboo management in south-central China. *Oryx* 52, 442–451.

Qian S, Yang Y, Tang CQ, *et al.* (2016) Effective conservation measures are needed for wild *Cathaya argyrophylla* populations in China: insights from the population structure and regeneration characteristics. *Forest Ecology and Management* 361, 358–367.

Qiao X, Jabot F, Tang Z, Jiang M, Fang J (2015) A latitudinal gradient in tree community assembly processes evidenced in Chinese forests. *Global Ecology and Biogeography* 24, 314–323.

Quesada M, Stoner KE, Rosas-Guerrero V, Palacios-Guevara C, Lobo JA (2003) Effects of habitat disruption on the activity of nectarivorous bats (Chiroptera: Phyllostomidae) in a dry tropical forest: implications for the reproductive success of the neotropical tree *Ceiba grandiflora. Oecologia* 135, 400–406.

Quintana-Ascencio PF, Menges ES, Weekley CW (2003) A fire-explicit population viability analysis of *Hypericum cumulicola* in Florida rosemary scrub. *Conservation Biology* 17, 433–449.

Rabinowitz D, Cairns S, Dillon T (1986) Seven forms of rarity and their frequency in the flora of the British Isles. In *Conservation Biology: The Science of Scarcity and Diversity*, Soulé ME (ed.). Sunderland, MA: Sinauer Associates, pp. 182–204.

Radeloff VC, Williams JW, Bateman BL, *et al.* (2015) The rise of novelty in ecosystems. *Ecological Applications* 25, 2051–2068.

Raes N, Roos MC, Slik JWF, van Loon EE, ter Steege H (2009) Botanical richness and endemicity patterns of Borneo derived from species distribution models. *Ecography* 32, 180–192.

Randall JM (2011) Protected areas. In *Encyclopaedia of Biological Invasions*, Simberloff D, Rejmánek M (eds). Berkeley/Los Angeles, CA: University of California Press, pp. 563–567.

Rasanen K, Hendry AP (2008) Disentangling interactions between adaptive divergence and gene flow when ecology drives diversification. *Ecology Letters* 11, 624–636.

Rathcke BJ (1983) Competition and facilitation among plants for pollination. In *Pollination Biology*, Real L (ed.). Orlando, FL: Academic Press, pp. 305–329.

Ratnamhin A, Elliott S, Wangpakapattanawong P (2011) Vegetative propagation of rare tree species for forest restoration. *Chiang Mai Journal of Science* 38, 306–310.

Raven PH (1981) Research in botanical gardens. *Botanische Jahrbücher fur Systematik, Pflanzengeschichte und Pflanzengeographie* 102, 53–72.

Ray GJ, Brown BJ (1995) Restoring Caribbean dry forests: evaluation of tree propagation techniques. *Restoration Ecology* 3, 86–94.

Reay SD, Norton DA (1999) Assessing the success of restoration plantings in a temperate New Zealand forest. *Restoration Ecology* 7, 298–308.

Rebele F (2013) Differential succession towards woodland along a nutrient gradient. *Applied Vegetation Science* 16, 365–378.

Redford KH (1992) The empty forest. *BioScience* 42, 412–422.

Reed AW, Kaufman GA, Kaufman DW (2006) Effect of plant litter on seed predation in three prairie types. *American Midland Naturalist* 155, 278–285.

Reed DH, Frankham R (2001) How closely correlated are molecular and quantitative measures of genetic variation? A meta-analysis. *Evolution* 55, 1095–1103.

Reed DH, Frankham R (2003) Correlation between fitness and genetic diversity. *Conservation Biology* 17, 230–237.

Rees M, Condit R, Crawley M, Pacala S, Tilman D (2001) Long-term studies of vegetation dynamics. *Science* 293, 650–655.

Rehfeldt GE, Crookston NL, Warwell MV, Evans JS (2006) Empirical analyses of plant-climate relationships for the western United States. *International Journal of Plant Sciences* 167, 1123–1150.

Reid JL, Holl KD, Zahawi RA (2015) Seed dispersal limitations shift over time in tropical forest restoration. *Ecological Applications* 25, 1072–1082.

Reid WV (1998) Biodiversity hotspots. *Trends in Ecology and Evolution* 13, 275–280.

Reinhardt JR, Nagel LM, Swanston CW, Keough H (2017) Community-level impacts of management and disturbance in western Michigan oak savannas. *American Midland Naturalist* 177, 112–125.

Rejmanek M, Richardson DM (2013) Trees and shrubs as invasive alien species – 2013 update of the global database. *Diversity and Distributions* 19, 1093–1094.

Requena N, Perez-Solis E, Azcón-Aguilar C, Jeffries P, Barea J-M (2001) Management of indigenous plant-microbe symbioses aids restoration of desertified ecosystems. *Applied and Environmental Microbiology* 67, 495–498.

Reside AE, Welbergen JA, Phillips BL, *et al.* (2014) Characteristics of climate change refugia for Australian biodiversity. *Austral Ecology* 39, 887–897.

Rey Benayas JMR, Bullock JM, Newton AC (2008) Creating woodland islets to reconcile ecological restoration, conservation, and agricultural land use. *Frontiers in Ecology and the Environment* 6, 329–336.

Rey Benayas JMR, Camacho-Cruz A (2004) Performance of *Quercus ilex* saplings planted in abandoned Mediterranean cropland after long-term interruption of their management. *Forest Ecology and Management* 194, 223–233.

Rey PJ, Alcantara JM (2000) Recruitment dynamics of a fleshy-fruited plant (*Olea europaea*): connecting patterns of seed dispersal to seedling establishment. *Journal of Ecology* 88, 622–633.

Rey PJ, Siles G, Alcantara JM (2009) Community-level restoration profiles in Mediterranean vegetation: nurse-based vs. traditional reforestation. *Journal of Applied Ecology* 46, 937–945.

Reyes-Betancort JA, Santos Guerra A, Guma IR, Humphries CJ, Carine MA (2008) Diversity, rarity, and the evolution and conservation of the Canary Islands endemic flora. *Anales del Jardín Botánico de Madrid* 65, 25–45.

Richardson DM, Brown PJ (1986) Invasion of mesic mountain fynbos by *Pinus radiata*. *South African Journal of Botany* 52, 529–536.

Richardson DM, Hellmann JJ, McLachlan JS, *et al.* (2009) Multidimensional evaluation of managed relocation. *Proceedings of the National Academy of Sciences of the United States of America* 106, 9721–9724.

Richardson DM, Rejmanek M (2011) Trees and shrubs as invasive alien species – a global review. *Diversity and Distributions* 17, 788–809.

Richardson ML, Watson MLJ, Peterson CL (2013) Influence of community structure on the spatial distribution of critically endangered *Dicerandra immaculata* var. *immaculata* (Lamiaceae) at wild, introduced, and extirpated locations in Florida scrub. *Plant Ecology* 214, 443–453.

Ricketts TH (1999) *Terrestrial Ecoregions of North America: A Conservation Assessment.* Washington, DC: Island Press.

Ricketts TH, Dinerstein E, Boucher T, *et al.* (2005) Pinpointing and preventing imminent extinctions. *Proceedings of the National Academy of Sciences of the United States of America* 102, 18497–18501.

Rigg LS, Enright NJ, Jaffre T (1998) Stand structure of the emergent conifer *Araucaria laubenfelsii*, in maquis, and rainforest, Mont Do, New Caledonia. *Australian Journal of Ecology* 23, 528–538.

Rigg LS, Enright NJ, Jaffre T, Perry GLW (2010) Contrasting population dynamics of the endemic New Caledonian conifer *Araucaria laubenfelsii* in maquis and rain forest. *Biotropica* 42, 479–487.

Rigg LS, Enright NJ, Perry GLW, Miller BP (2002) The role of cloud combing and shading by isolated trees in the succession from maquis to rain forest in New Caledonia. *Biotropica* 34, 199–210.

Ripple WJ, Beschta RL (2004) Wolves and the ecology of fear: can predation risk structure ecosystems? *BioScience* 54, 755–766.

Ripple WJ, Newsome TM, Wolf C, *et al.* (2015) Collapse of the world's largest herbivores. *Science Advances* 1, e1400103.

Riswan S, Kenworthy JB, Kartawinata K (1985) The estimation of temporal processes in tropical rain forest: a study of primary mixed dipterocarp forest in Indonesia. *Journal of Tropical Ecology* 1, 171–182.

Rivera LW, Zimmerman JK, Aide TM (2000) Forest recovery in abandoned agricultural lands in a karst region of the Dominican Republic. *Plant Ecology* 148, 115–125.

Robertson AW, Trass A, Ladley JJ, Kelly D (2006) Assessing the benefits of frugivory for seed germination: the importance of deinhibition effect. *Functional Ecology* 20, 58–66.

Robiansyah I, Davy AJ (2015) Population status and habitat preferences of critically endangered *Dipterocarpus littoralis* in West Nusakambangan, Indonesia. *Makara Journal of Science* 19, 150–160.

Robichaux RH, Moriyasu PY, Enoka JH, *et al.* (2017) Silversword and lobeliad reintroduction linked to landscape restoration on Mauna Loa and Kilauea, and its implications for plant adaptive radiation in Hawai'i. *Biological Conservation* 213, 59–69.

Robinson JG (2006) Conservation biology and real-world conservation. *Conservation Biology* 20, 658–669.

Rodrigues ASL, Brooks TM, Gaston KJ (2005) Integrating phylogenetic diversity in the selection of priority areas for conservation: does it make a difference? In *Phylogeny and Conservation*, Purvis A, Gittleman JL, Brooks TM (eds). Cambridge: Cambridge University Press, pp. 101–119.

Rodrigues ASL, Gaston KJ (2002) Maximising phylogenetic diversity in the selection of networks of conservation areas. *Biological Conservation* 105, 103–111.

Rodrigues RR, Gandolfi S (2007) Restoration actions. In *High Diversity Forest Restoration in Degraded Areas: Methods and Projects in Brazil*, Rodrigues RR, Martins SV, Gandolfi S (eds). New York: Nova Science Publishers, pp. 77–101.

Rodrigues RR, Gandolfi S, Nave AG, *et al.* (2011) Large-scale ecological restoration of high-diversity tropical forests in SE Brazil. *Forest Ecology and Management* 261, 1605–1613.

Rodrigues RR, Lima RAF, Gandolfi S, Nave AG (2009) On the restoration of high diversity forests: 30 years of experience in the Brazilian Atlantic Forest. *Biological Conservation* 142, 1242–1251.

Rodriguez-Calcerrada J, Mutke S, Alonso J, *et al.* (2008) Influence of overstory density on understory light, soil moisture, and survival of two underplanted oak species in a Mediterranean montane Scots pine forest. *Investigacion Agraria: Sistemas y Recursos Forestales* 17, 31–38.

Rodriguez-Quilon I, Santos-del-Blanco L, Jesus Serra-Varela M, *et al.* (2016) Capturing neutral and adaptive genetic diversity for conservation in a highly structured tree species. *Ecological Applications* 26, 2254–2266.

Roll JR, Mitchell J, Cabin RJ, Marshall CR (1997) Reproductive success increases with local density of conspecifics in a desert mustard (*Lesquerella fendleri*). *Conservation Biology* 11, 738–746.

Roman-Danobeytia FJ, Levy-Tacher SI, Aronson J, Rodrigues RR, Castellanos-Albores J (2012) Testing the performance of fourteen native tropical tree species in two abandoned pastures of the Lacandon rainforest region of Chiapas, Mexico. *Restoration Ecology* 20, 378–386.

Romell E, Hallsby G, Karlsson A, Garcia C (2008) Artificial canopy gaps in a *Macaranga* spp. dominated secondary tropical rain forest: effects on survival and above ground increment of four under-planted dipterocarp species. *Forest Ecology and Management* 255, 1452–1460.

Roncal J, Maschinski J, Schaffer B, Gutierrez SM, Walters D (2012) Testing appropriate habitat outside of historic range: the case of *Amorpha herbacea* var. *crenulata* (Fabaceae). *Journal for Nature Conservation* 20, 109–116.

Rosauer D, Laffan SW, Crisp MD, Donnellan SC, Cook LG (2009) Phylogenetic endemism: a new approach for identifying geographical concentrations of evolutionary history. *Molecular Ecology* 18, 4061–4072.

Rose CL, Marcot BG, Mellen TK, *et al.* (2001) Decaying wood in Pacific Northwest forests: concepts and tools for habitat management. In *Wildlife-habitat Relationships in Oregon and Washington*. Corvallis, OR: Oregon State University Press, pp. 580–623.

Rose M, Hermanutz L (2004) Are boreal ecosystems susceptible to alien plant invasion? Evidence from protected areas. *Oecologia* 139, 467–477.

Rother DC, Rodrigues RR, Pizo MA (2016) Bamboo thickets alter the demographic structure of *Euterpe edulis* population: a keystone, threatened palm species of the Atlantic Forest. *Acta Oecologica* 70, 96–102.

Rousset O, Lepart J (1999) Shrub facilitation of *Quercus humilis* regeneration in succession on calcareous grasslands. *Journal of Vegetation Science* 10, 493–502.

Rousset O, Lepart J (2000) Positive and negative interactions at different life stages of a colonizing species (*Quercus humilis*). *Journal of Ecology* 88, 401–412.

Rousset O, Lepart J (2003) Neighbourhood effects on the risk of an unpalatable plant being grazed. *Plant Ecology* 165, 197–206.

Rovzar C, Gillespie TW, Kawelo K (2016) Landscape to site variations in species distribution models for endangered plants. *Forest Ecology and Management* 369, 20–28.

Rowland EL, Davison JE, Graumlich LJ (2011) Approaches to evaluating climate change impacts on species: a guide to initiating the adaptation planning process. *Environmental Management* 47, 322–337.

Rucinska A, Puchalski J (2011) Comparative molecular studies on the genetic diversity of an ex situ garden collection and its source population of the critically endangered Polish endemic plant *Cochlearia polonica* E. Frohlich. *Biodiversity and Conservation* 20, 401–413.

Rueger N, Huth A, Hubbell SP, Condit R (2009) Response of recruitment to light availability across a tropical lowland rain forest community. *Journal of Ecology* 97, 1360–1368.

Ruiz-Jaen MC, Aide TM (2005a) Restoration success: how is it being measured? *Restoration Ecology* 13, 569–577.

Ruiz-Jaen MC, Aide TM (2005b) Vegetation structure, species diversity, and ecosystem processes as measures of restoration success. *Forest Ecology and Management* 218, 159–173.

Rull V (2009) Microrefugia. *Journal of Biogeography* 36, 481–484.

Rünk K, Pihkva K, Zobel K (2014) Desirable site conditions for introduction sites for a locally rare and threatened fern species *Asplenium septentrionale* (L.) Hoffm. *Journal for Nature Conservation* 22, 272–278.

Runkle JR (1982) Patterns of disturbance in some old-growth mesic forests of eastern North America. *Ecology* 63, 1533–1546.

Rusch GM, Pausas JG, Leps J (2003) Plant functional types in relation to disturbance and land use: introduction. *Journal of Vegetation Science* 14, 307–310.

Rutten G, Ensslin A, Hemp A, Fischer M (2015) Forest structure and composition of previously selectively logged and non-logged montane forests at Mt. Kilimanjaro. *Forest Ecology and Management* 337, 61–66.

Ryder OA (1986) Species conservation and systematics: the dilemma of subspecies. *Trends in Ecology and Evolution* 1, 9–10.

Sáenz-Romero C, Rehfeldt GE, Duval P, Lindig-Cisneros RA (2012) *Abies religiosa* habitat prediction in climatic change scenarios and implications for monarch butterfly conservation in Mexico. *Forest Ecology and Management* 275, 98–106.

Saha S, Kuehne C, Bauhus J (2013) Tree species richness and stand productivity in low-density cluster plantings with oaks (*Quercus robur* L. and *Q. petraea* (Mattuschka) Liebl.). *Forests* 4, 650–665.

Saha S, Kuehne C, Bauhus J (2014) Intra- and interspecific competition differently influence growth and stem quality of young oaks (*Quercus robur* L. and *Quercus petraea* (Mattuschka) Liebl.). *Annals of Forest Science* 71, 381–393.

Saha S, Kuehne C, Bauhus J (2017) Lessons learned from oak cluster planting trials in central Europe. *Canadian Journal of Forest Research* 47, 139–148.

Sainsbury AW, Vaughan-Higgins RJ (2012) Analyzing disease risks associated with translocations. *Conservation Biology* 26, 442–452.

Sakai A, Visaratana T, Vacharangkura, Ratana T-N, Nakamura S (2014) Growth performance of four dipterocarp species planted in a *Leucaena leucocephala* plantation and in an open site on degraded land under a tropical monsoon climate. *Japan Agricultural Research Quarterly* 48, 95–104.

Sakai C, Subiakto A, Nuroniah HS, Kamata N, Nakamura K (2002) Mass propagation method from the cutting of three dipterocarp species. *Journal of Forest Research* 7, 73–80.

Saldarriaga JG, West DC, Tharp ML (1986) Forest succession in the upper Rio Negro of Colombia and Venezuela. *Journal of Ecology* 76, 939–958.

Saldarriaga JG, West DC, Tharp ML, Uhl C (1988) Long-term chronosequence of forest succession in the upper Rio Negro of Colombia and Venezuela. *Journal of Ecology* 76, 938–958.

Sampaio AB, Holl KD, Scariot A (2007) Does restoration enhance regeneration of seasonal deciduous forests in pastures in central Brazil? *Restoration Ecology* 15, 462–471.

Samways MJ, Hitchins P, Bourquin O, Henwood J (2010a) *Tropical Island Recovery: Cousine Island, Seychelles*. Hoboken,NJ: John Wiley and Sons.

Samways MJ, Hitchins PM, Bourquin O, Henwood J (2010b) Restoration of a tropical island: Cousine Island, Seychelles. *Biodiversity and Conservation* 19, 425–434.

Sanchez-Velasquez LR, Quintero-Gardilla S, Aragon-Cruz F, Pineda-Lopez MR (2004) Nurses for *Brosimum alicastrum* reintroduction in secondary tropical dry forest. *Forest Ecology and Management* 198, 401–404.

Santos BA, Peres CA, Oliveira MA, *et al.* (2008) Drastic erosion in functional attributes of tree assemblages in Atlantic Forest fragments of northeastern Brazil. *Biological Conservation* 141, 249–260.

Santos SLD, Válio IFM (2002) Litter accumulation and its effect on seedling recruitment in a Southeast Brazilian Tropical Forest. *Brazilian Journal of Botany* 25, 89–92.

Sarkar S, Illoldi-Rangel P (2010) Systematic conservation planning: an updated protocol. *Natureza and Conservacao* 8, 19–26.

Sarkar S, Pressey RL, Faith DP, *et al.* (2006) Biodiversity conservation planning tools: present status and challenges for the future. *Annual Review of Environment and Resources* 31, 123–159.

Satterthwaite WH, Holl KD, Hayes GF, Barber AL (2007) Seed banks in plant conservation: case study of Santa Cruz tarplant restoration. *Biological Conservation* 135, 57–66.

Sautu A, Baskin JM, Baskin CC, Condit R (2006) Studies on the seed biology of 100 native species of trees in a seasonal moist tropical forest, Panama, Central America. *Forest Ecology and Management* 234, 245–263.

Savolainen O, Pyhajarvi T, Knurr T (2007) Gene flow and local adaptation in trees. *Annual Review of Ecology, Evolution, and Systematics* 38, 595–619.

Sax DF, Gaines SD (2008) Species invasions and extinction: the future of native biodiversity on islands. *Proceedings of the Natural Academy of Sciences of United States of America* 105, 11490–11497.

Schemske DW, Husband BC, Ruckelshaus MH, *et al.* (1994) Evaluating approaches to the conservation of rare and endangered plants. *Ecology* 75, 584–606.

Schirone B, Salis A, Vessella F (2011) Effectiveness of the Miyawaki method in Mediterranean forest restoration programs. *Landscape and Ecological Engineering* 7, 81–92.

Schirone B, Vessella F (2014) Adapting the Miyawaki method in Mediterranean forest reforestation practices. In *Genetic Considerations in Ecosystem Restoration Using Native Tree Species. State of the World's Forest Genetic Resources: Thematic Study*, Bozzano M, Jalonen R, Thomas E, *et al.* (eds). Rome: Food and Agriculture Organization and Bioversity International, pp. 140–144.

Schlawin J, Zahawi RA (2008) "Nucleating" succession in recovering neotropical wet forests: the legacy of remnant trees. *Journal of Vegetation Science* 19, 485–487.

Schmitt CB, Burgess ND, Coad L, *et al.* (2009) Global analysis of the protection status of the world's forests. *Biological Conservation* 142, 2122–2130.

Schmitz OJ, Lawler JJ, Beier P, *et al.* (2015) Conserving biodiversity: practical guidance about climate change adaptation approaches in support of land-use planning. *Natural Areas Journal* 35, 190–203.

Schneider T, Ashton MS, Montagnini F, Milan PP (2014) Growth performance of sixty tree species in smallholder reforestation trials on Leyte, Philippines. *New Forests* 45, 83–96.

Schnitzer SA, Bongers F (2002) The ecology of lianas and their role in forests. *Trends in Ecology and Evolution* 17, 223–230.

Schnitzer SA, Carson WP (2010) Lianas suppress tree regeneration and diversity in treefall gaps. *Ecology Letters* 13, 849–857.

Schnitzer SA, Dalling JW, Carson WP (2000) The impact of lianas on tree regeneration in tropical forest canopy gaps: evidence for an alternative pathway of gap-phase regeneration. *Journal of Ecology* 88, 655–666.

Schoen DJ, Brown ADH (2001) The conservation of wild plant species in seed banks. *BioScience* 51, 960–966.

Schönenberger W (2001) Cluster afforestation for creating diverse mountain forest structures: a review. *Forest Ecology and Management* 145, 121–128.

Schonewald-Cox CM, Chambers SM, MacBryde B, Thomas WL (1983) *Genetics and Conservation: A Reference for Managing Wild Animal and Plant Populations.* Menlo Park, CA: Benjamin Cummings.

Schreiber SG, Ding C, Hamann A, *et al.* (2013) Frost hardiness vs. growth performance in trembling aspen: an experimental test of assisted migration. *Journal of Applied Ecology* 50, 939–949.

Schroder A, Persson L, De Roos AM (2005) Direct experimental evidence for alternative stable states: a review. *Oikos* 110, 3–19.

Schulze M (2008) Technical and financial analysis of enrichment planting in logging gaps as a potential component of forest management in the eastern Amazon. *Forest Ecology and Management* 255, 866–879.

Schulze M, Grogan J, Uhl C, Lentini M, Vidal E (2008) Evaluating ipê (*Tabebuia,* Bignoniaceae) logging in Amazonia: sustainable management or catalyst for forest degradation? *Biological Conservation* 141, 2071–2085.

Schupp EW (1988) Factors affecting post-dispersal seed survival in a tropical forest. *Oecologia* 76, 525–530.

Schwartz G, Falkowski V, Pena-Claros M (2017) Natural regeneration of tree species in the Eastern Amazon: short-term responses after reduced-impact logging. *Forest Ecology and Management* 385, 97–103.

Schwartz G, Lopes JCA, Mohren GMJ, Pena-Claros M (2013) Post-harvesting silvicultural treatments in logging gaps: a comparison between enrichment planting and tending of natural regeneration. *Forest Ecology and Management* 293, 57–64.

Schwartz MW, Brigham CA, Hoeksema JD, *et al.* (2000) Linking biodiversity to ecosystem function: implications for conservation ecology. *Oecologia* 122, 297–305.

Schwartz MW, Hellmann JJ, McLachlan JM, *et al.* (2012) Managed relocation: integrating the scientific, regulatory, and ethical challenges. *BioScience* 62, 732–743.

Schwartz MW, Iverson LR, Prasad AM, Matthews SN, O'Connor RJ (2006) Predicting extinctions as a result of climate change. *Ecology* 87, 1611–1615.

Scowcroft PG, Yeh JT (2013) Passive restoration augments active restoration in deforested landscapes: the role of root suckering adjacent to planted stands of *Acacia koa.* *Forest Ecology and Management* 305, 138–145.

Sechrest W, Brooks TM, da Fonseca GAB, *et al.* (2002) Hotspots and the conservation of evolutionary history. *Proceedings of the National Academy of Sciences of the United States of America* 99, 2067–2071.

Seddon PJ (2010) From reintroduction to assisted colonization: moving along the conservation translocation spectrum. *Restoration Ecology* 18, 796–802.

Seddon PJ, Armstrong DP, Maloney RF (2007) Developing the science of reintroduction biology. *Conservation Biology* 21, 303–312.

Seddon PJ, Griffiths CJ, Soorae PS, Armstrong DP (2014) Reversing defaunation: restoring species in a changing world. *Science* 345, 406–412.

Seibold S, Baessler C, Brandl R, *et al.* (2015) Experimental studies of deadwood biodiversity: a review identifying global gaps in knowledge. *Biological Conservation* 191, 139–149.

Seiwa K, Etoh Y, Hisita M, *et al.* (2012) Roles of thinning intensity in hardwood recruitment and diversity in a conifer, *Criptomeria japonica* plantation: a 5-year demographic study. *Forest Ecology and Management* 269, 177–187.

SER (2002) *The SER Primer on Ecological Restoration.* Tucson, AZ: Science and Policy Working Group, Society for Ecological Restoration International.

SER (2004) *The SER Primer on Ecological Restoration. Version 2.* Tucson, AZ: Society for Ecological Restoration Science and Policy Working Group.

Sethi P, Howe HF (2009) Recruitment of hornbill dispersed trees in hunted and logged forests of the Indian Eastern Himalaya. *Conservation Biology* 23, 710–718.

Setsuko S, Nagamitsu T, Tomaru N (2013) Pollen flow and effects of population structure on selfing rates and female and male reproductive success in fragmented *Magnolia stellata* populations. *BMC Ecology* 13, 10.

Sexton JP, Strauss SY, Rice KJ (2011) Gene flow increases fitness at the warm edge of a species range. *Proceedings of the National Academy of Sciences of the United States of America* 108, 11704–11709.

Sgro CM, Lowe AJ, Hoffmann AA (2011) Building evolutionary resilience for conserving biodiversity under climate change. *Evolutionary Applications* 4, 326–337.

Shachak M, Sachs M, Moshe I (1998) Ecosystem management of desertified shrublands in Israel. *Ecosystems* 1, 475–483.

Shaffer M (1987) Minimum viable populations: coping with uncertainty. In *Viable Populations for Conservation*, Soulé ME (ed.). Cambridge: Cambridge University Press, pp. 69–86.

Shaffer ML (1981) Minimum population sizes for species conservation. *BioScience* 31, 131–134.

Shaw K, Nicholson M, Hardwick K (2015) Encouraging and enabling a science-based approach to ecological restoration: an introduction to the work of the Ecological Restoration Alliance of Botanic Gardens (ERA). *Sibbaldia* 13, 145–152.

Shen S-K, Wang Y-H (2011) Arbuscular mycorrhizal (AM) status and seedling growth response to indigenous AM colonisation of *Euryodendron excelsum* in China: implications for restoring an endemic and critically endangered tree. *Australian Journal of Botany* 59, 460–467.

Shimshi D (1979/80) Two ecotypes of *Iris atrofusca* Bak. and their relations to man-modified habitats. *Israel Journal of Botany* 28, 80–86.

Shono K, Cadaweng EA, Durst PB (2007a) Application of assisted natural regeneration to restore degraded tropical forestlands. *Restoration Ecology* 15, 620–626.

Shono K, Davies SJ, Chua YK (2007b) Performance of 45 native tree species on degraded lands in Singapore. *Journal of Tropical Forest Science* 19, 25–34.

Shono K, Davies SJ, Kheng CY (2006) Regeneration of native plant species in restored forests on degraded lands in Singapore. *Forest Ecology and Management* 237, 574–582.

Shoo LP, Freebody K, Kanowski J, Catterall CP (2016) Slow recovery of tropical old-field rainforest regrowth and the value and limitations of active restoration. *Conservation Biology* 30, 121–132.

Shoo LP, Storlie C, Vanderwal J, Little J, Williams SE (2011) Targeted protection and restoration to conserve tropical biodiversity in a warming world. *Global Change Biology* 17, 186–193.

Shriner SA, Wilson KR, Flather CH (2006) Reserve networks based on richness hotspots and representation vary with scale. *Ecological Applications* 16, 1660–1673.

Shultz LM (1993) Patterns of endemism in Utah flora. In *Southwestern Rare and Endangered Plants: Proceedings of the Conference; 1993. March 30–April 2, 1992. Santa Fe, New Mexico*, Sivinski RC, Lightfoot K (eds). Santa Fe, NM: New Mexico Forestry and Resources Conservation Division, pp. 249–269.

Shumway SW (2000) Facilitative effects of a sand dune shrub on species growing beneath the shrub canopy. *Oecologia* 124, 138–148.

Siitonen J (2001) Forest management, coarse woody debris and saproxylic organisms: Fennoscandian boreal forests as an example. *Ecological Bulletins* 49, 11–41.

Siitonen J, Martikainen P, Punttila P, Rauh J (2000) Coarse woody debris and stand characteristics in mature managed and old-growth boreal mesic forests in southern Finland. *Forest Ecological Management* 128, 211–225.

Silander JA (1978) Density-dependent control of reproductive success in *Cassia biflora*. *Biotropica* 10, 292–296.

Silander JA, Primack RB (1978) Pollination intensity and seed set in evening primrose (*Oenothera fruticosa*). *American Midland Naturalist* 100, 213–216.

Siles G, Alcantara JM, Rey PJ, Bastida JM (2010a) Defining a target map of native species assemblages for restoration. *Restoration Ecology* 18, 439–448.

Siles G, Rey PJ, Alcantara JM, Bastida JM, Herreros JL (2010b) Effects of soil enrichment, watering, and seedling age on establishment of Mediterranean woody species. *Acta Oecologica* 36, 357–364.

Siles G, Rey PJ, Alcantara JM, Ramirez JM (2008) Assessing the long-term contribution of nurse plants to restoration of Mediterranean forests through Markovian models. *Journal of Applied Ecology* 45, 1790–1798.

Simberloff D (2014) Eradication: pipe dream or real option? In *Plant Invasions in Protected Areas: Patterns, Problems and Challenges*, Foxcroft LC, Pyšek P, Richardson DM (eds). Dordrecht, The Netherlands: Springer, pp. 549–559.

Simberloff D, Murcia C, Aronson J (2015) "Novel ecosystems" are a Trojan horse for conservation. *Ensia*, https://ensia.com/voices/novel-ecosystems-are-a-trojan-horsefor-conservation/.

Simberloff DJ, Doak D, Groom M, *et al.* (1999) Regional and continental restoration. In *Continental Conservation: Scientific Foundations of Regional Reserve Networks*, Soulé´ ME, Terborgh J (eds). Washington, DC: Island Press, pp. 65–98.

Sinclair A, Catling PM (2003) Restoration of *Hydrastis canadensis* by transplanting with disturbance simulation: results of one growing season. *Restoration Ecology* 11, 217–222.

Sinclair E, Krauss S, Cheetham B, Hobbs R (2010) High genetic diversity in a clonal relict *Alexgeorgea nitens* (Restionaceae): implications for ecological restoration. *Australian Journal of Botany* 58, 206–213.

Singleton R, Gardescu S, Marks PL, Geber MA (2001) Forest herb colonization of postagricultural forests in central New York State, USA. *Journal of Ecology* 89, 325–338.

Sist P, Nguyen-The N (2002) Logging damage and the subsequent dynamics of a dipterocarp forest in East Kalimantan (1990–1996). *Forest Ecology and Management* 165, 85–103.

Sivinski RC, Knight PJ (1996) Narrow endemism in the New Mexico flora. In *Southwestern Rare and Endangered Plants: Proceedings of the Second Conference. 1995 September 11–14; Flagstaff, Arizona. General Technical Report RM-GTR-283. US*, Maschinski J, Hammond HD, Holter L (eds). Fort Collins, CO: US Department of Agriculture, Forest Service, Rocky Mountain Forest, and Range Experiment Station, pp. 286–296.

Skinner M, Pavlik B (1994) *California Native Plant Society's Inventory of Rare and Endangered Plants in California*, 5th edition. Sacramento, CA: California Native Plant Society.

Sloan S, Goosem M, Laurance SG (2016) Tropical forest regeneration following land abandonment is driven by primary rainforest distribution in an old pastoral region. *Landscape Ecology* 31, 601–618.

Sloan S, Jenkins CN, Joppa LN, Gaveau DLA, Laurance WF (2014) Remaining natural vegetation in the global biodiversity hotspots. *Biological Conservation* 177, 12–24.

Slocum MG (2001) How tree species differ as recruitment foci in a tropical pasture. *Ecology* 82, 2547–2559.

Smart NOE, Hatton JC, Spence DHN (1985) The effect of long-term exclusion of large herbivores on vegetation in Murchison Falls National Park, Uganda. *Biological Conservation* 33, 229–245.

Smit C, Den Ouden JAN, Muller-Scharer H (2006) Unpalatable plants facilitate tree sapling survival in wooded pastures. *Journal of Applied Ecology* 43, 305–312.

Smith DM, Larson BC, Kelty MJ, Ashton PMS (1997) *The Practice of Silviculture: Applied Forest Ecology.* Hoboken, NJ: John Wiley and Sons.

Smith DR, Brown JA, Lo NCH (2004) Application of adaptive sampling to biological populations. In *Sampling Rare or Elusive Species: Concepts, Designs, and Techniques for Estimating Population Parameters*, Thompson WL (ed.). Washington, DC: Island Press, pp. 77–122.

Smith PP (2014) The role of seed banks in habitat restoration. In *Genetic Considerations in Ecosystem Restoration Using Native Tree Species. State of the World's Forest Genetic Resources: Thematic Study*, Bozzano M, Jalonen R, Thomas E, *et al.* (eds). Rome: Food and Agriculture Organization and Bioversity International, pp. 106–107.

Smith R, Olff H (1998) Woody species colonisation in relation to habitat productivity. *Plant Ecology* 139, 203–209.

Smith RJ, Di Minin E, Linke S, Segan DB, Possingham HP (2010) An approach for ensuring minimum protected area size in systematic conservation planning. *Biological Conservation* 143, 2525–2531.

Smith ZF, James EA, McDonnell MJ, McLean CB (2009) Planting conditions improve translocation success of the endangered terrestrial orchid *Diuris fragrantissima* (Orchidaceae). *Australian Journal of Botany* 57, 200–209.

Sommers KP, Elswick M, Herrick GI, Fox GA (2011) Inferring microhabitat characteristics of *Lilium catesbaei* (Liliaceae). *American Journal of Botany* 98, 819–828.

Soulé M (1980) Thresholds for survival: maintaining fitness and evolutionary potential. In *Conservation Biology: An Evolutionary-ecological Perspective*, Soulé M, Wilcox BA (eds). Sunderland, MA: Sinauer Associates, pp. 151–169.

Soulé ME (1985) What is conservation biology? *BioScience* 35, 727–734.

Soulé ME (1986) *Conservation Biology, the Science of Scarcity, and Diversity.* Sunderland, MA: Sinauer Associates.

Soulé ME (1987) *Viable Populations for Conservation.* Cambridge: Cambridge University Press.

Soulé ME, Terborgh J (1999) Conserving nature at regional and continental scales – a scientific program for North America. *BioScience* 49, 809–817.

Soulé ME, Wilcox BA (1980) *Conservation Biology: An Evolutionary-ecological Perspective.* Sunderland, MA: Sinauer Associates.

Souza AF (2007) Ecological interpretation of multiple population size structures in trees: the case of *Araucaria angustifolia* in South America. *Austral Ecology* 32, 524–533.

Souza AF, Forgiarini C, Longhi SJ, Brena DA (2008) Regeneration patterns of a long-lived dominant conifer and the effects of logging in southern South America. *Acta Oecologica* 34, 221–232.

Sovu S, Tigabu M, Savadogo P, Oden PD, Xayvongsa L (2010) Enrichment planting in a logged-over tropical mixed deciduous forest of Laos. *Journal of Forestry Research* 21, 273–280.

Spies TA, Franklin JF, Thomas TB (1988) Coarse woody debris in Douglas-fir forests of western Oregon and Washington. *Ecology* 69, 1689–1702.

Sprugel DG (1991) Disturbance, equilibrium, and environmental variability: what is "natural" vegetation in a changing environment? *Biological Conservation* 58, 1–18.

Sprugel DG, Bormann FH (1981) Natural disturbance and the steady state in high-altitude balsam fir forests. *Science* 211, 390–393.

Stacy EA (2001) Cross-fertility in two tropical tree species: evidence of inbreeding depression within populations and genetic divergence among populations. *American Journal of Botany* 88, 1041–1051.

Stacy EA, Hamrick JL, Nason JD, *et al.* (1996) Pollen dispersal in low-density populations of three neotropical tree species. *American Naturalist* 148, 275–298.

Standards Petitions Working Group (2017) *Guidelines for Using the IUCN Red List Categories and Criteria: Version 13.* Gland, Switzerland: IUCN.

Stanturf JA, Palik BJ, Dumroese RK (2014) Contemporary forest restoration: a review emphasizing function. *Forest Ecology and Management* 331, 292–323.

Staudinger MD, Carter SL, Cross MS, *et al.* (2013) Biodiversity in a changing climate: a synthesis of current and projected trends in the US. *Frontiers in Ecology and the Environment* 11, 465–473.

Ste-Marie C, Nelson EA, Dabros A, Bonneau M-E (2011) Assisted migration: introduction to a multifaceted concept. *Forestry Chronicle* 87, 724–730.

Steffan-Dewenter I, Schiele S (2008) Do resources or natural enemies drive bee population dynamics in fragmented habitats? *Ecology* 89, 1375–1387.

Steiner KC, Westbrook JW, Hebard FV, *et al.* (2017) Rescue of American chestnut with extraspecific genes following its destruction by a naturalized pathogen. *New Forests* 48, 317–336.

Steininger MK (2000) Secondary forest structure and biomass following short and extended land-use in central and southern Amazonia. *Journal of Tropical Ecology* 16, 689–708.

Stephens EL, Castro-Morales L, Quintana-Ascencio PF (2012) Post-dispersal seed predation, germination, and seedling survival of five rare Florida scrub species in intact and degraded habitats. *American Midland Naturalist* 167, 223–239.

Stephens SL, Martin RE, Clinton NE (2007) Prehistoric fire area and emissions from California's forests, woodlands, shrublands, and grasslands. *Forest Ecology and Management* 251, 205–216.

Stephens SL, Ruth LW (2005) Federal forest-fire policy in the United States. *Ecological Applications* 15, 532–542.

Steward G, Beveridge A (2010) A review of New Zealand kauri (*Agathis australis* (D. Don) Lindl.): its ecology, history, growth, and potential for management for timber. *New Zealand Journal of Forestry Science* 40, 33–59.

Stewart JR, Lister AM, Barnes I, Dalen L (2010) Refugia revisited: individualistic responses of species in space and time. *Proceedings of the Royal Society B: Biological Sciences* 277, 661–671.

Still SM, Frances AL, Treher AC, Oliver L (2015) Using two climate change vulnerability assessment methods to prioritize and manage rare plants: a case study. *Natural Areas Journal* 35, 106–121.

Stockwell DRB, Peters D (1999) The GARP modeling system: problems and solutions to automated spatial prediction. *International Journal of Geographic Information Sience* 13, 143–158.

Stoner KE, Vulinec K, Wright SJ, Peres CA (2007) Hunting and plant community dynamics in tropical forests: a synthesis and future directions. *Biotropica* 39, 385–392.

Stott G, Gill D (2014) *How to Design and Manage a Basic Tree Nursery.* Cambridge: Global Trees Campaign, Fauna and Flora International.

Strandgaard H (1972) The roe deer (*Capreolus capreolus*) population at Kalø and the factors regulating its size. *Danish Review of Game Biology* 7, 1–205.

Strong TF, Erdmann GG (2000) Effects of residual stand density on growth and volume production in even-aged red maple stands. *Canadian Journal of Forest Research* 30, 372–378.

Subiakto A, Rachmat HH, Sakai C (2016) Choosing native tree species for establishing man-made forest: a new perspective for sustainable forest management in changing world. *Biodiversitas* 17, 620–625.

Suding KN (2011) Toward an era of restoration in ecology: successes, failures, and opportunities ahead. *Annual Review of Ecology, Evolution, and Systematics* 42, 465–487.

Suding KN, Gross KL, Houseman GR (2004) Alternative states and positive feedbacks in restoration ecology. *Trends in Ecology and Evolution* 19, 46–53.

Suding KN, Hobbs RJ (2009) Threshold models in restoration and conservation: a developing framework. *Trends in Ecology and Evolution* 24, 271–279.

Suding KN, Lavorel S, Chapin FS, *et al.* (2008) Scaling environmental change through the community-level: a trait-based response-and-effect framework for plants. *Global Change Biology* 14, 1125–1140.

Sutton RF (1993) Mounding site preparation: a review of European and North American experience. *New Forests* 7, 151–192.

Suzan H, Nabhan GP, Patten DT (1996) The importance of *Olneya tesota* as a nurse plant in the Sonoran Desert. *Journal of Vegetation Science* 7, 635–644.

Svoboda M, Fraver S, Janda P, Bače R, Zenáhlíková J (2010) Natural development and regeneration of a Central European montane spruce forest. *Forest Ecology and Management* 260, 707–714.

Swaine MD, Hall JB (1983) Early succession on cleared forest land in Ghana. *Journal of Ecology* 71, 601–627.

Swanson FJ, Jones JA, Wallin DO, Cissel JH (1994) Natural variability: implications for ecosystem management. Volume II: ecosystem management principles and applications. In *Eastside Forest Ecosystem Health Assessment*, Jensen ME, Bourgeron PS (eds). Washington, DC: US Forest Service General Technical Report PNW-GTR-318, pp. 80–94.

Swetnam TW (1993) Fire history and climate change in giant sequoia groves. *Science* 262, 885–889.

Swetnam TW, Allen CD, Betancourt JL (1999) Applied historical ecology: using the past to manage for the future. *Ecological Applications* 9, 1189–1206.

Swinfield T, Afriandi R, Antoni F, Harrison RD (2016) Accelerating tropical forest restoration through the selective removal of pioneer species. *Forest Ecology and Management* 381, 209–216.

Szewczyk J, Szwagrzyk J (1996) Tree regeneration on rotten wood and on soil in old-growth stand. *Vegetatio* 122, 37–46.

Tabarelli M, Lopes AV, Peres CA (2008) Edge-effects drive forest fragments towards an early successional system. *Biotropica* 40, 657–661.

Tabarelli M, Mantovani W (2000) Gap-phase regeneration in a tropical montane forest: the effects of gap structure and bamboo species. *Plant Ecology* 148, 149–155.

Takahashi M, Sakai Y, Ootomo R, Shiozaki M (2000) Establishment of tree seedlings and water-soluble nutrients in coarse woody debris in old-growth *Picea abies* forest in Hokkaido, northern Japan. *Canadian Journal of Forestry Research* 30, 1148–1155.

Talamo A, Barchuk A, Cardozo S, *et al.* (2015) Direct versus indirect facilitation (herbivore mediated) among woody plants in a semiarid Chaco forest: a spatial association approach. *Austral Ecology* 40, 573–580.

Tallmon DA, Jules ES, Radke NJ, Mills LS (2003) Of mice and men and trillium: cascading effects of forest fragmentation. *Ecological Applications* 13, 1193–1203.

Tambosi LR, Martensen AC, Ribeiro MC, Metzger JP (2014) A framework to optimize biodiversity restoration efforts based on habitat amount and landscape connectivity. *Restoration Ecology* 22, 169–177.

Tang CQ, He L-Y, Gao Z, *et al.* (2011a) Habitat fragmentation, degradation, and population status of endangered *Michelia coriacea* in Southeastern Yunnan, China. *Mountain Research and Development* 31, 343–350.

Tang CQ, Ohsawa M (2002) Tertiary relic deciduous forests on a humid subtropical mountain, Mt. Emei, Sichuan, China. *Folia Geobotanica* 37, 93–106.

Tang CQ, Peng M-C, He L-Y, *et al.* (2013) Population persistence of a Tertiary relict tree *Tetracentron sinense* on the Ailao Mountains, Yunnan, China. *Journal of Plant Research* 126, 651–659.

Tang CQ, Werger MJA, Ohsawa M, Yang Y (2014) Habitats of Tertiary relict trees in China. In *Endemism in Vascular Plants. Plants and Vegetation*, Hobohm C (ed.). Dordrecht, The Netherlands: Springer, pp. 289–308.

Tang CQ, Yang Y, Ohsawa M, *et al.* (2011b) Population structure of relict *Metasequoia glyptostroboides* and its habitat fragmentation and degradation in south-central China. *Biological Conservation* 144, 279–289.

Tang CQ, Yang Y, Ohsawa M, *et al.* (2015) Community structure and survival of Tertiary relict *Thuja sutchuenensis* (Cupressaceae) in the subtropical Daba Mountains, Southwestern China. *PLoS ONE* 10, e0125307.

Tang L, Shao G, Piao Z, *et al.* (2010) Forest degradation deepens around and within protected areas in East Asia. *Biological Conservation* 143, 1295–1298.

Tani N, Tsumura Y, Kado T, *et al.* (2009) Paternity analysis-based inference of pollen dispersal patterns, male fecundity variation, and influence of flowering tree density and general flowering magnitude in two dipterocarp species. *Annals of Botany* 104, 1421–1434.

Taylor AH, Huang JY, Zhou SQ (2004) Canopy tree development and under-growth bamboo dynamics in old-growth *Abies-Betula* forests in Southwestern China: a 12-year study. *Forest Ecology and Management* 200, 347–360.

Taylor AH, Qin ZS (1988) Regeneration patterns in old-growth *Abies-Betula* forests in the Wolong Natural Reserve, Sichuan, China. *Journal of Ecology* 76, 1204–1218.

Taylor AH, Zisheng Q, Jie L (1996) Structure and dynamics of subalpine forests in the Wang Lang Natural Reserve, Sichuan, China. *Vegetatio* 124, 25–38.

Tchouto MGP, Yemefack M, De Boer WF, *et al.* (2006) Biodiversity hotspots and conservation priorities in the Campo-Ma'an rain forests, Cameroon. *Biodiversity and Conservation* 15, 1219–1252.

Temperton VM, Hobbs RJ, Nuttle T, Halle S (2004) *Assembly Rules and Restoration Ecology*. Washington, DC: Island Press.

Terborgh J, Nunez-Iturri G, Pitman NCA, *et al.* (2008) Tree recruitment in an empty forest. *Ecology* 89, 1757–1768.

Tewksbury JJ, Lloyd JD (2001) Positive interactions under nurse-plants: spatial scale, stress gradients, and benefactor size. *Oecologia* 127, 425–434.

The Long N, Hoelscher D (2014) The fate of five rare tree species after logging in a tropical limestone forest (Xuan Son National Park, northern Vietnam). *Tropical Conservation Science* 7, 336–341.

Theodoropoulos D (2013) *Invasion Biology: Critique of a Pseudoscience*. Blythe, CA: Avvar Books.

Thomas CD (2011) Translocation of species, climate change, and the end of trying to recreate past ecological communities. *Trends in Ecology and Evolution* 26, 216–221.

Thomas CD, Bodsworth EJ, Wilson RJ, *et al.* (2001) Ecological and evolutionary processes at expanding range margins. *Nature* 411, 577–581.

Thomas E, Jalonen R, Loo J, *et al.* (2014) Genetic considerations in ecosystem restoration using native tree species. *Forest Ecology and Management* 333, 66–75.

Thompson SK, Seber GAF (1996) *Adaptive Sampling*. Hoboken, NJ: John Wiley and Sons.

Thorne RF (1999) Eastern Asia as a living museum for archaic angiosperms and other seed plants. *Taiwania* 44, 413–422.

Thorpe AS, Stanley AG (2011) Determining appropriate goals for restoration of imperilled communities and species. *Journal of Applied Ecology* 48, 275–279.

Thorpe HC, Thomas SC, Caspersen JP (2007) Residual-tree growth responses to partial stand harvest in the black spruce (*Picea mariana*) boreal forest. *Canadian Journal of Forest Research* 37, 1563–1571.

Thuiller W, Lavorel S, Araujo MB, Sykes MT, Prentice IC (2005) Climate change threats to plant diversity in Europe. *Proceedings of the National Academy of Sciences of the United States of America* 102, 8245–8250.

Thysell DR, Carey AB (2001) Manipulation of density of *Pseudotsuga menziesii* canopies: preliminary effects on understory vegetation. *Canadian Journal of Forest Research* 31, 1513–1525.

Tilman D, Lehman CL (1997) Habitat destruction and species extinction. In *Spatial Ecology: The Role of Space in Population Dynamics and Interspecific Interactions*, Tilman D, Kareiva P (eds). Princeton, NJ: Princeton Univesity Press, pp. 233–249.

Tilman D, May RM, Lehman CL, Nowak MA (1994) Habitat destruction and the extinction debt. *Nature* 371, 65–66.

Timyan JC, Reep SF (1994) Conservation status of *Attalea crassipatha* (Mart.) Burret, the rare, and endemic oil palm of Haiti. *Biological Conservation* 68, 11–18.

Tobon W, Urquiza-Haas T, Koleff P, *et al.* (2017) Restoration planning to guide Aichi targets in a megadiverse country. *Conservation Biology* 31, 1086–1097.

Toh I, Gillespie M, Lamb D (1999) The role of isolated trees in facilitating tree seedling recruitment at a degraded sub-tropical rainforest site. *Restoration Ecology* 7, 288–297.

Tones RC, Renison D (2016) Indirect facilitation becomes stronger with seedling age in a degraded seasonally dry forest. *Acta Oecologica* 70, 138–143.

Torroba-Balmori P, Zaldivar P, Alday JG, Fernandez-Santos B, Martinez-Ruiz C (2015) Recovering *Quercus* species on reclaimed coal wastes using native shrubs as restoration nurse plants. *Ecological Engineering* 77, 146–153.

Towns DR (2002) Korapuki Island as a case study for restoration of insular ecosystems in New Zealand. *Journal of Biogeography* 29, 593–607.

Tozer MG, Mackenzie BDE, Simpson CC (2012) An application of plant functional types for predicting restoration outcomes. *Restoration Ecology* 20, 730–739.

Traveset A (1998) Effect of seed passage through vertebrate frugivores' guts on germination: a review. *Perspectives in Plant Ecology, Evolution and Systematics* 1/2, 151–190.

Traveset A, Gonzalez-Varo JP, Valido A (2012) Long-term demographic consequences of a seed dispersal disruption. *Proceedings of the Royal Society B: Biological Sciences* 279, 3298–3303.

Traveset A, Richardson DM (2006) Biological invasions as disruptors of plant reproductive mutualisms. *Trends in Ecology and Evolution* 21, 208–216.

Traveset A, Riera N (2005) Disruption of a plant-lizard seed dispersal system and its ecological effects on a threatened endemic plant in the Balearic Islands. *Conservation Biology* 19, 421–431.

Traveset A, Verdu M (2002) A meta-analysis of gut treatment on seed germination. In *Seed Dispersal and Frugivory: Ecology, Evolution and Conservation,* Levey DJ, Silva WR, Galetti M (eds). Wallington, UK: CAB International, pp. 339–350.

Tribsch A (2004) Areas of endemism of vascular plants in the Eastern Alps in relation to Pleistocene glaciation. *Journal of Biogeography* 31, 747–760.

Trubat R, Cortina J, Vilagrosa A (2011) Nutrient deprivation improves field performance of woody seedlings in a degraded semi-arid shrubland. *Ecological Engineering* 37, 1164–1173.

Tsuyuzaki S, Titus JH, Moral R (1997) Seedling establishment patterns on the pumice plain, Mount St. Helens, Washington. *Journal of Vegetation Science* 8, 727–734.

Tu M, Robison RA (2014) Overcoming barriers to the prevention and management of alien plant invasions in protected areas: a practical approach. In *Plant Invasions in Protected Areas: Patterns, Problems and Challenges,* Foxcroft LC, Pyšek P, Richardson DM (eds). Dordrecht, The Netherlands: Springer, pp. 529–547.

Tucker NIJ, Murphy TM (1997) The effects of ecological rehabilitation on vegetation recruitment: some observations from the wet tropics of North Queensland. *Forest Ecology and Management* 99, 133–152.

Tuomela K, Kuusipalo J, Vesa L, *et al.* (1996) Growth of dipterocarp seedlings in artificial gaps: an experiment in a logged-over rainforest in South Kalimantan, Indonesia. *Forest Ecology and Management* 81, 95–100.

Turesson G (1922) The genotypical response of the plant species to the habitat. *Hereditas* 3, 211–350.

Turnau K, Haselwandter K (2002) Arbuscular mycorrhizal fungi, an essential component of soil microflora in ecosystem restoration. In *Mycorrhizal Technology in Agriculture: From Genes to Bioproducts,* Gianinazzi S, Schüepp H, Barea JM, Haselwandter K (eds). Basel: Birkhäuser, pp. 137–149.

Turnbull LA, Crawley MJ, Rees M (2000) Are plant populations seed-limited? A review of seed sowing experiments. *Oikos* 88, 225–238.

Turner IM (2001) *The Ecology of Trees in the Tropical Rain Forest.* Cambridge: Cambridge University Press.

Tylianakis JM, Laliberté E, Nielsen A, Bascompte J (2010) Conservation of species interaction networks. *Biological Conservation* 143, 2270–2279.

Tymen B, Rejou-Mechain M, Dalling JW, *et al.* (2016) Evidence for arrested succession in a liana-infested Amazonian forest. *Journal of Ecology* 104, 149–159.

Tzedakis PC, Lawson IT, Frogley MR, Hewitt GM, Preece RC (2002) Buffered tree population changes in a Quaternary refugium: evolutionary implications. *Science* 297, 2044–2047.

Uhl C, Buschbacher R, Serrao EAS (1988) Abandoned pastures in eastern Amazonia. I. Patterns of plant succession. *Journal of Ecology* 76, 663–681.

Uhl C, Jordan CF (1984) Succession and nutrient dynamics following forest cutting and burning in Amazonia. *Ecology* 65, 1476–1490.

Uhl C, Kauffman JB (1990) Deforestation, fire susceptibility, and potential tree responses to fire in the eastern Amazon. *Ecology* 71, 437–449.

Uhl C, Vieira ICG (1989) Ecological impacts of selective logging in the Brazilian Amazon: a case study from the Paragominas region of the state of Pará. *Biotropica* 21, 98–106.

Valdecantos A, Fuentes D, Smanis A, *et al.* (2014) Effectiveness of low-cost planting techniques for improving water availability to *Olea europaea* seedlings in degraded drylands. *Restoration Ecology* 22, 327–335.

Valiente-Banuet A, Ezcurra E (1991) Shade as a cause of the association between the cactus *Neobuxbaumia tetetzo* and the nurse plant *Mimosa luisana* in the Tehuacan Valley, Mexico. *Journal of Ecology* 79, 961–971.

Vallauri DR, Aronson J, Barbero M (2002) An analysis of forest restoration 120 years after reforestation on badlands in the Southwestern Alps. *Restoration Ecology* 10, 16–26.

Vallauri DR, Aronson J, Dudley N, Vallejo R (2005) Monitoring and evaluating forest restoration success. In *Forest Restoration in Landscapes: Beyond Planting Trees*, Mansourian S, Vallauri D, Dudley N (eds). New York: Springer, pp. 150–158.

Vallee L, Hogbin T, Monks L, *et al.* (2004) *Guidelines for the Translocation of Threatened Plants in Australia*, 2nd edition. Canberra, Australia: Australian Network for Plant Conservation.

van Andel J (1998) Intraspecific variability in the context of ecological restoration projects. *Perspectives in Plant Ecology, Evolution and Systematics* 1, 221–237.

Van Gemerden BS, Shu GN, Olff H (2003) Recovery of conservation values in Central African rain forest after logging and shifting cultivation. *Biodiversity and Conservation* 12, 1553–1570.

van Wilgen BW, Richardson DM (2014) Challenges and trade-off in the management of invasive alien trees. *Biological Invasions* 16, 721–734.

van Zonneveld MJ, Gutierrez JR, Holmgren M (2012) Shrub facilitation increases plant diversity along an arid scrubland-temperate rain forest boundary in South America. *Journal of Vegetation Science* 23, 541–551.

Vander Mijnsbrugge K (2014) Continuity of local genetic diversity as an alternative to importing foreign provenances. In *Genetic Considerations in Ecosystem Restoration Using Native Tree Species. State of the World's Forest Genetic Resources: Thematic Study*, Bozzano M, Jalonen R, Thomas E, *et al.* (eds). Rome: Food and Agriculture Organization and Bioversity International, pp. 38–46.

Vander Wall SB (1994) Seed fate pathways of antelope bitterbrush: dispersal by seed-caching yellow pine chipmunks. *Ecology* 75, 1911–1926.

Vandergast AG, Bohonak AJ, Hathaway SA, Boys J, Fisher RN (2008) Are hotspots of evolutionary potential adequately protected in southern California? *Biological Conservation* 141, 1648–1664.

Vane-Wright RI, Humphries CJ, Williams PH (1991) What to protect? Systematics and the agony of choice. *Biological Conservation* 55, 235–254.

Vane-Wright RI, Smith CR, Kitching IJ (1994) Systematic assessment of taxic diversity by summation. In *Systematics and Conservation Evaluation. Systematics Association Special Volume 50*, Forey PL, Humphries CJ, Vane-Wright RI (eds). Oxford, UK: Clarendon Press, pp. 309–326.

Vanha-Majamaa I, Lilja S, Ryömä R, *et al.* (2007) Rehabilitating boreal forest structure and species composition in Finland through logging, dead wood creation and fire: the EVO experiment. *Forest Ecology and Management* 250, 77–88.

Vanthomme H, Bellé B, Forget PM (2010) Bushmeat hunting alters recruitment of large-seeded plant species in Central Africa. *Biotropica* 42, 672–679.

Veblen TT, Veblen AT, Schlegel FM (1979) Understorey patterns in mixed evergreen-deciduous Nothofagus forests in Chile. *Journal of Ecology* 67, 809–823.

Vellend M, Verheyen K, Jacquemyn H, *et al.* (2006) Extinction debt of forest plants persists for more than a century following habitat fragmentation. *Ecology* 87, 542–548.

Verburg R, van Eijk-Bos C (2003) Effects of selective logging on tree diversity, composition, and plant functional type patterns in a Bornean rain forest. *Journal of Vegetation Science* 14, 99–110.

Verdu M, Garcia-Fayos P (1996) Nucleation processes in a Mediterranean bird-dispersed plant. *Functional Ecology* 10, 275–280.

Vergeer P, van den Berg LJL, Roelofs JG, Ouborg NJ (2005) Single-family versus multi-family introduction. *Plant Biology* 7, 509–515.

Verschuyl J, Riffell S, Miller D, Wigley TB (2011) Biodiversity response to intensive biomass production from forest thinning in North American forests: a meta-analysis. *Forest Ecology and Management* 261, 221–232.

Viani RA, Holl KD, Padovezi A, *et al.* (2017) Protocol for monitoring tropical forest restoration: perspectives from the Atlantic Forest Restoration Pact in Brazil. *Tropical Conservation Science* 10, 1–8.

Vieira ICG, Uhl C, Nepstad D (1994) The role of the shrub *Cordia multispicata* Cham. as a "succession facilitator" in an abandoned pasture, Paragominas, Amazonia. *Vegetatio* 115, 91–99.

Vila M, Vayreda J, Comas L, *et al.* (2007) Species richness and wood production: a positive association in Mediterranean forests. *Ecology Letters* 10, 241–250.

Vilagrosa A, Cortina J, Gil-Pelegrin E, Bellot J (2003) Suitability of drought-preconditioning techniques in Mediterranean climate. *Restoration Ecology* 11, 208–216.

Villar-Salvador P, Puértolas J, Peñuelas J, Peñuelas R (2009) Morphological and physiological plant quality in woodland restoration: a Mediterranean perspective. In *Land Restoration to Combat Desertification: Innovative Approaches, Quality Control and Project Evaluation*, Bautista S, Aronson J, Vallejo R (eds). Valencia, Spain: Fundación Centro de Estudios Ambientales de Mediterráneo (CEAM), pp. 103–120.

Villard M-A, Metzger JP (2014) Beyond the fragmentation debate: a conceptual model to predict when habitat configuration really matters. *Journal of Applied Ecology* 51, 309–318.

Villegas Z, Pena-Claros M, Mostacedo B, *et al.* (2009) Silvicultural treatments enhance growth rates of future crop trees in a tropical dry forest. *Forest Ecology and Management* 258, 971–977.

Vitt P, Belmaric PN, Book R, Curran M (2016) Assisted migration as a climate change adaptation strategy: lessons from restoration and plant reintroductions. *Israel Journal of Plant Sciences* 63, 250–261.

Vitt P, Havens K (2004) Integrating quantitative genetics into ex situ conservation and restoration practices. In *Ex Situ Plant Conservation: Supporting Species Survival in the Wild*, Guerrant EO, Havens K, Maunder M (eds). Washington, DC: Island Press, pp. 286–304.

Vitt P, Havens K, Kramer AT, Sollenberger D, Yates E (2010) Assisted migration of plants: changes in latitudes, changes in attitudes. *Biological Conservation* 143, 18–27.

Vlam M, Baker PJ, Bunyavejchewin S, Mohren GMJ, Zuidema PA (2014) Understanding recruitment failure in tropical tree species: insights from a tree-ring study. *Forest Ecology and Management* 312, 108–116.

Vogler AP, Desalle R (1994) Diagnostic units of conservation management. *Conservation Biology* 8, 354–363.

Volis S (2011) Adaptive genetic differentiation in a predominantly self-pollinating species analyzed by transplanting into natural environment, crossbreeding, and $Q_{ST}$–$F_{ST}$ test. *New Phytologist* 192, 237–248.

Volis S (2016a) Conservation-oriented restoration: how to make it a success? *Israel Journal of Plant Sciences* 63, 276–296.

Volis S (2016b) How to conserve threatened Chinese species with extremely small populations? *Plant Diversity* 38, 53–62.

Volis S (2016c) Species-targeted plant conservation: time for conceptual integration. *Israel Journal of Plant Sciences* 63, 232–249.

Volis S (2017a) Complementarities of two existing intermediate conservation approaches. *Plant Diversity* 39, 379–382.

Volis S (2017b) Conservation utility of botanic garden living collections: setting a strategy and appropriate methodology. *Plant Diversity* 39, 365–372.

Volis S (2017c) Plant conservation in the Anthropocene: definitely not win–win but maybe not lose–lose? In *Reference Module in Earth Systems and Environmental Sciences*. New York: Elsevier.

Volis S (2018) Securing a future for China's plant biodiversity through an integrated conservation approach. *Plant Diversity* 40, 91–105.

Volis S, Blecher M (2010) Quasi *in situ* – a bridge between ex situ and *in situ* conservation of plants. *Biodiversity and Conservation* 19, 2441–2454.

Volis S, Blecher M, Sapir Y (2010) Application of complex conservation strategy to *Iris atrofusca* of the Northern Negev, Israel. *Biodiversity and Conservation* 19, 3157–3169.

Volis S, Bohrer G, Oostermeijer G, van Tienderen P (2005) Regional consequences of local population demography and genetics in relation to habitat management in *Gentiana pneumonanthe*. *Conservation Biology* 19, 357–367.

Volis S, Mendlinger S, Ward D (2002) Adaptive traits of wild barley plants of Mediterranean and desert origin. *Oecologia* 133, 131–138.

Volis S, Ormanbekova D, Shulgina I (2016a) Role of selection and gene flow in population differentiation at the edge vs. interior of the species range differing in climatic conditions. *Molecular Ecology* 25, 1449–1464.

Volis S, Ormanbekova D, Yermekbayev K, Song M, Shulgina I (2015) Multi-approaches analysis reveals local adaptation in the emmer wheat (*Triticum dicoccoides*) at macro- but not micro-geographical scale. *PLoS ONE* 10, e0121153.

Volis S, Ormanbekova D, Yermekbayev K, Song M, Shulgina I (2016b) The conservation value of peripheral populations and a relationship between quantitative trait and molecular variation. *Evolutionary Biology* 43, 26–36.

Volis S, Zhang Y-H, Dorman M, Blecher M (2016c) *Iris atrofusca* genetic and phenotypic variation, the role of habitat-specific selection in this variation structuring, and conservation implications using quasi *in situ* guidelines. *Israel Journal of Plant Sciences* 63, 347–354.

Vos CC, Berry P, Opdam P, *et al.* (2008) Adapting landscapes to climate change: examples of climate-proof ecosystem networks and priority adaptation zones. *Journal of Applied Ecology* 45, 1722–1731.

Vovides AP, Pérez-Farrera MA, Iglesias C (2010) Cycad propagation by rural nurseries in Mexico as an alternative conservation strategy: 20 years on. *Kew Bulletin* 65, 603–611.

Vucetich JA, Waite TA (2003) Spatial patterns of demography and genetic processes across the species' range: null hypotheses for landscape conservation genetics. *Conservation Genetics* 4, 639–645.

Wadsworth FH, Zweede JC (2006) Liberation: acceptable production of tropical forest timber. *Forest Ecology and Management* 233, 45–51.

Wagner M, Bullock JM, Hulmes L, *et al.* (2016) Creation of micro-topographic features: a new tool for introducing specialist species of calcareous grassland to restored sites? *Applied Vegetation Science* 19, 89–100.

Wagner S, Collet C, Madsen P, *et al.* (2010) Beech regeneration research: from ecological to silvicultural aspects. *Forest Ecology and Management* 259, 2172–2182.

Walker BH (1992) Biodiversity and ecological redundancy. *Conservation Biology* 6, 18–23.

Walters CJ, Holling CS (1990) Large scale management experiments and learning by doing. *Ecology* 71, 2060–2068.

Wandelli EV, Fearnside PM (2015) Secondary vegetation in central Amazonia: land-use history effects on aboveground biomass. *Forest Ecology and Management* 347, 140–148.

Wang B, Chen G, Li C, Sun W (2017) Floral characteristics and pollination ecology of *Manglietia ventii* (Magnoliaceae), a plant species with extremely small populations (PSESP) endemic to South Yunnan of China. *Plant Diversity* 39, 52–59.

Wang T, O'Neill GA, Aitken SN (2010) Integrating environmental and genetic effects to predict responses of tree populations to climate. *Ecological Applications* 20, 153–163.

Wang W, Franklin SB, Ren Y, Ouellette JR (2006) Growth of bamboo *Fargesia qinlingensis* and regeneration of trees in a mixed hardwood-conifer forest in the Qinling Mountains, China. *Forest Ecology and Management* 234, 107–115.

Waring RH, Pitman GB (1985) Modifying lodgepole pine stands to change susceptibility to mountain pine beetle attack. *Ecology* 66, 889–897.

Watson GW, Heywood V, Crowley W (1993) North American botanic gardens. *Horticultural Reviews* 15, 1–62.

Way MJ (2003) Collecting seed from non-domesticated plants for long-term conservation. In *Seed Conservation. Turning Science into Practice*, Smith RD, Dickie JB, Linington SH, Pritchard HW, Probert RJ (eds). Richmond, UK: The Royal Botanic Gardens Kew, pp. 163–201.

Wearn OR, Reuman DC, Ewers RM (2012) Extinction debt and windows of conservation opportunity in the Brazilian Amazon. *Science* 337, 228–232.

Weaver PL (1987) Enrichment planting in tropical America. In *Management of the Forests of Tropical America: Prospects and Technologies*, Figueroa Colon JC, Wadsworth FH, Branham S (eds). Río Piedras, Puerto Rico: Institute of Tropical Forestry, USDA Forest Service and University of Puerto Rico, pp. 258–278.

Webb CO, Peart DR (2000) Habitat associations of trees and seedlings in a Bornean rain forest. *Journal of Ecology* 88, 464–478.

Weekley CW, Kubisiak TL, Race TM (2002) Genetic impoverishment and cross-incompatibility in remnant genotypes of *Ziziphus celata* (Rhamnaceae), a rare shrub endemic to the Lake Wales Ridge, Florida. *Biodiversity and Conservation* 11, 2027–2046.

Weeks AR, Sgro CM, Young AG, *et al.* (2011) Assessing the benefits and risks of translocations in changing environments: a genetic perspective. *Evolutionary Applications* 4, 709–725.

Weiher E, Keddy PA (1999) *Ecological Assembly Rules: Perspectives, Advances, Retreats.* Cambridge: Cambridge University Press.

Weiher E, van der Werf A, Thompson K, *et al.* (1999) Challenging Theophrastus: a common core list of plant traits for functional ecology. *Journal of Vegetation Science* 10, 609–620.

Weller SG (1994) The relationship of rarity to plant reproductive biology. In *Restoration of Endangered Species: Conceptual Issues, Planning and Implementation*, Bowles ML, Whelan CJ (eds). Cambridge: Cambridge University Press, pp. 90–117.

Wendelberger KS, Fellows MQN, Maschinski J (2008) Rescue and restoration: experimental translocation of *Amorpha herbacea* Walter var. *crenulata* (Rybd.) Isley into a novel urban habitat. *Restoration Ecology* 16, 542–552.

Wendelberger KS, Maschinski J (2009) Linking geographical information systems and observational and experimental studies to determine optimal seedling microsites of an endangered plant in a subtropical urban fire-adapted ecosystem. *Restoration Ecology* 17, 845–853.

Wendelberger KS, Maschinski J (2016) Assessing microsite and regeneration niche preferences through experimental reintroduction of the rare plant *Tephrosia angustissima* var. *corallicola*. *Plant Ecology* 217, 155–167.

Wheeler CE, Omeja PA, Chapman CA, *et al.* (2016) Carbon sequestration and bio-diversity following 18 years of active tropical forest restoration. *Forest Ecology and Management* 373, 44–55.

White PS, Walker JL (1997) Approximating nature's variation: selecting and using reference information in restoration ecology. *Restoration Ecology* 5, 338–349.

White TL, Adams WT, Neale DB (2007) *Forest Genetics.* Wallington, UK: CAB International.

Whitmore TC (1975) *Tropical Rain Forests of the Far East.* Oxford, UK: Clarendon Press.

Whitmore TC (1989) Canopy gaps and the 2 major groups of forest trees. *Ecology* 70, 536–538.

Whitmore TC (1991) Tropical rain forest dynamics and its implications for man-agement. In *Rain Forest Regeneration and Management,* Gomez-Pompa A, Whitmore TC, Hadley M (eds). Lancaster, UK: Parthenon Publishing, pp. 67–89.

Whitmore TC, Brown ND (1996) Dipterocarp seedling growth in rain forest canopy gaps during six and a half years. *Philosophical Transactions of the Royal Society of London Series B: Biological Sciences* 351, 1195–1203.

Whittaker RH (1965) Dominance and diversity in land plant communities: numer-ical relations of species express the importance of competition in community function and evolution. *Science* 147, 250–260.

Whittaker RJ, Maria Fernandez-Palacios J, Matthews TJ, Borregaard MK, Triantis KA (2017) Island biogeography: taking the long view of nature's laboratories. *Science* 357, eaam8326.

Whitworth A, Downie R, von May R, Villacampa J, MacLeod R (2016) How much potential biodiversity and conservation value can a regenerating rainforest pro-vide? A "best-case scenario" approach from the Peruvian Amazon. *Tropical Conservation Science* 9, 224–245.

Wiens JA, Hobbs RJ (2015) Integrating conservation and restoration in a changing world. *BioScience* 65, 302–312.

Wilcove DS, Giam X, Edwards DP, Fisher B, Koh LP (2013) Navjot's nightmare revisited: logging, agriculture, and biodiversity in Southeast Asia. *Trends in Ecology and Evolution* 28, 531–540.

Williams JN, Seo C, Thorne J, *et al.* (2009) Using species distribution models to predict new occurrences for rare plants. *Diversity and Distributions* 15, 565–576.

Williams JW, Jackson ST (2007) Novel climates, no-analog communities, and eco-logical surprises. *Frontiers in Ecology and the Environment* 5, 475–482.

Williams MC, Wardle GM (2005) The invasion of two native *Eucalypt* forests by *Pinus radiata* in the Blue Mountains, New South Wales, Australia. *Biological Conservation* 125, 55–64.

Williams MI, Dumroese RK (2013) Preparing for climate change: forestry and assisted migration. *Journal of Forestry* 111, 287–297.

Williams NM (2011) Restoration of nontarget species: bee communities and pollination function in riparian forests. *Restoration Ecology* 19, 450–459.

Williams P, Gibbons D, Margules C, *et al.* (1996) A comparison of richness hotspots, rarity hotspots, and complementary areas for conserving diversity of British birds. *Conservation Biology* 10, 155–174.

Williams SE, Shoo LP, Isaac JL, Hoffmann AA, Langham G (2008) Towards an integrated framework for assessing the vulnerability of species to climate change. *PLoS Biology* 6, 2621–2626.

Wilson K, Pressey RL, Newton A, *et al.* (2005) Measuring and incorporating vulnerability into conservation planning. *Environmental Management* 35, 527–543.

Winder R, Nelson EA, Beardmore T (2011) Ecological implications for assisted migration in Canadian forests. *Forestry Chronicle* 87, 731–744.

Wishnie MH, Dent DH, Mariscal E, *et al.* (2007) Initial performance and reforestation potential of 24 tropical tree species planted across a precipitation gradient in the Republic of Panama. *Forest Ecology and Management* 243, 39–49.

Wong M, Wright SJ, Hubbell SP, Foster RB (1990) The spatial pattern and reproductive consequences of outbreak defoliation in *Quararibea asterolepis*, a tropical tree. *Journal of Ecology* 78, 579–588.

Wotton DM, Kelly D (2011) Frugivore loss limits recruitment of large-seeded trees. *Proceedings of the Royal Society B: Biological Sciences* 278, 3345–3354.

Wright A, Tobin M, Mangan S, Schnitzer SA (2015) Unique competitive effects of lianas and trees in a tropical forest understory. *Oecologia* 177, 561–569.

Wright SJ (2002) Plant diversity in tropical forests: a review of mechanisms of species coexistence. *Oecologia* 130, 1–14.

Wright SJ, Hernandez A, Condit R (2007) The bushmeat harvest alters seedling banks by favoring lianas, large seeds, and seeds dispersed by bats, birds, and wind. *Biotropica* 39, 363–371.

Wunderle JM (1997) The role of animal seed dispersal in accelerating native forest regeneration on degraded tropical lands. *Forest Ecology and Management* 99, 223–235.

Wyse Jackson P (1997) Botanic gardens and the convention on biological diversity. *Botanic Gardens Conservation News* 2, 26–30.

Wyse Jackson PS (1999) Experimentation on a large scale – an analysis of the holdings and resources of botanic gardens. *Botanic Gardens Conservation News* 3, 27–32.

Wyse Jackson PS, Kennedy K (2009) The global strategy for plant conservation: a challenge and opportunity for the international community. *Trends in Plant Science* 14, 578–580.

Xia J, Lu J, Wang ZX, *et al.* (2013) Pollen limitation and Allee effect related to population size and sex ratio in the endangered *Ottelia acuminata* (Hydrocharitaceae): implications for conservation and reintroduction. *Plant Biology* 15, 376–383.

Xu H, Li Y, Liu S, *et al.* (2015) Partial recovery of a tropical rain forest a half-century after clear-cut and selective logging. *Journal of Applied Ecology* 52, 1044–1052.

Yamada T, Hosaka T, Okuda T, Kassim AR (2013) Effects of 50 years of selective logging on demography of trees in a Malaysian lowland forest. *Forest Ecology and Management* 310, 531–538.

Yang J, Zhao LL, Yang J, Sun W (2015) Genetic diversity and conservation evaluation of a critically endangered endemic maple, *Acer yangbiense*, analyzed using microsatellite markers. *Biochemical Systematics and Ecology* 60, 193–198.

Yang JM, Yang XY, Liang H (2004) The discovery of buried *Metasequoia* wood in Lichuan, Hubei, China, and its significance. *Acta Palaeontologica Sinica* 43, 124–131 (in Chinese).

Yang X, Bauhus J, Both S, *et al.* (2013) Establishment success in a forest biodiversity and ecosystem functioning experiment in subtropical China (BEF-China). *European Journal of Forest Research* 132, 593–606.

Yarranton GA, Morrison RG (1974) Spatial dynamics of a primary succession: nucleation. *Journal of Ecology* 62, 417–428.

Ying CC, Yanchuk AD (2006) The development of British Columbia's tree seed transfer guidelines: purpose, concept, methodology, and implementation. *Forest Ecology and Management* 227, 1–13.

York RA, Battles JJ, Heald RC (2007) *Gap based silviculture in a Sierran mixed conifer forest: effects of gap size on early survival and 7-year seedling growth.* USDA Forest Service General Technical Report. PSW-GTR-203. Washington, DC: US Forest Service, pp. 181–191.

York RA, Fuchs D, John J. Battles JJ, Stephens SL (2010) Radial growth responses to gap creation in large, old *Sequoiadendron giganteum*. *Applied Vegetation Science* 13, 498–509.

York RA, O'Hara KL, Battles JJ (2013) Density effects on giant sequoia (*Sequoiadendron giganteum*) growth through 22 years: implications for restoration and plantation management. *Western Journal of Applied Forestry* 28, 30–36.

Yoshioka A, Akasaka M, Kadoya T (2014) Spatial prioritization for biodiversity restoration: a simple framework referencing past species distributions. *Restoration Ecology* 22, 185–195.

Young AG, Pickup M (2010) Low S-allele numbers limit mate availability, reduce seed set, and skew fitness in small populations of a self-incompatible plant. *Journal of Applied Ecology* 47, 541–548.

Young BE, Hall KR, Byers E, *et al.* (2012) Rapid assessment of plant and animal vulnerability to climate change. In *Wildlife Conservation in a Changing Climate*, Brodie J, Post E, Doak D (eds). Chicago, IL: University of Chicago Press, pp. 129–152.

Young KR, Ewel JJ, Brown BJ (1987) Seed dynamics during forest succession in Costa Rica. *Vegetatio* 71, 157–173.

Young TP (2000) Restoration ecology and conservation biology. *Biological Conservation* 92, 73–83.

Young TP, Petersen DA, Clary JJ (2005) The ecology of restoration: historical links, emerging issues, and unexplored realms. *Ecology Letters* 8, 662–673.

Youngblood AP (1991) Radial growth after a shelterwood seed cut in a mature stand of white spruce in interior Alaska. *Canadian Journal of Forest Research* 21, 410–413.

Zahawi RA, Augspurger CK (1999) Early plant succession in abandoned pastures in Ecuador. *Biotropica* 31, 540–552.

Zahawi RA, Augspurger CK (2006) Tropical forest restoration: tree islands as recruitment foci in degraded lands of Honduras. *Ecological Applications* 16, 464–478.

Zahawi RA, Holl KD (2009) Comparing the performance of tree stakes and seedlings to restore abandoned tropical pastures. *Restoration Ecology* 17, 854–864.

Zahawi RA, Holl KD (2014) Evaluation of different tree propagation methods in ecological restoration in the neotropics. In *Genetic Considerations in Ecosystem Restoration Using Native Tree Species. State of the World's Forest Genetic Resources: Thematic Study*, Bozzano M, Jalonen R, Thomas E, *et al.* (eds). Rome: Food and Agriculture Organization and Bioversity International, pp. 85–96.

Zahawi RA, Holl KD, Cole RJ, Reid JL (2013) Testing applied nucleation as a strategy to facilitate tropical forest recovery. *Journal of Applied Ecology* 50, 88–96.

Zamith LR, Scarano FR (2010) Restoration of a coastal swamp forest in southeast Brazil. *Wetlands Ecology and Management* 18, 435–448.

Zandavalli RB, Dillenburg LR, de Souza PVD (2004) Growth responses of *Araucaria angustifolia* (Araucariaceae) to inoculation with the mycorrhizal fungus *Glomus clarum*. *Applied Soil Ecology* 25, 245–255.

Zavodna M, Abdelkrim J, Pellissier V, Machon N (2015) A long-term genetic study reveals complex population dynamics of multiple-source plant reintroductions. *Biological Conservation* 192, 1–9.

Zeide B (2001) Thinning and growth: a full turnaround. *Journal of Forestry* 99, 20–25.

Zeigler SL, Che-Castaldo JP, Neel MC (2013) Actual and potential use of population viability analyses in recovery of plant species listed under the US Endangered Species Act. *Conservation Biology* 27, 1265–1278.

Zender S (2014) Assessing the ecological restoration of Motutapu Island, New Zealand. *Eukaryon* 10, 41–44.

Zeng FP, Peng WX, Song TQ, *et al.* (2007) Changes in vegetation after 22 years' natural restoration in the karst disturbed area in northwest Guangxi. *Acta Ecologica Sinnica* 27, 5110–5119.

Zenner EK (2004) Does old-growth condition imply high live-tree structural complexity? *Forest Ecology and Management* 195, 243–258.

Zerbe S (2002) Restoration of natural broad-leaved woodland in Central Europe on sites with coniferous forest plantations. *Forest Ecology and Management* 167, 27–42.

Zermeno-Hernandez I, Mendez-Toribio M, Siebe C, Benitez-Malvido J, Martinez-Ramos M (2015) Ecological disturbance regimes caused by agricultural land uses and their effects on tropical forest regeneration. *Applied Vegetation Science* 18, 443–455.

Zhang S, Shi F, Yang W, *et al.* (2015) Autotoxicity as a cause for natural regeneration failure in *Nyssa yunnanensis* and its implications for conservation. *Israel Journal of Plant Sciences* 62, 187–197.

Zhu JJ, Matsuzaki T, Lee FQ, Gonda Y (2003) Effect of gap size created by thinning on seedling emergency, survival, and establishment in a coastal pine forest. *Forest Ecology and Management* 182, 339–354.

Zhu YL, Comita LS, Hubbell SP, Ma K (2015) Conspecific and phylogenetic density-dependent survival differs across life stages in a tropical forest. *Journal of Ecology* 103, 957–966.

Zielonka T (2006) When does dead wood turn into a substrate for spruce replacement? *Journal of Vegetation Science* 17, 739–746.

Zimmer HC, Offord CA, Auld TD, Baker PJ (2016) Establishing a wild, ex situ population of a critically endangered shade-tolerant rainforest conifer: a translocation experiment. *PLoS ONE* 11, e0157559.

Zimmerman BL, Kormos CF (2012) Prospects for sustainable logging in tropical forests. *BioScience* 62, 479–487.

Zimmerman JK, Pascarella JB, Aide TM (2000) Barriers to forest regeneration in an abandoned pasture in Puerto Rico. *Restoration Ecology* 8, 350–360.

Zobel M (1997) The relative role of species pools in determining plant species richness: an alternative explanation of species coexistence? *Trends in Ecology and Evolution* 12, 266–269.

Zobel M, Otsus M, Liira J, Moora M, Mols T (2000) Is small-scale species richness limited by seed availability or microsite availability? *Ecology* 81, 3274–3282.

Zobel M, van der Maarel E, Dupre C (1998) Species pool: the concept, its determination, and significance for community restoration. *Applied Vegetation Science* 1, 55–66.

Zohary D, Hopf M, Weiss E (2012) *Domestication of Plants in the Old World: The Origin and Spread of Domesticated Plants in Southwest Asia, Europe, and the Mediterranean Basin.* Oxford, UK: Oxford University Press.

Zubek S, Turnau K, Tsimilli-Michael M, Strasser RJ (2009) Response of endangered plant species to inoculation with arbuscular mycorrhizal fungi and soil bacteria. *Mycorrhiza* 19, 113–123.

Zysk-Gorczynska E, Jakubiec Z, Wuczynski A (2015) Brown bears (*Ursus arctos*) as ecological engineers: the prospective role of trees damaged by bears in forest ecosystems. *Canadian Journal of Zoology* 93, 133–141.

# Index

Page numbers in **bold** refer to figures; those in *italic* to tables